FLUOROCARBON REFRIGERANTS HANDBOOK

RALPH C. DOWNING

PRENTICE HALL, Englewood Cliffs, New Jersey 07632

Library of Congress Cataloging-in-Publication Data

Downing, Ralph C.
Fluorocarbon refrigerants handbook.

Bibliography: p.
Includes index.
1. Refrigerants. 2. Fluorocarbons. I. Title.
TP492.7.D63 1988 621.5′64 87-14572
ISBN 0-13-322504-6

Editorial/production supervision and
interior design: Theresa A. Soler
Cover design: Edsal Enterprises
Manufacturing buyer: Peter Havens

Printed in the United States of America

10 9 8 7 6 5 4 3 2 1

ISBN 0-13-322504-6

Prentice-Hall International (UK) Limited, *London*
Prentice-Hall of Australia Pty. Limited, *Sydney*
Prentice-Hall Canada Inc., *Toronto*
Prentice-Hall Hispanoamericana, S.A., *Mexico*
Prentice-Hall of India Private Limited, *New Delhi*
Prentice-Hall of Japan, Inc., *Tokyo*
Simon & Schuster Southeast Asia Pte. Ltd., *Singapore*
Editora Prentice-Hall do Brasil, Ltda., *Rio de Janeiro*

CONTENTS

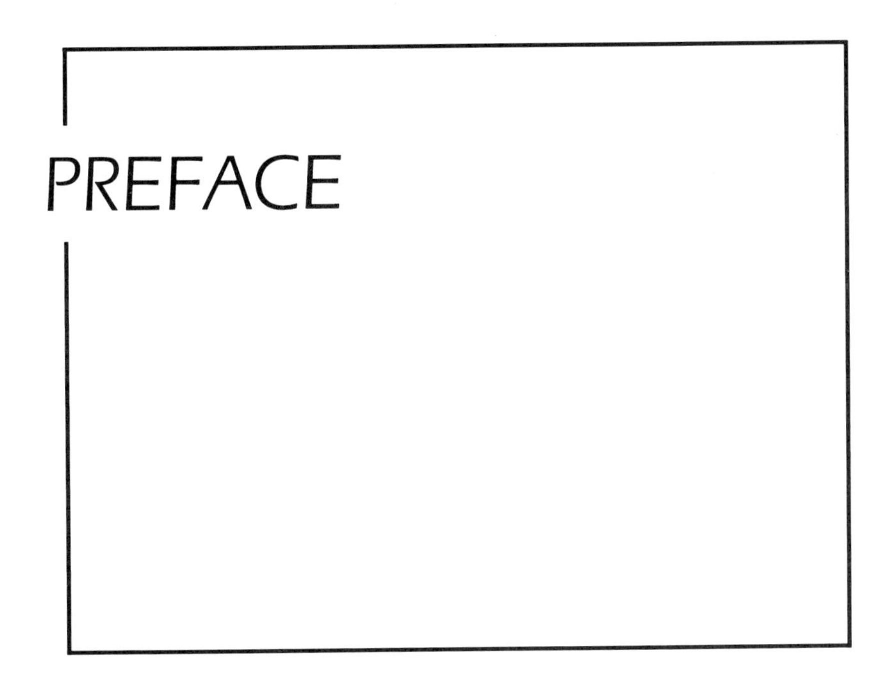

PREFACE

The fluorocarbon refrigerants were first proposed, developed, and evaluated by Midgley, Henne, and McNary of the General Motors Corporation. Early tests were encouraging and larger quantities than could be made in the laboratory were needed [P–1].

The DuPont Company was asked to develop processes for the manufacture of these promising new refrigerants. A new corporation, Kinetic Chemicals, jointly owned by DuPont and General Motors, was organized to promote and market the products. Later, DuPont purchased General Motors' interest in Kinetic Chemicals and it became a part of the DuPont Company, first as the Kinetic Chemicals Division and finally as the Freon Products Division.

Over the years, the DuPont Company has spent a great deal of time and effort in the development of good manufacturing procedures, in the study of physical and chemical properties, in the introduction of new and useful fluorinated refrigerants, and in the development of other applications as solvents, foam blowing agents, fire extinguishing agents, aerosol propellants, and so on. During the past 50 and more years a very extensive amount of information has been created and accumulated by the DuPont Company and by others interested in the fluorinated refrigerants.

The objective of this handbook is to record in one place properties of some fluorinated products especially pertinent to their use in refrigeration and air conditioning. I am very grateful to the DuPont Company for permission to use information from their files, especially information that may not have been previously published. The list of people who have added so much to the store of knowledge about the fluorinated refrigerants is too long to include here, but their contributions have been great and are truly recognized and acknowledged. The use of duplicating and other facilities at the Freon Products Laboratory is also greatly appreciated.

The support, encouragement, forbearance, solicitude, and sufferance of my wife, Rosalie, have been vital to the completion of this volume and are much appreciated.

Ralph C. Downing

REFERENCE

P–1. R. C. Downing, "Development of Chlorofluorocarbon Refrigerants," *ASHRAE Tran.*, 90, pt 2, (1984), 8.

FLUOROCARBON
REFRIGERANTS
HANDBOOK

1

FLUOROCARBON REFRIGERANTS

For the sake of simplicity, organic compounds containing a few carbon atoms (usually four or less) and one or more fluorine atoms are sometimes called *fluorocarbon refrigerants* when used in air conditioning and refrigeration applications. However, they may also contain hydrogen, chlorine, or bromine atoms. Other descriptive names include *halogenated compounds, chlorofluorocompounds* (CFC), and *fluorinated compounds*.

HISTORY [*1–1*]

Throughout the early history of mechanical refrigeration in the latter part of the nineteenth century and the first part of the twentieth century, the nature of the refrigerant was a problem. In 1962, Gosney reported that by then more than 50 refrigerants had been tested and many more proposed [1–2]. This large number of candidate refrigerants reflects the need for better products. Most of the early refrigerants were thermodynamically suitable and performed well when tested. Some of them, such as methyl chloride, sulfur dioxide, carbon dioxide, ammonia, and some hydrocarbons, became commercially important. Others, such as ether, acetone, and alcohol, were discarded after a few tests.

All of the early refrigerants suffered from one or more properties that restricted their general use and inhibited the development of the refrigerating and air-conditioning industry as we know it today. Some were toxic or flammable—and sometimes both. Others were lacking in stability or operated at very high pressures. These deficiencies were generally recognized—among others by Willis Carrier, the father of centrifugal refrigeration [1–3]. After a long search he finally chose dichloroethylene, commonly called "dilene," as the refrigerant for his new centrifugal compressor. In 1926, methylene chloride (R-30) replaced dilene and was the basic centrifugal refrigerant for a few years. When R-11 became available, its superior properties were quickly recognized and it was adopted by Carrier. Later, R-113 and other fluorinated refrigerants were also used.

In 1928, the search for better refrigerants was formalized and intensified. Thomas Midgley, then employed by the Frigidaire Division of General Motors, was asked to lead the search for a nonflammable refrigerant with good stability, low toxicity, and an atmospheric boiling point between −40 and 32°F [1–4]. At first it seemed unlikely that a single compound with all of these properties could be found, but three days after receiving the assignment, Midgley and his associates, Henne and McNary, had synthesized the first fluorocarbon refrigerant and demonstrated on guinea pigs that it had unusually low toxicity [1–5,1–6].

Midgley is especially noted for his use of the periodic table arrangement of chemical elements to find solutions for industrial problems. One very successful development was tetraethyllead for improving the octane rating of gasoline, and another of special interest to the refrigeration industry was his recognition of the desirable properties of the fluorocarbon compounds. While studying the periodic table, Midgley and his associates found that the commercial refrigerants then available contained relatively few chemical elements and that these elements were closely related in structure. A small portion of the periodic table of elements has the following arrangement:

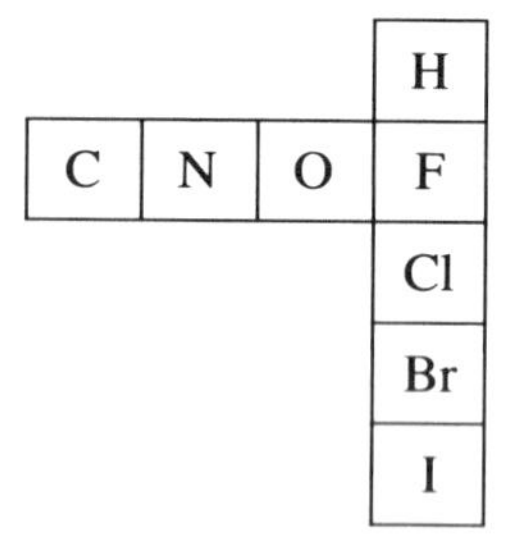

The chemical formulas of some older refrigerants are:

Ammonia	NH_3
Carbon dioxide	CO_2
Sulfur dioxide	SO_2
Hydrocarbons	C_nH_m
Methyl chloride	CH_3Cl
Water	H_2O

In studying the properties of compounds containing the elements in the portion of the periodic table shown earlier, attention was directed toward fluorine.

Flammability:

Hydrocarbons: very flammable
Ammonia: flammable
Water: not flammable
Hydrofluoric acid (HF): not flammable

Stability: The heats of formation of compounds containing carbon and halogens increase in the following order:

least stable   iodine < bromine < chlorine < fluorine   most stable

Toxicity: This was a difficult question. Many inorganic fluorides are toxic. Hydrofluoric acid was well known to be a dangerous material. Fluorine itself is very reactive and toxic. However, there were some clues. Compounds with good stability tend to be less toxic. Elements with low molecular weight tend to be less toxic than heavier ones. The toxicity of the methyl halides is inversely related to their stability. The American Conference of Industrial Hygienists has assigned *threshold limit values* (TLVs) for many chemical compounds [1–7]. The TLV is the parts per million by volume in air for a compound that is considered the safe upper limit for exposure during a working day. The TLVs listed below are all in the toxic range but indicate that methyl fluoride would be the least toxic. It is now known that the toxicity decreases rapidly with increasing fluorine content.

	TLV (ppm)
Methyl fluoride (CH_3F)	Not listed
Methyl chloride (CH_3Cl)	50
Methyl bromide (CH_3Br)	5
Methyl iodide (CH_3I)	2

Midgley and his fellow workers prepared all 15 possible one-carbon combinations of hydrogen, chlorine, and fluorine and evaluated their properties. They finally selected one, dichlorodifluoromethane (CCl_2F_2, or R-12) as the compound most nearly meeting the criteria of the perfect refrigerant which they had sought. The new candidate was announced at the April 1930 meeting of the American Chemical Society in Atlanta, Georgia [1–8]. In recognition of its importance, all other sections of the society adjourned their meetings. At the conclusion of his talk, Midgley gave a convincing demonstration of two of the properties of the new refrigerant by inhaling a lung full of the gas and using it to extinguish a lighted candle.

Historical Highlights [*1–1*]

Some highlights in the progress of refrigeration and the development of refrigerants are outlined below.

1600 A mixture of snow and salt was found to lower the temperature enough to freeze water.

1775	Professor Cullen at the University of Edinburgh used an air pump to reduce the pressure and lower the evaporating temperature of water.
1810	Sir John Leslie in Scotland used sulfuric acid to absorb water vapor and help maintain a low pressure.
1834	The first practical refrigerating machine using the compression cycle was built by Jacob Perkins in London with ether as the refrigerant.
1845	The expansion of compressed air was used to freeze water by John Gorrie and others.
1850	A small, practical absorption machine was developed by E. Carre using water and sulfuric acid.
1857	James Harrison successfully applied the Perkins machine to produce refrigeration for breweries and cooling for meat and perishable produce.
1859	An ammonia and water absorption system was built by Ferdinand Carre.
1869	By this time, several plants selling artificial ice were in existence.
1873	Carl Linde introduced an ammonia vapor compressor.
1876	Raoul Pictet built a sulfur dioxide compressor.
1876	Methyl ether was used as a refrigerant in shipping meat from Argentina to France.
1877	Developmental work on air refrigeration machinery was started, leading to the Bell–Coleman compressors.
1877	The *Ice Trade Journal* began publication.
1880–1890	Several different refrigerants were used successfully, including methyl chloride and carbon dioxide.
1891	*Ice and Refrigeration* began publication.
1930	The development of fluorocarbon refrigerants was announced by Midgley and Henne.
1931	R-12 was introduced as a commercial refrigerant.
1932	Commercial introduction of R-11.
1933	Commercial introduction of R-114.
1934	Commercial introduction of R-113.
1936	Commercial introduction of R-22. Large-scale use, however, did not develop until after World War II.
1943	Use of R-11 and R-12 as aerosol propellants was developed by Goodhue and Sullivan of the Department of Agriculture.
1945	Commercial introduction of R-13.
1949	DuPont purchased General Motors' interest in Kinetic Chemicals, Inc.
1952	Manufacture of fluorocarbon refrigerants started by the Allied Chemical Corporation.
1955	Commercial introduction of R-14.

1956 Refrigerant numbering system released by DuPont for general use.

1957 Manufacture of fluorocarbon refrigerants started by Pennsalt Chemicals Corporation.

1958 Manufacture of fluorocarbon refrigerants started by Union Carbide Corporation.

1961 Commercial introduction of R-502.

1963 Manufacture of fluorocarbon refrigerants started by Kaiser Aluminum and Chemical Company.

1963 Manufacture of fluorocarbon refrigerants started by the Racon Company.

NOMENCLATURE

The chemical names for the fluorinated refrigerants are long and cumbersome and a numbering system was developed to identify the different products. The number was originally part of the registered trade name but was later donated to the industry by the DuPont Company to avoid confusion and the proliferation of different numbers for the same product.

The number assigned to each refrigerant is related to its chemical composition and the system has been formalized as ASHRAE Standard 34 [1–9]. The numbering rules follow.

1. The first digit on the right is the number of fluorine (F) atoms.
2. The second digit from the right is one more than the number of hydrogen (H) atoms.
3. The third digit from the right is one less than the number of carbon (C) atoms. When this digit is zero it is omitted.
4. The number of chlorine (Cl) atoms is found by subtracting the sum of the F and H atoms from the total number that can be attached to carbon. With one carbon, the total number of attached atoms is four. With two carbons, the total number is six.
5. In some cases, part or all of the chlorine atoms are replaced with bromine and the letter B is used in the number.
6. Azeotropic mixtures are assigned numbers in the 500 series in order of their commercial introduction (e.g., R-500, R-501, R-502, etc.).
7. Ordinary (nonazeotropic) mixtures are assigned numbers in the 400 series.
8. Hydrocarbon refrigerants are identified by the regular numbering system except for butane and isobutane, which are included in the 600 series.
9. Miscellaneous organic refrigerants are assigned arbitrary numbers in the 600 series.
10. Inorganic refrigerants are in the 700 series, using the molecular weight as part of the number.

The chemical name, formula, refrigerant number, and chemical abstract registry number for some common refrigerants are given in Table 1–1.

The numbers can be used generally, such as R-12 or refrigerant 12, or with a specific trademark, such as Freon 12 or Genetron 12.

The nameplate on refrigeration compressors specifies the refrigerant that should be used in it. The designation may be the standard refrigerant number, the chemical name, or sometimes the chemical formula. At one time the term "mono" was used when only one atom of a halogen species was present: for example, monochlorodifluoromethane for R-22 or trichloromonofluoromethane for R-11. The modern practice is to omit "mono."

Perhaps a simpler method of relating refrigerant number to chemical composition is the addition of 90 to the refrigerant number. The result gives, in order, the number of carbon, hydrogen, and fluorine atoms. The number of chlorine atoms is found in the same way, by subtracting from the total number of possible atoms attached to the carbon atom(s). For example, with R-12, adding 90 gives 102; the R-12 molecule contains one carbon, no hydrogen, and two fluorine atoms.

TABLE 1–1 REFRIGERANT NAMES AND NUMBERS

Refrigerant number	Chemical name	Chemical formula	Molecular weight	Chemical abstracts registry number
11	Trichlorofluoromethane	CCl_3F	137.4	75–69–4
12	Dichlorodifluoromethane	CCl_2F_2	120.9	75–71–8
13	Chlorotrifluoromethane	$CClF_3$	104.5	75–72–9
13B1	Bromotrifluoromethane	$CBrF_3$	148.9	75–63–8
14	Tetrafluoromethane	CF_4	88.0	75–73–0
22	Chlorodifluoromethane	$CHClF_2$	86.5	75–45–6
23	Trifluoromethane	CHF_3	70.0	75–46–7
40	Methyl chloride	CH_3Cl	50.5	74–87–3
113	Trichlorotrifluoroethane	CCl_2FCClF_2	187.4	76–13–1
114	Dichlorotetrafluoroethane	$CClF_2CClF_2$	170.9	76–14–2
115	Chloropentafluoroethane	$CClF_2CF_3$	154.5	76–15–3
116	Hexafluoroethane	CF_3CF_3	138.0	76–16–4
124a	Chlorotetrafluoroethane	$CClF_2CHF_2$	136.5	354–25–6
142b	Chlorodifluoroethane	$CClF_2CH_3$	100.5	75–68–3
152a	Difluoroethane	CHF_2CH_3	66.0	75–37–6
170	Ethane	CH_3CH_3	30.0	74–84–0
290	Propane	$CH_3CH_2CH_3$	44.0	74–98–6
500	R-12/R-152a (73.8/26.2 wt %)		99.3	
501	R-22/R-12 (75/25 wt %)		93.1	
502	R-22/R-115 (48.8/51.2 wt %)		111.6	
503	R-23/R-13 (40.1/59.9 wt %)		87.5	
504	R-32/R-115 (48.2/51.8 wt %)		79.2	
717	Ammonia	NH_3	17.0	7664–41–7
718	Water	H_2O	18.0	7732–18–5
744	Carbon dioxide	CO_2	44.0	124–38–9
764	Sulfur dioxide	SO_2	64.0	7446–09–5
1150	Ethylene	$CH_2{=}CH_2$	28.0	74–84–0
1270	Propylene	$CH_3CH{=}CH_2$	42.1	115–07–1

When the Carrier Corporation began manufacturing centrifugal compressors for refrigeration and air conditioning, they used the trade name Carrene, followed by a number for the refrigerants. This system has been abandoned for many years but some of the old numbers are occasionally seen.

Refrigerant number	Carrene number
30 (methylene chloride)	1
11	2
113	3
500	7

Before DuPont released the numbering system described earlier, the Allied Corporation began manufacturing fluorocarbon refrigerants using different numbers. References to the Allied numbers appear now and then.

Refrigerant number	Allied number
11	11
12	12
22	141
113	226
114	316
114a	320
142b	101
152a	100

Another method of identifying products containing halogen atoms was developed by the U.S. Corps of Engineers during studies of their effectiveness as fire extinguishing agents [1–10]. Starting at the left, the first digit is the number of carbon (C) atoms, the second digit is the number of fluorine atoms (F), the third digit is the number of chlorine atoms (Cl), and the fourth digit the number of bromine atoms (Br). For fire extinguishant applications, these products are known as Halon, followed by the appropriate number.

Refrigerant number	Chemical formula	Halon number
12	CCl_2F_2	122
12B1	$CBrClF_2$	1211
12B2	CBr_2F_2	1202
13B1	$CBrF_3$	1301
114B2	$CBrF_2CBrF_2$	2402

MANUFACTURE

In the fifteenth century it was discovered that the addition of a common mineral during the processing of ores, ceramics, and glasses lowered the melting point and improved the product [1–1]. This material was called "fluores" from the Latin *fluere*, to flow. From this beginning, the words "fluorspar" for the mineral and "fluorine" for its principal element have evolved. Fluorspar is essentially calcium fluoride and is still widely used in the metallurgical and glass industries. Fluorine is estimated to be the thirteenth most abundant element on the earth and is also found in other minerals, such as cryolite and apatite.

During the latter part of the nineteenth century and the first part of the twentieth century, there was considerable interest in the preparation of organic fluorine compounds. A number of derivatives of the lower aliphatic hydrocarbons were prepared and their properties determined. Much of the experimental work was conducted by Swarts at the University of Ghent in Belgium; Meslans, Lebeau, and Damiens in France; and Ruff and Fredenhagen in Germany. In this early work, the fluorination was generally accomplished by using metal fluorides, such as those of silver, antimony, tin, mercury, and lead. In most cases the reaction was very slow and incomplete. About 1890, a milestone in the development of the organic fluorine chemical industry was discovered by Swarts. He found that the addition of a small amount of pentavalent antimony to antimony trifluoride had a marked influence on the reaction. The exchange of fluorine for other halogens occurred rapidly and completely. The *Swarts reaction* is still the basis for many commercial processes for making organic fluorine products [1–5,1–11).

Since most of the fluorinated compounds suitable for refrigerant use had been reported in the chemical literature, they were not patentable as such. However, application and process patents were obtained. An essential part of the manufacturing process involved making and handling anhydrous hydrogen fluoride. The aqueous solution had been used for many years for etching glass and for other purposes, but the anhydrous acid was not then a commercial product. Methods of making, drying, storing, and handling hydrogen fluoride were successfully developed. All of the reactants had to be dry and of high quality.

The starting materials for the Swarts reaction are carbon tetrachloride for the one-carbon compounds not containing hydrogen, chloroform for those with one hydrogen, and tetrachloroethylene for the two-carbon products. The source of fluorine is hydrofluoric acid and the catalyst is antimony pentachloride. The reaction begins with a rapid reaction between the HF and the $SbCl_5$.

$$SbCl_5 + HF \longrightarrow SbCl_4F + HCl$$

The $SbCl_4F$ is a powerful fluorinating agent and reacts immediately with the chlorinated compound.

$$2CCl_4 + 3SbCl_4F \longrightarrow CCl_3F + CCl_2F_2 + 3SbCl_5$$

The reaction is carried out continuously. A small amount of chlorine gas is added together with the carbon tetrachloride and hydrofluoric acid to keep the catalyst at a high level of activity. Both R-11 and R-12 are formed in the reaction in ratios that can be controlled by the temperature and pressure in the reaction pot. The reaction vessel is equipped with a reflux column to separate the more volatile gases from the higher-boiling products. The R-12 in this case is withdrawn from the top of the column and the R-11 at a point lower down in the column. The R-11 can be collected or returned to the pot for further fluorination. A small amount of R-13 ($CClF_3$) may also be formed.

Starting with chloroform ($CHCl_3$), a similar series of steps produces first R-21 ($CHCl_2F$) and then R-22 ($CHClF_2$). R-113 and R-114 are manufactured in the same way from tetrachloroethylene ($CCl_2{=}CCl_2$). Sufficient chlorine is added to make a saturated compound during the reaction.

As the fluorine content of the molecule increases, the replacement of chlorine atoms with additional fluorine atoms becomes more difficult. For the more highly fluorinated products, manufacturing processes are usually different from those used to make less fluorinated products. For example, under certain catalytic conditions, R-12 can undergo disproportionation to produce R-11 and R-13. Gas-phase reactions with specific catalysts are used in some cases. An interesting example is R-114. The two isomers, $CClF_2CClF_2$ and CCl_2FCF_3, are both formed during the fluorination, have similar properties, and boil within a few degrees of each other. In the liquid-phase fluorination, the symmetric compound is the principal product and commercial R-114 contains about 6 to 8% of the other isomer. In vapor-phase fluorination the mixture contains perhaps 60% of asymmetric isomer. In refrigeration either mixture could be used since the boiling points are so close together. The asymmetric isomer, however, is less stable toward hydrolysis, and in some applications the mixture with a high concentration of this isomer might be less suitable.

The requirements of the modern refrigeration industry demand a degree of refrigerant purity not often found in organic compounds produced on a large scale. Organic purity generally exceeds 99.9%. Water is limited to 10 ppm by weight. The air specification for the commercial products is less than 1.5 vol % in the vapor phase. Since air is not very soluble in the liquid, the total air concentration is exceedingly small. There must by no acidity or detectable chloride ion concentration. These specifications are met by distillation and highly developed water- and acid-removal techniques.

REFERENCES

1–1. R. C. Downing, "Development of Chlorofluorocarbon Refrigerants," *ASHRAE Trans.*, 90, pt 2 (1984), 8.

1–2. W. B. Gosney, "A Survey of the Newer Refrigerants," *J. Refrig.*, 5 (Sept.–Oct. 1962), 113.

1–3. W. A. Pennington, "Progress in Refrigerants: 1," *World Refrig.*, 8 (Feb.–Mar. 1957), 85.

1–4. T. Midgley, Jr., "From the Periodic Table to Production," *Ind. Eng. Chem.*, 29 (1937), 239.

1–5. J. M. Hamilton, Jr., "The Organic Fluorochemicals Industry," *Advances in Fluorine Chemistry*. London: Butterworth & Company (Publishers) Ltd., 3 (1962), 117.

1–6. R. C. Downing, ''History of the Organic Fluorine Industry,'' in *Kirk-Othmer*: *Encyclopedia of Chemical Technology*, 2nd ed., Vol. 9. New York: John Wiley & Sons, Inc. (1966), p. 704.

1–7. American Conference of Governmental Industrial Hygienists, Threshold Limit Values for Chemical Substances in the Work Environment, Cincinnati, Ohio: ACGIH (1983–1984).

1–8. T. Midgley, Jr., and A. L. Henne, ''Organic Fluorides as Refrigerants,'' *Ind. Eng. Chem.*, 22 (1930), 542.

1–9. American Society of Heating, Refrigerating and Air-Conditioning Engineers, ''Number Designation of Refrigerants,'' ASHRAE Standard 34. Atlanta, Ga.: ASHRAE (1978).

1–10. R. C. Downing, B. J. Eiseman, Jr., and J. E. Malcolm, ''Halogenated Extinguishing Agents,'' *Nat. Fire Prot. Assoc.* (Oct. 1951).

1–11. J. S. Sconce, ed. *Chlorine*, *Its Manufacture*, *Properties*, *and Uses*, Am. Chem. Soc. Monograph 154, New York: Reinhold Publishing Corporation (1962).

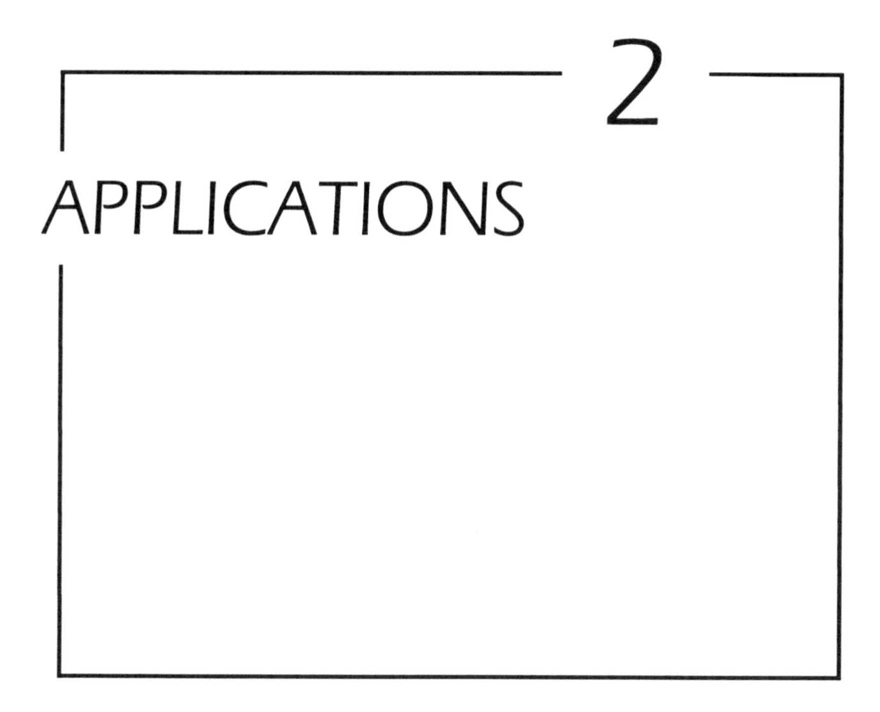

The use of the fluorocarbons as refrigerants is based on their stability, nonflammability, and low toxicity. These properties are also responsible for a number of other minor but interesting applications. The choice of a particular fluorocarbon for a specific application depends on a number of factors, including boiling point, effect on elastomers and plastics, degree of stability, historical inertia, cost, solubility in oil, efficiency, and perhaps others. In some cases controversial, and as yet unproven, environmental effects may be a factor.

INDUSTRIAL COOLING AND LARGE-TONNAGE AIR CONDITIONING

R-11 in centrifugal compressors is used in much of the larger-tonnage industrial and air-conditioning equipment. These compressors are generally available in only a few standard sizes since they are expensive to build. For applications requiring more or less capacity than provided with R-11, R-114 or R-113 can be used as the refrigerant without changing the compressor except for providing a properly sized motor. The comparison shown in Table 2–1 assumes a constant volumetric output from the compressor with evaporation at 40°F and condensing at 100°F.

Since the evaporating pressure for these refrigerants is generally below atmospheric, leakage, if any, will be inward, introducing air and moisture. A means of removing these unwelcome impurities is usually provided. Hermetic compressors

TABLE 2-1. RELATIVE CAPACITY OF R-11, R-113, and R-114

Refrigerant	Relative capacity
R-114	1.74
R-11	1.00
R-113	0.41

eliminate shaft seals as a source of leakage, but for these low-pressure refrigerants, the shell is ordinarily hinged and the two halves are sealed with a gasket. Inward leakage is reduced but not completely stopped. The intrusion of air and moisture still occurs at a slower rate.

Air can be automatically vented when the internal pressure exceeds a given limit. A small amount of refrigerant is vented with the air, but over the years, improvements in design and operation have reduced refrigerant loss to relatively insignificant levels. Water is more difficult to remove but probably more harmful in the system. The solubility of water in liquid refrigerant decreases with decreasing temperature. This change is used in some removal schemes. Liquid from the condenser is cooled by liquid from the evaporator. Excess water separates as a layer on top and the two liquid phases are manually separated. This procedure leaves the refrigerant saturated with water at evaporating temperatures but less than saturated at condensing temperatures. There is some evidence that dissolved water is less harmful than excess water. The water level can be further reduced by cooling liquid from the condenser to lower temperatures with a separate small refrigeration system.

R-11, R-113, and R-114 have all been used for many years with very good results. However, if the reaction of refrigerant with water should be a problem, R-114 is generally more stable toward hydrolysis than R-11 or R-113.

The capacity developed by a given compressor is directly related to the boiling point of the refrigerant, assuming that sufficient power is available. The lower-boiling refrigerants, such as R-12, R-500, R-22, and R-502, are also used with centrifugal compressors. They must be completely hermetic since in most applications, suction pressures will be higher than atmospheric. Any leakage would be outward, with resulting loss of refrigerant. With these refrigerants, much smaller compressors are needed than the large units required to produce the same capacity with R-11, R-113, or R-114.

REFRIGERANTS

R-12 versus R-22 versus R-502

R-12 was the first fluorocarbon-type refrigerant developed and used commercially and is still a standard for comparing other refrigerants. It has many properties especially suitable for use in refrigeration, such as good stability, little effect on elastomers and plastics, good solubility in lubricating oil, and reasonable compression ratio. In spite of much higher cost, R-12 was able to compete with and in most cases supplant other refrigerants, including sulfur dioxide, methyl chloride, hydrocarbons, and ammonia, on the basis of safety. Nonflammability, low toxicity, inertness, and stability were also recognized as essential refrigerant properties. Over a period of years, equipment manufacturers developed an understanding of the properties of R-12 and systems were designed to optimize its usefulness.

R-22 was introduced a few years after R-12 and the general reaction was that it was just another fluorocarbon refrigerant with a lower boiling point than R-12. It gradually became apparent that this was not the case. R-22 is a different chemical compound and has different properties, which at first caused problems in several areas.

1. With open compressors, shaft seals adequate for R-12 permitted leakage with R-22. In studying the problem it was found that elastomers may become enlarged and weakened in contact with R-22 but are little affected by R-12. The swelling is caused by some of the R-22 dissolving in the elastomer. In some applications the swelling may not be harmful, but when an elastomer is used in a seal, the rate of leakage can be approximately correlated with the degree of swelling. Refrigerant is apparently dissolved at the inner surface of the elastomer and evaporated from the surface exposed to the air. It also developed that some types of lubricating oils combined with the refrigerant-affected elastomers more than would be expected from results with the oil and refrigerant separately. Eventually, suitable shaft seals were found and the leakage problem with R-22 was solved.

2. Another example of the difference between these refrigerants was found in the hermetically sealed compressors that were beginning to become popular. R-12 had been used in the new compressors for several years with complete satisfaction. When R-22 was first used, motor failures and burnouts rose to unacceptable levels. After some study it was found that R-22 partially dissolved in the magnet wire coating of the motor, causing blistering and softening of the coating and eventually, arcing and failure. New coatings were quickly developed and curing times and temperatures were increased to provide materials not affected by R-22.

3. The solubility of refrigerants in lubricating oil is important at low evaporating temperatures. Oil by itself becomes viscous and tends to accumulate in the evaporator and low-pressure piping. When refrigerant dissolves in the oil it stays thin and is more easily pushed along by the flowing refrigerant gas. Good solubility usually means no oil-return problems, and poor solubility often means poor oil return. In this property, R-12 is best, R-22 is fair, and R-502 is worst. For example, R-12 is completely miscible with Suniso 3GS refrigeration oil down to at least −100°F, solutions of the oil with R-22 separate into two liquid phases at about 35°F, and R-502 does not completely dissolve at any temperature below the critical (180°F). These temperatures are for mixtures of refrigerant and oil containing approximately 50% of each. For each refrigerant a considerable amount is dissolved in oil even at low temperatures. In many cases a sufficient amount is dissolved for good oil return, even with R-502. In applications with R-22 or R-502 where good oil return is a problem, a small amount of R-12 is sometimes added because it will be

more soluble in the oil and thereby keep it fluid enough to flow well at low temperatures.

A synthetic lubricating oil based on alkylated benzene has recently been developed and is finding applications where oil return has been difficult [2–1]. Both R-12 and R-22 are completely miscible with the new oil to temperatures below −100°F. R-502 and the synthetic oil are miscible above −25°F and considerable solubility remains at lower temperatures. The solubility of other refrigerants, such as R-13 and R-503, is also much greater than in so-called "natural" oils. The alkylated benzene oil is especially valuable in very-low-temperature applications.

It is rather curious that R-12 is a better solvent for oil than is R-22 since in most other cases R-22 is the better solvent.

4. Compressor discharge temperatures are much higher with R-22 than with either R-12 or R-502, due partly to differences in the heat of compression and partly to differences in the flow rate.

5. The refrigerating capacity when used in the same compressor running at the same speed is about 60% greater with R-22 than with R-12 at any evaporating temperature. With R-502, the capacity is about 30% greater than with R-22 at −40°F but about the same or perhaps a little less than R-22 at +40°F evaporation.

A simplified comparison of these three refrigerants might be that R-12 has low capacity and low discharge temperatures, R-22 has high capacity and high discharge temperatures, and R-502 has high capacity and low discharge temperatures.

R-11

The centrifugal compressor developed by Carrier many years ago used methylene chloride as the refrigerant and a few are still in operation. However, R-11 has long since replaced it as the refrigerant in systems using large centrifugal compressors. These systems provide air conditioning for large office buildings, hotels, warehouses, industrial plants, and other locations where large amounts of air conditioning are needed. Industrial process water and other cooling applications are also served by R-11.

Before the ozone controversy and chlorofluorocarbon ban, R-11 was used in mixtures with lower-boiling products as aerosol propellants. Small amounts are still used in essential or specialized applications.

A minor use is in an instrument for measuring temperatures at higher altitudes based on the vapor pressure.

R-12

As a refrigerant, R-12 is used in household and commercial refrigeration and to some extent in air conditioning. It is used in home freezers, in water coolers, in vegetable display cases, in walk-in freezers, and in industrial applications.

R-12 is used as an aerosol propellant, either alone or in mixtures with higher-boiling products.

Other applications include freezing tissue for eye operations, wind tunnel studies, the extraction of low-boiling components in a dehydration process, heat pipes, and other similar heat-transfer operations.

It is generally used in reciprocating compressors ranging in size from fractional to 800 hp, in rotary-type compressors in the smaller sizes, and to some extent in large, hermetically-sealed centrifugal compressors.

R-13

R-13 is used in very-low-temperature specialty applications in cascade systems with R-12, R-22, or R-502 in the higher stage.

A minor application is maintaining a constant temperature by slow evaporation for unattended radar stations.

R-13B1

R-13B1 is used to some extent in medium-low-temperature applications. It is an excellent refrigerant and would undoubtedly be more widely used if it were more competitive with other fluorocarbon refrigerants in cost. It has an unusually low compression ratio and can be used in single-stage compressors at quite low evaporating temperatures.

The principal application of this product is as a fire extinguishing agent. It provides excellent protection without damage to material objects at concentrations low enough to be safe in the event of human inhalation.

R-14

Systems for producing ultralow temperatures use R-14 in the bottom stage, with R-13 or R-503 in the middle stage and R-12, R-22, or R-502 in the upper stage.

R-14 has also been used in wind tunnel studies and in the guidance of cold rockets. Mixtures with R-23 have been used to maintain constant temperatures in heat-seeking rocket missiles.

A rapidly increasing use of R-14 is in the removal of the oxide from silicon wafers.

R-21

In spite of good physical properties, R-21 has never been used as a refrigerant, primarily because of excessive attack on elastomers and plastics and decomposition in the presence of water. It was used as a heat transfer agent in the first space shuttle but was replaced by R-114 in later versions. Recent long-term studies have revealed serious physiological effects and it is no longer manufactured in the United States.

R-22

R-22 is used in practically all residential and light commercial air conditioning and to some extent in very-large-tonnage applications with centrifugal compressors. It is also used for some industrial low-temperature requirements and for medium- and low-temperature display cases in supermarkets.

R-22 is the starting material for making tetrafluoroethylene, used in the manufacture of fluorinated plastics such as Teflon.

R-113

R-113 is used in commercial and industrial air conditioning and process water and brine cooling with centrifugal compression. A given compressor designed for use with R-11 can be used with R-113 to produce less capacity or with R-114 to produce more capacity.

Trichlorotrifluoroethane is used by itself or in mixtures with other liquids in many solvent cleaning applications. As a solvent, trade names other than R-113 are used. The basic chemical is the same, but in solvent applications, quality specifications are more restrictive, especially for particulate matter and residue on evaporation.

R-114

R-114 is sometimes used in small refrigeration systems with rotary compressors and in large industrial process cooling and air-conditioning systems with centrifugal compressors.

It is used in a number of heat transfer applications such as the separation of uranium isotopes in the gaseous diffusion process, in space vehicles, in heat pipes, in solar collectors, and so on.

There is some limited usage as an aerosol propellant and in skin freezing for cosmetic surgery.

Other Refrigerants

R-115. R-115 is a component of R-502 and a small amount is used in stabilizing aerosol whipped cream.

R-116. R-116 has some application as an electrical insulating gas.

R-500. R-500 is used in home and commercial air conditioning, refrigeration, water coolers, and so on, using reciprocating compressors in smaller tonnages and hermetic centrifugal compressors in larger tonnages.

R-502. The primary use of R-502 is in the storage and display of frozen-food products. It is also used to some extent for industrial low-temperature cooling.

R-503. R-503 is used in very-low-temperature cascade systems, offering more capacity than R-13.

REFERENCE

2–1. W. D. Cooper, R. C. Downing, and J. B. Gray, "Alkyl Benzene as a Compressor Lubricant," *Proceedings of the 1974 Purdue Compressor Technology Conference*, July 10–12, p. 88.

3 REFRIGERATION

In refrigeration it is the duty of the refrigerant to carry heat from some place where it is not wanted to a place where it can be discarded. This cycle consists of compression of the refrigerant gas, condensation to a liquid, and evaporation at a lower pressure. While all of these steps are important, the compression step is a very crucial part of the cycle and some relationships describing what happens during compression are briefly reviewed here. The following equations are from Dodge [3–1]. For a more complete discussion of gas relationships, see Dodge or another reference work on thermodynamics, such as the discussion by Suttle [3–2].

An ideal gas is defined as one obeying the following equation:

$$PV = RT \qquad \text{(for 1 mol)} \tag{3–1}$$

Differentiation gives

$$\left(\frac{dP}{dT}\right)_v = \frac{R}{V} \tag{3–2}$$

$$\left(\frac{dV}{dT}\right)_p = \frac{R}{P} \tag{3–3}$$

$$\left(\frac{d^2P}{dT^2}\right)_v = 0 \tag{3–4}$$

$$\left(\frac{d^2V}{dT^2}\right)_p = 0 \tag{3–5}$$

Other relationships include the following:

$$dE = C_v\, dT \tag{3–6}$$

$$dH = C_p\, dT \tag{3–7}$$

$$H = E + PV \tag{3–8}$$

$$\left(\frac{dC_v}{dV}\right)_T = 0 \tag{3–9}$$

$$\left(\frac{dC_p}{dP}\right)_T = 0 \tag{3–10}$$

$$C_p - C_v = R \tag{3–11}$$

$$\Delta H = Q = \int_{T_1}^{T_2} C_p\, dT \tag{3–12}$$

$$dS = C_v \frac{dT}{T} + R \frac{dV}{V} \tag{3–13}$$

$$dS = C_p \frac{dT}{T} - R \frac{dP}{P} \tag{3–14}$$

where P = pressure, psia
V = volume, ft^3
T = temperature, °F + 459.67
R = gas constant (10.7318 when P is psia, V is ft^3, and T is °R
E = internal energy, Btu/lb
H = enthalpy or heat content, Btu/lb
S = entropy, Btu/lb-°R
C_v = heat capacity at constant volume, Btu/lb·°F
C_p = heat capacity at constant pressure, Btu/lb·°F
Q = amount of heat transferred
W = work

For perfect gases, the entropy, enthalpy, and heat capacity at constant volume and at constant pressure are all functions of temperature. Change in pressure or volume at constant temperature will not change their values.

Tables of thermodynamic properties often list values for enthalpy (H), specific volume, and entropy (S) at constant pressure at several temperatures for the superheated gas. The change in enthalpy with temperature can easily be obtained from such tables. Since the tables are at constant pressure, values for PV can also be obtained (with some interpolation) and a value for E can be calculated from equation (3–8). Values of H and E are not absolute, but that is of little consequence since it is the changes in H and E that are of interest in refrigeration.

On the other hand, Q, the heat transferred, and W, the work done on or by a gas, are definite quantities, although the value depends on how the process is operated (i.e., isothermally, adiabatically, etc.).

Dodge lists the following general equations for changes of state for any fluid:

$$dQ = T\,dS = C_v\,dT + T\left(\frac{dP}{dT}\right)_v dV \tag{3–15}$$

$$dQ = T\,dS = C_p\,dT - T\left(\frac{dV}{dT}\right)_p dP \tag{3–16}$$

$$dQ = T\,dS = C_p\left(\frac{dT}{dV}\right)_p dV + C_v\left(\frac{dT}{dP}\right)_v dP \tag{3–17}$$

For the special case of an ideal gas and using the equation of state, $PV = RT$ (per mole), these equations reduce to

$$T\,dS = C_v\,dT + P\,dV \tag{3–18}$$

$$T\,dS = C_p\,dT - \frac{RT\,dP}{P} \tag{3–19}$$

$$T\,dS = \frac{C_pT\,dV}{\mathrm{V}} + \frac{C_vT\,dP}{P} \tag{3–20}$$

ISOTHERMAL CHANGE

For an isothermal (constant temperature) change, the work done with an ideal gas is

$$W = \int_{V_1}^{V_2} P\,dV = RT\int_{V_1}^{V_2}\frac{dV}{V} = -RT\int_{P_1}^{P_2}\frac{dP}{P} \tag{3–21}$$

or

$$W = -RT\ln\frac{P_2}{P_1} = -2.3026RT\log\frac{P_2}{P_1} \quad \text{(per mole)} \tag{3–22}$$

For a nonideal gas, the work can be calculated in the same way if an equation of state is known, by substituting it for the pressure. This calculation can be simple if a short equation of state is available, but as a rule equations for the fluorocarbon refrigerants tend to be more complicated, in an effort to better fit experimental data. One of the earliest attempts to represent the behavior of gases by an equation was by van der Waals in about 1873. He proposed the following equation:

$$P = \frac{RT}{V - B} - \frac{a}{V} \tag{3–23}$$

Using this equation, the work of compression becomes

$$W = \int_{V_1}^{V_2} P\,dV = RT\int_{V_1}^{V_2}\frac{dV}{V-b} - a\int_{V_1}^{V_2}\frac{dV}{V} \tag{3–24}$$

$$W = RT\ln\frac{V-b}{V-b} - a\left(\frac{1}{V_1} - \frac{1}{V_2}\right) \tag{3–25}$$

Isothermal work is the absolute minimum that must be used to compress a gas over a given pressure range. However, it is usually not practical to remove the heat fast enough to approach isothermal conditions. Rather, only a relatively small amount of heat is lost and compression is nearly adiabatic.

ADIABATIC CHANGE

In adiabatic compression there is no change in entropy and

$$T\,dS = 0$$

All of the heat developed in compressing the gas is retained and goes into raising the temperature of the gas. Although compression is not exactly at constant entropy, it is usually close. Some heat is lost by conduction through the walls of the compressor and some is gained from mechanical friction and the passage of the gas through small valve openings. If there is a departure from constant entropy, it is often in the direction of more heat lost than gained, so real gas discharge temperatures may be a bit lower than calculated assuming constant entropy.

The work of adiabatic compression can be calculated with the following relationships:

$$W = \frac{P_1 V_1}{k-1}\left(1 - \frac{P_2}{P_1}\right)^{(k-1)/k} \tag{3–26}$$

$$W = \frac{RT}{k-1}\left(1 - \frac{P_2}{P_1}\right)^{(k-1)/k} \tag{3–27}$$

In these equations, k is the heat capacity ratio, C_p/C_v, and is assumed to be constant during the compression. Although the ratio does vary with conditions, the change is usually quite small and an average value can be used.

The increase in temperature during compression can be calculated as follows:

$$T_2 = T_1\left(\frac{P_2}{P_1}\right)^{(k-1)/k} \tag{3–28}$$

It should be noted that the work of compression depends on the pressure ratio and not on the pressure level. For example, the work will be the same whether compressing from 100 psia to 1000 psia or from 0.1 psia to 1 psia.

The relationships above are based on the *ideal gas law*, $PV = RT$. Actual performance of a given refrigerant will be somewhat different and can be determined exactly only by measurement. For the ideal gas, $PV/RT = 1$, but for a real gas, $PV/RT = z$, where z is called the compressibility factor. The ideal quantities are useful as standards of performance and approximations of the real quantities. They can be converted to actual values by using experimental correction factors such as the compressibility.

CHARTS AND TABLES [3–3]

When evaluating a refrigerant, many general properties must be considered, such as toxicity, flammability, stability, relationships with water and oil, electrical properties, and so on. All of these properties have a direct bearing on the suitability of a fluid for use as a refrigerant. However, from the standpoint of performance, five physical or thermodynamic properties are especially important: temperature, volume, pressure, enthalpy, and entropy. Since the refrigeration cycle involves a change of state from liquid to gas and back to liquid, knowledge of these properties for both liquid and gas is important. "Vapor" and "gas" are sometimes used interchangeably to describe the same physical state. At other times, "vapor" is reserved for "gas" that is in equilibrium with a liquid phase, and "gas" is used for conditions where a liquid phase is not possible.

Temperature is a measure of how hot or cold an object is, but is not a measure of how much heat the object will hold or how much heat is needed to change from one temperature to another. Two temperature scales are in general use. The Celsius (formerly centigrade) scale is used in many parts of the world and in most "scientific" work. It is the official temperature unit in the SI system of measurement and in earlier practice with metric units. It is based on 0 degrees at the freezing point of water and 100 degrees at the atmospheric boiling point of water. The Fahrenheit scale is named after the German chemist Gabriel Fahrenheit and is based on 0 degrees for the coldest temperature then available with a mixture of salt and water and 100 degrees for normal body temperature. However, he must have had a slight fever on the day he fixed the upper temperature since normal body temperature is 98.6°F. It is perhaps fortuitous that a Celsius degree is exactly 1.8 times larger than a Fahrenheit degree.

Volume is a measure of the space occupied by a specific amount of liquid or gas with units of cubic feet per pound, cubic meters per kilogram, and so on. It is the reciprocal of the density.

Three different types of pressure may be encountered in refrigeration applications: vapor, gas, and hydrostatic. *Vapor pressure* is exerted when both liquid and gas are present and in equilibrium. It is affected only by changes in temperature, not by the amounts of liquid and gas present. It is thus said to have one *degree of freedom*. For a particular refrigerant there can be only one value for the vapor pressure at a given temperature. In a refrigeration system the pressure is determined by the temperature of a liquid phase somewhere. For example, the temperature of the liquid in the condenser governs the pressure in the compressor discharge lines subject to some pressure drop in the piping. Liquid temperatures in the evaporator determine the pressure on the suction side of the compressor.

When no liquid is present, the gas is said to have two degrees of freedom and both temperature and volume affect the pressure. At a given temperature *gas pressure* may have several different values, depending on the specific volume of the gas.

When a cylinder, container, pipe, or any enclosed space becomes completely filled with liquid, *hydrostatic pressure* develops. This pressure changes rapidly with temperature and can be extremely hazardous. For example, with R-12 the hydrostatic pressure increases about 40 psi for each 1-degree rise in temperature at room temperature. Great care should be used to avoid overfilling cylinders, trapping liquid between two shutoff valves in a line, or other conditions where a liquid-full condition could develop.

Enthalpy or *heat content* is a measure of how much heat or cold a fluid can hold and how much heat is needed to change the temperature. Absolute values of enthalpy are not usually known and are not necessary since only changes in enthalpy are important in the refrigeration cycle. An arbitrary value for the enthalpy is assigned at a given temperature—often a value of zero for the liquid at −40°F. Changes in enthalpy with changes in temperature, pressure, and volume can be calculated with thermodymically exact equations. One problem with a −40-degree reference point is the presence of negative enthalpies at temperatures lower than −40°F. This concept does not affect the mathematical solution of refrigeration problems but does add a little confusion to the understanding of refrigeration. In more recent tables of thermodynamic properties, the reference point has been selected to avoid the use of negative numbers.

Entropy is more difficult to define. It can be regarded as a measure of that portion of the heat energy transferred which is unavailable for work [3–1]. It is the ratio of the heat transferred to the absolute temperature. Entropy stays the same during compression if heat is not added to or taken from the gas. The compression is called *adiabatic* when the entropy is constant. Absolute values for the entropy are not precisely known and for refrigeration purposes are unnecessary. An arbitrary reference value is selected—usually zero for the liquid at −40 degrees and values at other conditions are calculated using thermodynamically developed equations. In order to eliminate negative numbers at temperatures below −40 degrees, current practice is to assign a value for entropy at a reference point so that negative numbers are not involved.

For refrigeration use, these five properties of fluids are provided in tables and charts as illustrated in Chapter 16.

Richard Mollier, a German professor after whom charts of this type are named, had a great interest in displaying as much information as possible on one chart. Most of his work was with the properties of steam. He is best known for an arrangement with enthalpy and entropy shown along the axes and other properties as families of curves, although his name is often associated with other, similar arrangements.

The chart most often used in refrigeration work has pressure on the vertical axis and enthalpy on the horizontal axis, as shown in Figure 3–1. This chart is for R-12 and will be used to illustrate the information that can be obtained from it. The curved line at the left in Figure 3–1 represents liquid at the saturation temperature. The vapor pressure can be read on the vertical scale and the heat content or enthalpy on the bottom scale at a number of different temperatures. The curve at the right represents saturated vapor in equilibrium with the liquid, such as in a cylinder, in a condenser, or in a flooded evaporator. All of the area to the right of this curve is for superheated gas where no liquid can be present.

Use of the chart to obtain the compression ratio is illustrated in Figure 3–1. An evaporation temperature of 0°F and a condensing temperature of 120°F are as-

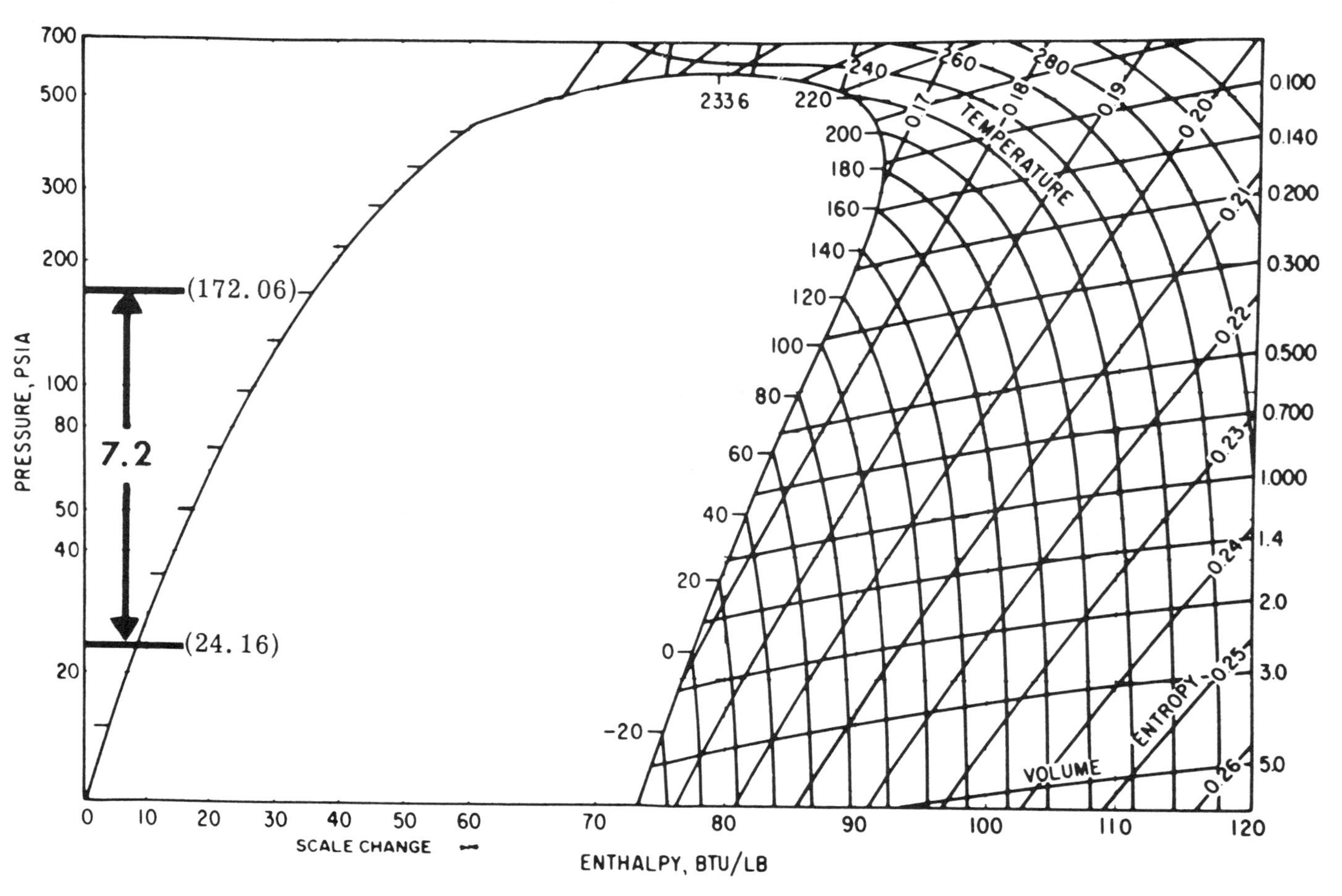

Figure 3–1 Pressure–enthalpy chart for R-12. (Reprinted by permission of the Refrigeration Service Engineers Society.)

sumed. The pressures corresponding to these temperatures can be read on the vertical scale. Dividing the higher pressure by the lower pressure will give the compression ratio. Note that the pressures must be absolute. In this case the ratio is a little more than 7 and can be handled easily by reciprocating compressors. Compression ratios of 12 to 15 and higher will be found with reciprocating compressors, especially in small units where operating costs may not be a large factor. In larger equipment, when ratios higher than about 10 are called for, a second stage of compression is often recommended. With rotary compressors, compression ratios up to about 5 are used and with centrifugal compressors up to about 3.

A typical refrigeration cycle is drawn on a pressure–enthalpy chart in Figure 3–2. An evaporating temperature of 0°F and condensing temperature of 120°F have been used in this illustration and refer to actual refrigerant temperatures rather than temperatures in the vicinity of the condensor or evaporator.

The latent heat or heat of vaporization is the amount of heat needed to change a liquid to a vapor at the same temperature. It is the difference between the enthalpy of the saturated vapor and the saturated liquid. These numbers can be read from the chart or with greater accuracy from tables of thermodynamic properties. In refrigeration the change from liquid to gas takes place in the evaporator and is the only place where any cooling occurs. All of the rest of the equipment is needed to turn the gas back into a liquid and get it back into the evaporator. In Figure 3–2 the latent heat at 0°F for R-12 is the enthalpy change represented by the line *AC* and amounts to 68.8 Btu/lb. It can be seen that when evaporation takes place at higher temperatures the latent heat becomes smaller and vanishes completely at the critical temperature of 233.6°F, where the liquid line and vapor line come together. No liquid phase can exist above the critical temperature. In the area above the critical temperature, to the right of the saturated vapor line and to the left of the saturated liquid line, only one phase exists. At higher pressures, higher than the critical pressure, the refrigerant could be called a light liquid or a dense gas. It does not really make any difference what the fluid is called—especially in refrigeration, where pressures never get that high (or at least should not).

Not all of the latent heat of vaporization is available for cooling the food in the refrigerator or the air in the air conditioner. Part of it must be used to reduce the temperature of the hot liquid coming from the condenser. Cooling of the liquid is represented by *EB* in Figure 3–2. Of course, heat changes in the evaporator do not occur in such neat, consecutive steps. The net change is found by the heat content of the vapor leaving the evaporator at *C* minus the heat content of the liquid entering the evaporator at *E* or leaving the condenser, assuming no heat change in the liquid line. The amount of heat to be extracted from the liquid to cool it from 120°F to 0°F is shown by the segment *AB*, or 27.5 Btu/lb. The rest of the latent heat, *BC*, or 41.3 Btu/lb, is available for cooling work and is called the *net refrigerating effect*.

Lines of constant volume for the gas are shown on the chart as nearly horizontal lines. Point *C* is shown as the outlet of the evaporator and the inlet to the compressor. In a real machine the temperature of the gas entering the compressor would be considerably higher than shown at *C*. However, the pressure would be substantially constant and would be the vapor pressure at 0°F. The volume of the gas actually

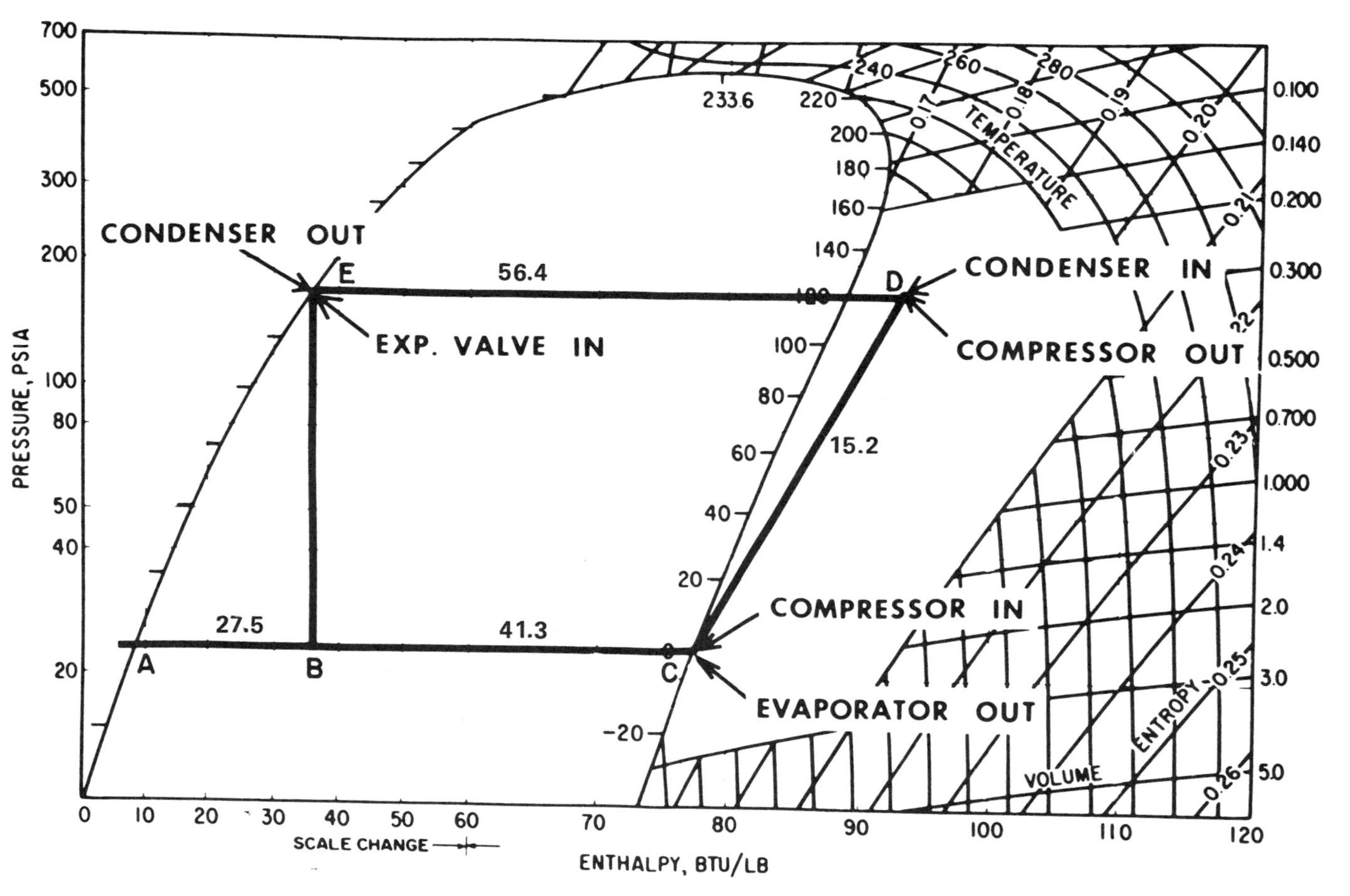

Figure 3–2 Pressure–enthalpy diagram for R-12 showing refrigeration cycle. (Reprinted by permission of the Refrigeration Service Engineers Society.)

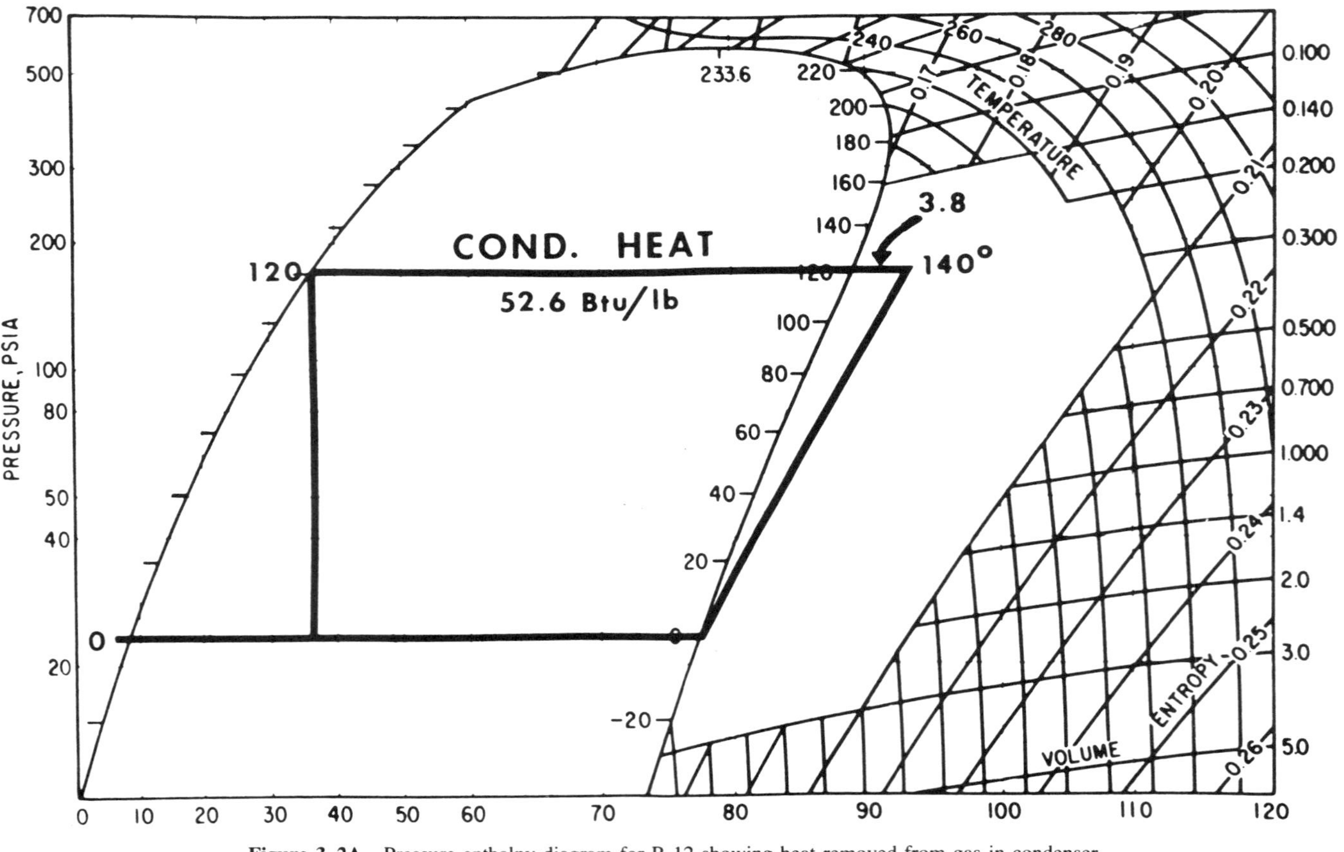

Figure 3–2A Pressure-enthalpy diagram for R-12 showing heat removed from gas in condenser.

entering the cylinder of the compressor depends on the temperature of the gas at that point. This temperature is difficult to measure since it is in the interior of the compressor and is considerably higher than the temperature at the entrance to the compressor shell. However, the volume change is small for small changes in temperature. The compressor inlet temperature would probably be between 80 and 100°F, but for this simplified cycle it will be considered as at point *C*. The volume of the saturated vapor at 0°F is about 1.6 Btu/lb.

The specific volume of the refrigerant gas entering the compressor cylinder can be used to calculate the load if the compressor displacement is known or the compressor size needed for a given capacity. For example, suppose that it is desired to develop 1 ton of refrigeration—or 12,000 Btu/hr or 200 Btu/min. The amount of cooling available is 41.3 Btu/lb, so to get 200 Btu/min, the refrigerant must go around the system at a rate of about 4.8 lb/min. With a specific volume for the refrigerant of 1.6 ft^3/lb, the volumetric displacement of the compressor must be about 7.8 ft^3/min or more.

Assuming that the refrigerant gas enters the cylinder of the compressor at *C* and that the compression is adiabatic or at constant entropy, the increase in enthalpy of the gas is represented by *CD* and is 15.2 Btu/lb. The point *D* is located by the pressure in the condenser and is the vapor pressure of the liquid at 120°F. The intersection of the constant-pressure line and the constant-entropy line fixes point *D*. This analysis ignores pressure drop in the discharge line from the compressor. A correction can be made by measuring the pressure at the compressor outlet or by calculating the pressure drop as described elsewhere. A correction can be made for departure from constant entropy by measuring the temperature of the gas at the compressor discharge. Without these corrections the temperature of the gas as it leaves the compressor at point *D* is 140°F.

The heat added to the gas during compression is given in Btu per pound in Figure 3–2. To convert to power, a time factor must be added. Multiplying the compressor heat in Btu/lb by the refrigerant circulated in lb/min, the power required to compress the gas is found to be 73.6 Btu/min. Some other units for expressing power are shown below. These values are the theoretical power needed to produce 1 ton of refrigeration with R-12 at 0°F evaporating and 120°F condensing temperatures.

73.6 Btu/min
1.73 hp
1293 W
954 ft·lb/sec

During the condensing of the refrigerant, shown in Figure 3–2 as *DE*, the pressure remains at the vapor pressure of the liquid at 120°F. Some heat must be removed from the gas before condensation begins; in the example, this is 3.8 Btu/lb. As more heat is removed, more and more liquid is formed until the liquid line is reached at *E*. The latent heat of condensation is 52.6 Btu/lb. The amount of superheat in the gas going to the condenser is important because heat transfer from a gas is much poorer than with a condensing liquid. Several times as much condenser surface must be provided for removing superheat as for condensing liquid. If the condenser is on the small side, the condensing temperature and pressure will be higher and more power will be required to compress the refrigerant.

One of the uses of the pressure–enthalpy chart is to compare different operating conditions. Suppose that the condenser is at 100°F instead of at 120°F, as illustrated in Figure 3–3. Perhaps the condenser has been cleaned or the ambient air temperature is lower or a larger condenser is used. What will happen to the capacity and the horsepower needed for compression? Although the latent heat of evaporation remains the same, the net refrigeration available is increased because the liquid entering the evaporator requires less cooling. Less compressor heat and power are needed because the compression ratio is lower. The discharge temperature would be lower and the amount of superheat in the gas would be slightly less. The total amount of heat removed in the condenser, however, would be a little greater.

The effect of subcooling the liquid before it enters the evaporator is illustrated in Figure 3–4. Subcooling means removing heat from the liquid without a change in pressure. If the condenser is large enough, subcooling may occur in the bottom rows of the condenser. In order to be helpful, the heat must be removed by some means other than evaporation of the liquid. In the example in Figure 3–4, reducing the temperature of the liquid by 20°F will increase the net refrigeration and also the capacity by more than 10%. The compression power, however, will stay the same since the pressure in the condenser will remain at the vapor pressure at 120°F.

Earlier it was assumed that saturated refrigerant vapor left the evaporator and immediately entered the compressor cylinder. However, this is not ordinarily a very realistic or desirable condition. Some superheat is needed to ensure that liquid refrigerant does not enter the compressor. The effect of superheat is illustrated in Figure 3–5. Assume that the refrigerant is superheated 65°F—in this case the gas will be at 65°F since the evaporating temperature is 0°F. The pressure does not change because it is controlled by the vapor pressure of the evaporating liquid. The increase in enthalpy is represented by *CK* in Figure 3–5. If all of the enthalpy change occurs outside the evaporator, there will be no change in the net refrigerating effect and the capacity will remain the same. However, if part of the increase in enthalpy takes place inside the evaporator, that amount of enthalpy would be added to the net effect and the capacity would be increased proportionally. In either case, the gas entering the compressor will be at a different entropy and somewhat more heat and power will be needed for compression, as illustrated in Figure 3–5. The discharge temperature will also be considerably higher. Often, some heat is taken from the hot liquid and added to the cold suction vapor with direct benefit from subcooling the liquid. This can be done with a formal heat exchanger or by soldering together the liquid and suction lines for a short distance. Of course, adding heat to the suction gas leads to an increase in power requirements and higher discharge temperatures, so the gain must be balanced against the loss. The nature of the refrigerant is also important. For example, more will be gained by using a liquid–vapor heat exchanger with R-502 than with R-12 or R-22.

Pressure–enthalpy charts are tools to use in better understanding refrigeration

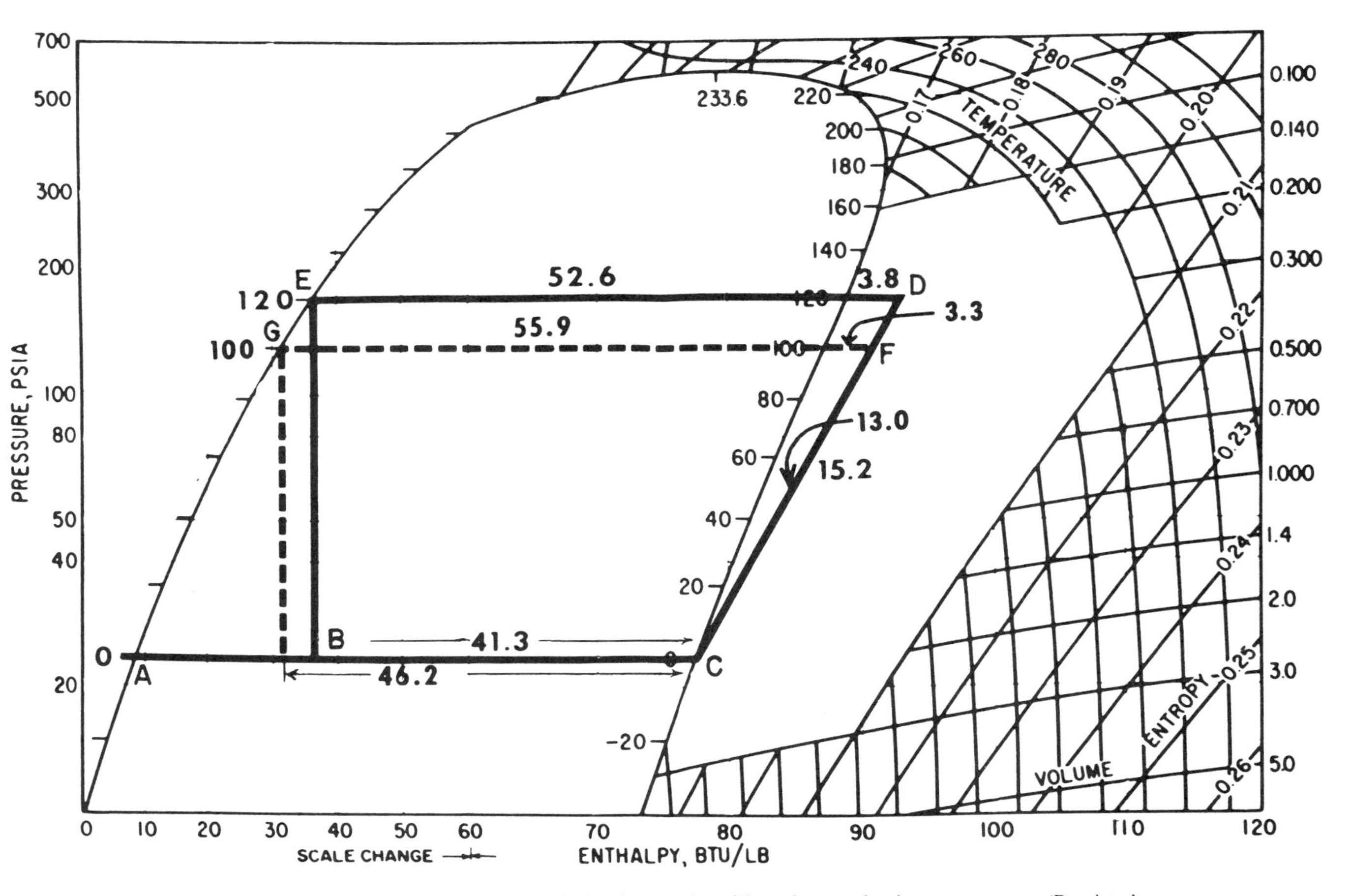

Figure 3–3 Diagram for R-12 showing results of lowering condensing temperature. (Reprinted by permission of the Refrigeration Service Engineers Society.)

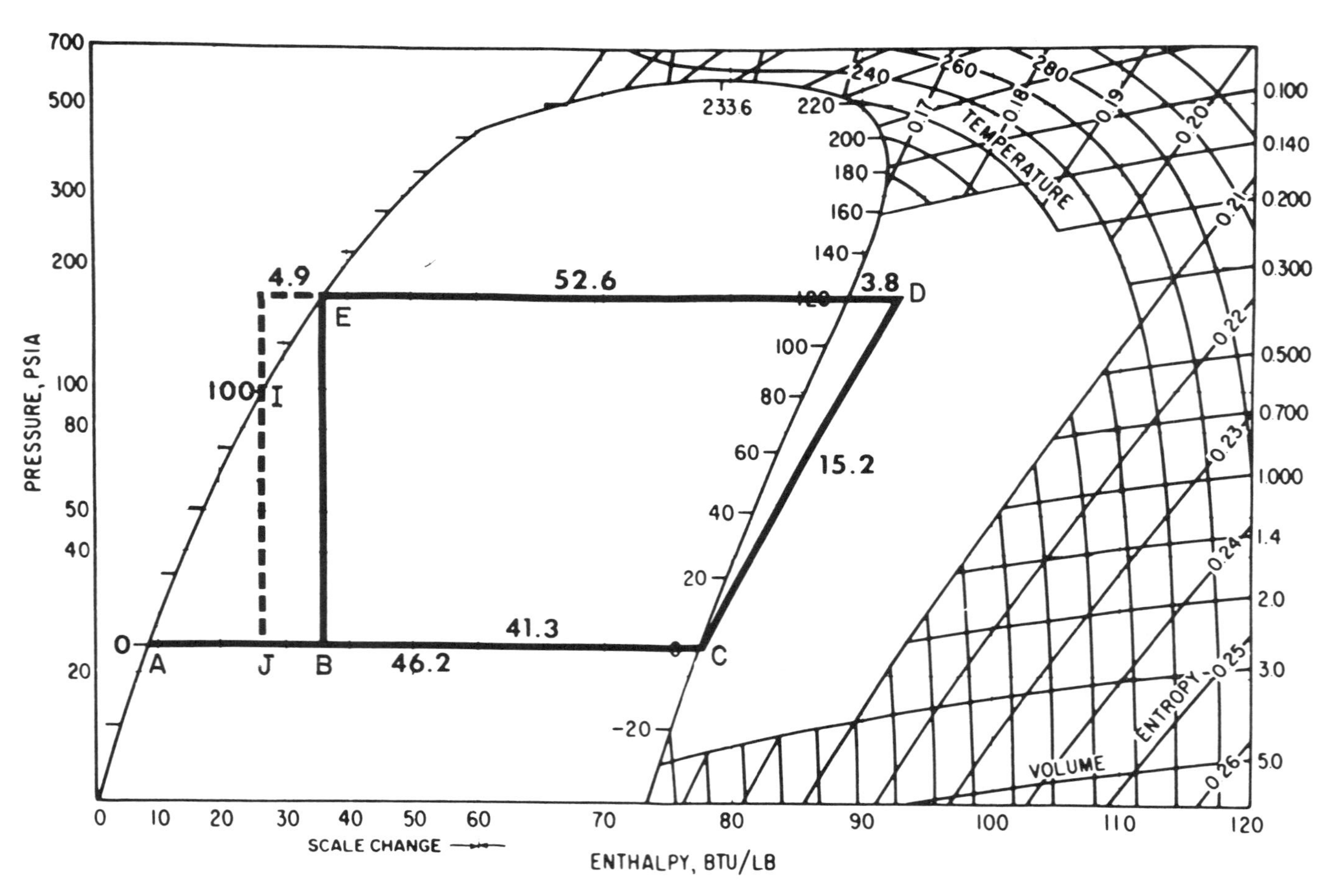

Figure 3–4 Pressure–enthalpy diagram for R-12 showing effect of liquid subcooling. (Reprinted by permission of the Refrigeration Service Engineers Society.)

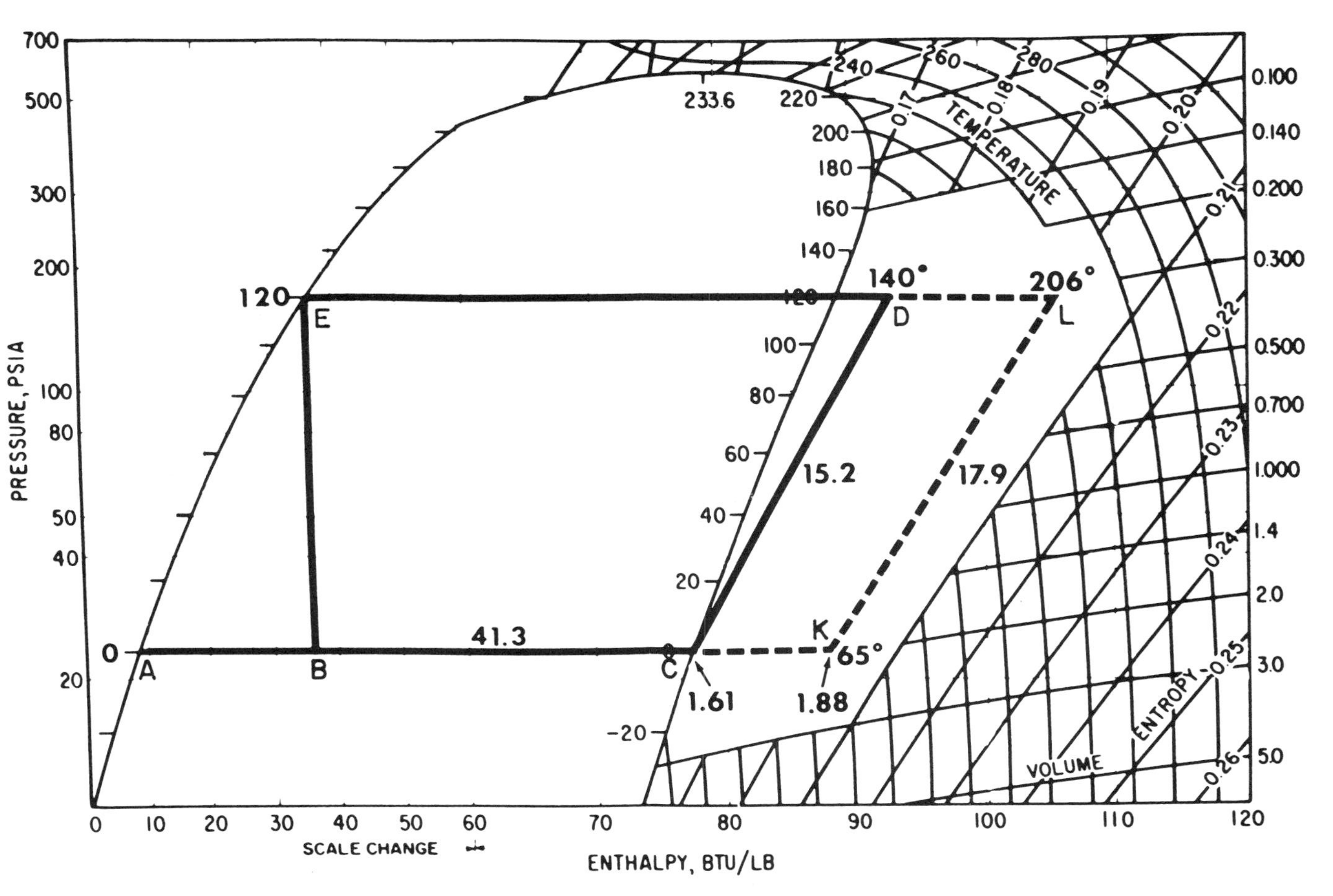

Figure 3–5 Pressure–enthalpy diagram for R-12 showing effect of suction superheating. (Reprinted by permission of the Refrigeration Service Engineers Society.)

cycles, for judging whether the operation is normal, and for estimating the effect of changes in operation. Learning how to read the charts and to do the calculations involved will pay rich dividends in all phases of refrigeration work. Examples of refrigeration calculations are shown in Table 3–1.

TABLE 3–1 REFRIGERATION CALCULATIONS

1. Net refrigerating effect (Btu/lb)
$$= \begin{bmatrix}\text{heat content of}\\ \text{vapor leaving}\\ \text{evaporator (Btu/lb)}\end{bmatrix} - \begin{bmatrix}\text{heat content of}\\ \text{liquid entering}\\ \text{evaporator (Btu/lb)}\end{bmatrix}$$

2. Net refrigerating effect (Btu/lb)
$$= \begin{bmatrix}\text{latent heat of}\\ \text{vaporization}\\ \text{(Btu/lb)}\end{bmatrix} - \begin{bmatrix}\text{change in heat content}\\ \text{of liquid from condens-}\\ \text{ing to evaporating}\\ \text{temperature (Btu/lb)}\end{bmatrix}$$

3. Net refrigerating effect (Btu/lb)
$$= \frac{\text{capacity (Btu/min)}}{\text{refrigerant circulated (lb/min)}}$$

4. Refrigerant circulated (lb/min)
$$= \frac{\text{load or capacity (Btu/min)}}{\text{net refrigerating effect (Btu/lb)}}$$

5. Compressor displacement (ft^3/min)
$$= \begin{bmatrix}\text{refrigerant}\\ \text{circulated}\\ \text{(lb/min)}\end{bmatrix} \times \begin{bmatrix}\text{volume of gas}\\ \text{entering com-}\\ \text{pressor (ft}^3\text{/lb)}\end{bmatrix}$$

6. Compressor displacement (ft^3/min)
$$= \frac{\begin{bmatrix}\text{capacity}\\ \text{(Btu/min)}\end{bmatrix} \times \begin{bmatrix}\text{volume of gas}\\ \text{entering com-}\\ \text{pressor (ft}^3\text{/lb)}\end{bmatrix}}{\text{net refrigerating effect (Btu/lb)}}$$

7. Heat of compression (Btu/lb)
$$= \begin{bmatrix}\text{heat content of}\\ \text{vapor leaving}\\ \text{compressor (Btu/lb)}\end{bmatrix} - \begin{bmatrix}\text{heat content of}\\ \text{vapor entering}\\ \text{compressor (Btu/lb)}\end{bmatrix}$$

8. Heat of compression (Btu/lb)
$$= \frac{\text{(42.418 Btu/min) (compression horsepower)}}{\text{refrigerant circulated (lb/min)}}$$

9. Compression work (Btu/min)
$$= \begin{bmatrix}\text{heat of com-}\\ \text{pression (Btu/lb)}\end{bmatrix} \times \begin{bmatrix}\text{refrigerant cir-}\\ \text{culated (lb/min)}\end{bmatrix}$$

10. Compression horsepower
$$= \frac{\text{compression work (Btu/min)}}{\text{conversion factor (42.418 Btu/min)}}$$

11. Compression horsepower
$$= \frac{\begin{bmatrix}\text{heat of com-}\\ \text{pression (Btu/lb)}\end{bmatrix} \times \begin{bmatrix}\text{capacity}\\ \text{(Btu/min)}\end{bmatrix}}{\text{(42.418 Btu/min)} \times \begin{bmatrix}\text{net refriger-}\\ \text{ating effect}\\ \text{(Btu/lb)}\end{bmatrix}}$$

12. Compression horsepower
$$= \frac{\text{capacity (Btu/min)}}{\text{(42.418 Btu/min)} \times \begin{bmatrix}\text{coefficient}\\ \text{of performance}\end{bmatrix}}$$

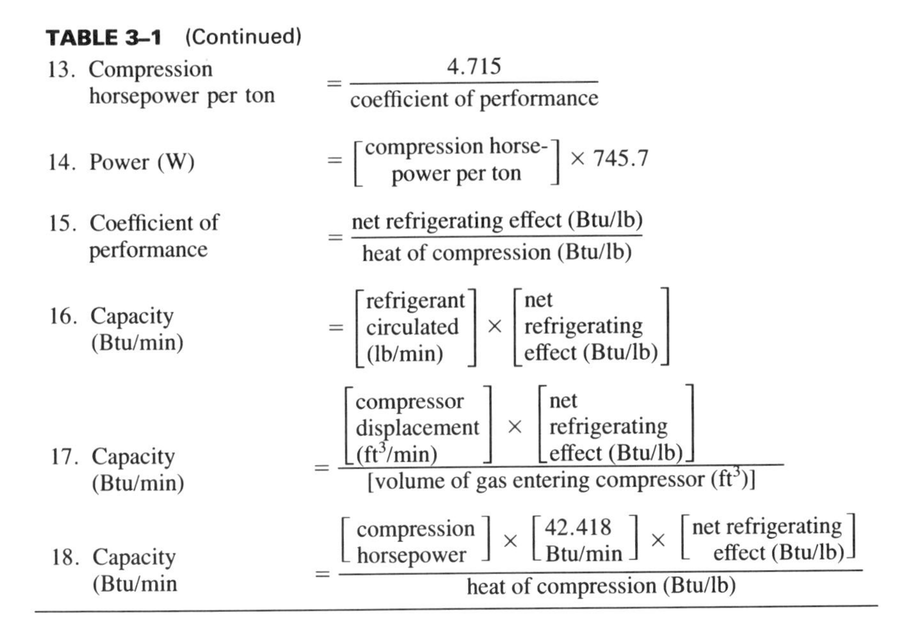

TABLE 3–1 (Continued)

13. Compression horsepower per ton $= \dfrac{4.715}{\text{coefficient of performance}}$

14. Power (W) $= \left[\begin{array}{c}\text{compression horse-}\\ \text{power per ton}\end{array}\right] \times 745.7$

15. Coefficient of performance $= \dfrac{\text{net refrigerating effect (Btu/lb)}}{\text{heat of compression (Btu/lb)}}$

16. Capacity (Btu/min) $= \left[\begin{array}{l}\text{refrigerant}\\ \text{circulated}\\ \text{(lb/min)}\end{array}\right] \times \left[\begin{array}{l}\text{net}\\ \text{refrigerating}\\ \text{effect (Btu/lb)}\end{array}\right]$

17. Capacity (Btu/min) $= \dfrac{\left[\begin{array}{l}\text{compressor}\\ \text{displacement}\\ (\text{ft}^3/\text{min})\end{array}\right] \times \left[\begin{array}{l}\text{net}\\ \text{refrigerating}\\ \text{effect (Btu/lb)}\end{array}\right]}{[\text{volume of gas entering compressor } (\text{ft}^3)]}$

18. Capacity (Btu/min $= \dfrac{\left[\begin{array}{l}\text{compression}\\ \text{horsepower}\end{array}\right] \times \left[\begin{array}{l}42.418\\ \text{Btu/min}\end{array}\right] \times \left[\begin{array}{r}\text{net refrigerating}\\ \text{effect (Btu/lb)}\end{array}\right]}{\text{heat of compression (Btu/lb)}}$

REFRIGERANTS

The operation of a refrigeration system depends on two principal parts: the compressor and the refrigerant. Other parts of the system, such as the condenser, evaporator, receiver, oil separator, and connecting piping, are also important and must be properly sized, located, and installed for best results. The compressor is the real heart of the system and the only part where mechanical movement occurs—other than automatic expansion valves. Compressor design must consider clearance volume,* heat losses by radiation and other means, compression ratios that might be encountered, expected life, cost, and similar factors, including the nature of the refrigerant.

For many applications, any of two or more refrigerants may be suitable. The decision regarding the refrigerant may be based on precedent, stability, toxicity, oil and water solubility, flammability, cost, and so on, and such criteria are indeed important and must be satisfied by the final refrigerant selection. However, it is also interesting, entertaining, and occasionally enlightening to compare refrigerants strictly on the basis of performance.

W. A. Pennington at the University of Maryland reviewed some refrigerant comparisons [3–5] and suggested a plot of evaporator pressure versus capacity as a correlating device. A straight line is obtained with a number of refrigerants, assuming the same volumetric displacement and the same evaporating and condensing temperatures. Such a comparison is shown in Figure 3–6. In this comparison both the capacity and evaporator pressure are given as percentages of R-12 since a direct

*In a reciprocating compressor the space between the top of the cylinder and the top of the piston at the apex of its stroke is called the clearance volume. Gas in this volume is compressed but not expelled from the cylinder, so compression work is required without any gain in refrigeration.

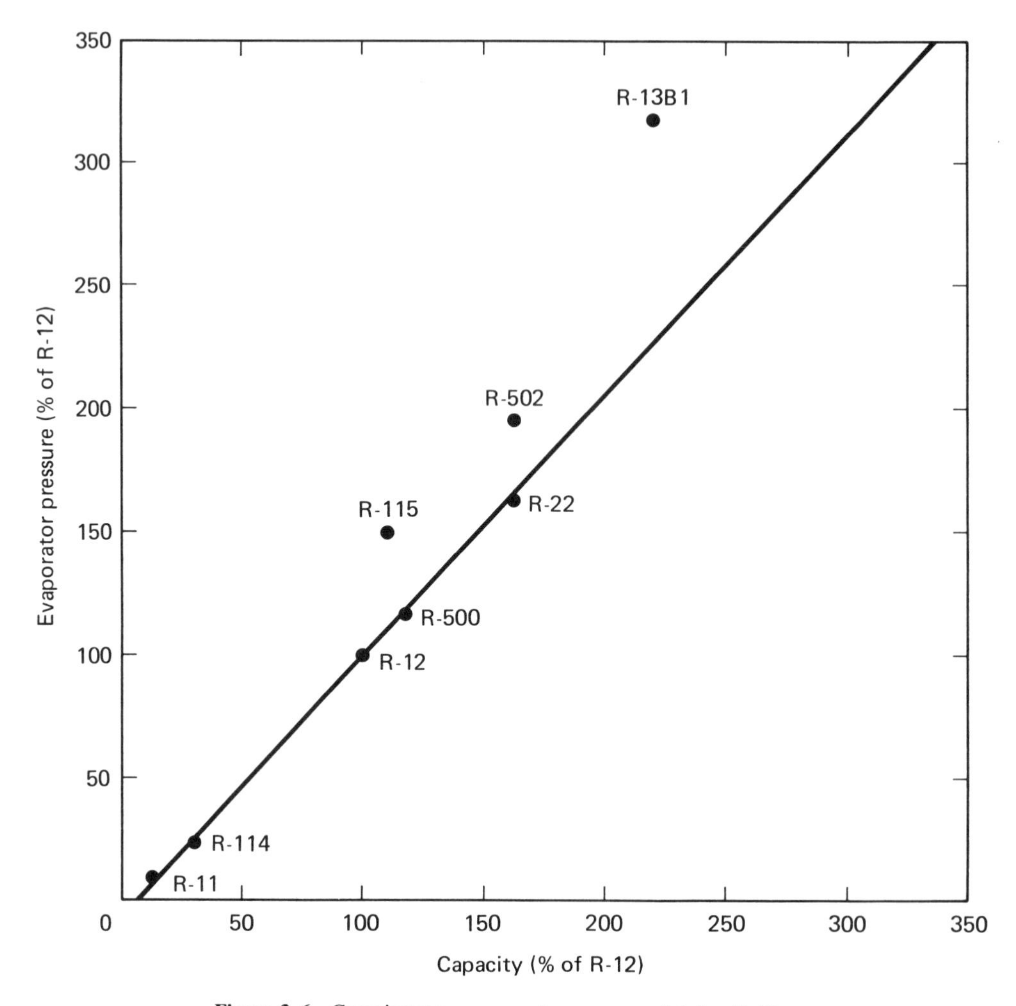

Figure 3–6 Capacity versus evaporator pressure related to R-12.

comparison would produce a different curve for each different refrigeration cycle. The line in Figure 3–6 is largely independent of a specific evaporator pressure and capacity. The points on the line are calculated with the following assumptions:

1. The evaporator temperature is −20°F.
2. The temperature of the gas leaving the evaporator and entering the compressor is 65°F.
3. The condensing temperature is 110°F.
4. Pressure drops in lines, and so on, are neglected.

Five refrigerants form a very nice straight line, but three are somewhat removed and produce less capacity than would be expected from the common line. A compressor displacement of 169 ft^3/hr is assumed for all refrigerants. Including superheated gas to 65°F in the evaporator gives a superheat of 85°F and is more than would ordinarily be found in an operating system. However, such a condition seems valid for the sake of comparison, although it may tend slightly to benefit gases with higher vapor heat capacity, such as R-115 and R-502. A temperature of 65°F for the gas at the beginning of compression is often assumed and seems to allow somewhat for departures from adiabatic compression and loss of heat by radiation or in other ways. The temperature of the gas actually entering the cylinder of the compressor is higher than 65°F but is seldom known since measurement is difficult. The following relationships are used in calculating the capacity:

$$\frac{\text{specific volume (ft}^3\text{/lb)}}{\text{net refrigeration effect (Btu/lb)}} = \text{ft}^3\text{/Btu}$$

$$\frac{\text{displacement (ft}^3\text{/hr)}}{\text{ft}^3\text{/Btu}} = \text{capacity (Btu/hr)}$$

To help explain the behavior of refrigerants that do not fit on the common line in Figure 3–6, Pennington invented an expression which he called the *latent heat efficiency*, L:

$$L = \frac{\text{net refrigeration effect} \times 100}{\text{total latent heat} + \text{superheat in the evaporator}}$$

An inspection of a pressure–enthalpy diagram shows that the latent heat of vaporization decreases as the temperature increases. The saturated liquid line and the saturated vapor line approach each other and join at the critical temperature. At this point the latent heat is zero. At temperatures near the critical, a significant portion of the latent heat is used up to cool the liquid, leaving less for useful refrigeration. When the latent heat efficiency is low, the refrigeration capacity available with a given refrigerant is less than would be expected based on the pressure in the evaporator. These comparisons are illustrated in Table 3–2.

TABLE 3–2 REFRIGERANT COMPARISONS[a]

R	Capacity (Btu/hr)	Latent heat[b] (Btu/lb)	Net refrigerating effect[b] (Btu/lb)	Latent heat efficiency	Critical properties	
					Temperature (°F)	Pressure (psia)
115	3114	65.85	33.73	51	176	453
13B1	6167	55.89	30.41	54	153	575
502	4553	84.34	48.40	57	180	591
114	828	76.33	46.16	60	294	473
12	2802	82.93	53.64	65	234	597
500	3294	99.81	64.77	65	222	642
22	4502	110.36	73.04	66	205	723
11	375	96.55	69.80	72	388	640

[a]Evaporator temperature, −20°F; condensing temperature, 110°F; evaporator outlet and compressor inlet temperature, 65°F.

[b]Including superheat to 65°F.

The latent heat efficiency is certainly a measure of difference in refrigerants and does help explain exceptions to linearity in Figure 3–6. However, it is probably not the whole story. Some anomalies, such as R-114 with a fairly high critical temperature but a rather low efficiency, and R-22 with the reverse, need further explanation.

Refrigerants that are not on the common line in Figure 3–6 are not necessarily "bad" refrigerants. The data for Figure 3–6 are based on measured properties of the refrigerants, but the relationships are calculated and actual performance may be somewhat different. Also, properties other than capacity may be of prime significance. For example, from Figure 3–6, R-502 would appear to produce about the same capacity as R-22 and less than would be indicated by its vapor pressure. However, in a calorimeter study [3–4], R-502 was found to give 12% more capacity than R-22 at the same operating conditions. R-502 is widely used in low-temperature refrigeration because low compressor discharge temperatures yield mild operating conditions, less maintenance, and longer compressor life. These factors are very important in grocery store applications, for example.

Pennington graphically demonstrated that the compression ratio for a given refrigeration cycle tends to increase linearly with the atmospheric boiling point of the refrigerant. He also showed the linear relationship of the latent heat of vaporization versus the boiling point, a restatement of *Trouton's rule*, which says that the molar latent heat of any compound divided by the atmospheric boiling point on the absolute scale is a constant. For the refrigerants this ratio is not exactly constant, but differences do not seem to help in explaining refrigeration performance. Trouton's rule does point out that molar quantities might be better for comparing refrigerants than amounts based on weight. For example, ammonia and R-500 have about the same boiling point, −28°F. The latent heat at the boiling point is 589 Btu/lb for ammonia and 86 Btu/lb for R-500, yet the capacity for the cycle used in Table 3–2 is 4509 Btu/hr for ammonia and 3294 Btu/hr for R-500. The latent heat efficiency for ammonia is quite high—76.9, compared with 60.1 for R-500.

The *coefficient of performance* (COP) has been a popular correlating property. Gosney [3–6] showed some regularity between COP and critical temperature for a number of refrigerants, without any clearly defined trends. Barger et al. [3–7] related the COP to the evaporator temperature in a series of curves with condenser temperature as a parameter and offer some suggestions for selecting the optimum refrigerant for a given cycle. The reader is referred to the reference for details since the explanation is a little involved. Pennington [3–5] found a broad correlation between normal boiling point and coefficient of performance.

Most comparisons of refrigerants are based on calculated data and some assumptions and are helpful in judging the effect of different operating conditions and estimating performance. However, real measurements would be better and some have been made. In 1962, McHarness and Chapman published the results of calorimeter studies with several refrigerants [3–4]. Similar data may be stored in the files of testing laboratories and equipment manufacturers but are not readily available to the public. A considerable amount of performance data, however, has been published and is exceedingly valuable in selecting compressors and other equipment and for refrigerant comparisons. However, the McHarness and Chapman study is rather unique in the number of refrigerants tested and measurements recorded.

A schematic diagram of the calorimeter showing where temperatures and pressures were measured is given in Figure 3–7. Of special interest are the temperature measurements at T5 and T6, where thermocouples were located in the suction and discharge gas streams just $\frac{1}{16}$ in. from the valve reeds. The compressor is described as semihermetic with a 1.5-hp 1750-rpm 230-V, 60-cycle single-phase motor and a two-cylinder compressor with a calculated displacement of 169 ft^3/hr. Tests were run at evaporating temperatures of −20°F, −10°F, and +40°F with a condensing temperature of 110°F in all cases. Some of the findings are recorded in Table 3–3.

The measured capacity is compared with calculated data in Table 3–4. The percentages are sometimes called the *compressor efficiency* and seem about right for a small compressor of this type. Obviously, the compressor is more efficient at higher evaporating temperatures. The suction temperature is that of the gas measured at T5 in Figure 3–7, just as it enters the cylinder of the compressor. The gas volume and entropy at this point were found from tables of thermodynamic properties assuming that the pressure is still the vapor pressure of the liquid at the evaporating temperature. Adiabatic compression is assumed. Since the actual temperature of the gas as compression begins is not easy to measure and is seldom known, a temperature of 65°F is often used for estimating properties during compression. Such an assumption in this case would predict a higher capacity than found. Whether or not this difference would be found with other compressors and other systems is uncertain. However, in the absence of other data, applying the correction factors shown in Table 3–4 would probably be reasonable.

Compressor discharge temperatures are listed in Table 3–5. The measured values are from a thermocouple in the gas stream $\frac{1}{16}$ in. from the valve reed. The calculated values assume adiabatic compression using entropy data from the thermodynamic tables at the vapor pressure of the evaporator—in one case at a temperature of 65°F and in the other at the temperature of the gas just before it enters the cylinder of the compressor. The calculated temperatures starting at 65°F are all below the measured data, indicating that (at least in this example) the assumption of 65°F as the temperature of the inlet gas is not valid. Using the measured inlet

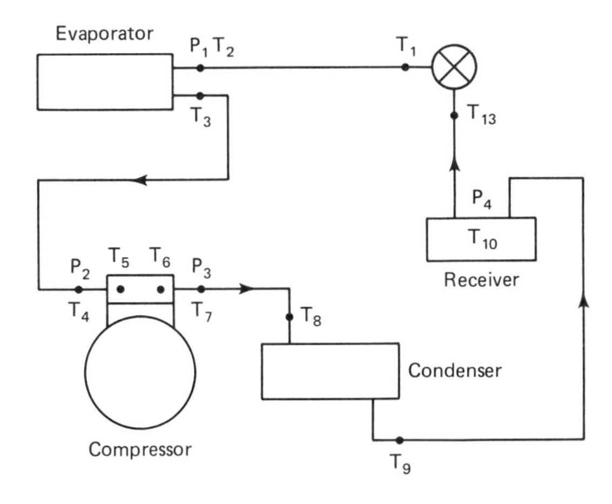

Figure 3–7 Schematic diagram of calorimeter.

TABLE 3–3 PERFORMANCE OF REFRIGERANTS IN CALORIMETER STUDIES[a]

	Evaporating temperature (°F)	Refrigerant				
		12	22	115	500	502
Measured refrigerating capacity (Btu/hr)	−20	1250	2360	1675	1525	2650
Coefficient of performance						
Actual		0.70	0.97	0.74	0.72	0.95
Indicated[b]		1.66	1.67	1.35	1.40	1.49
Theoretical		2.50	2.30	1.93	2.46	2.28
Compression ratio		9.88	9.74	9.15	10.07	8.88
Pressure (psia)						
Suction		15.3	25.0	22.8	17.8	29.8
Discharge		151	243	209	180	265
Temperature (°F)						
Suction		134	142	102	130	116
Discharge		236	282	194	238	238
Motor input (W)		520	715	665	620	820
Evaporator input (W)		366	692	491	447	776
Measured refrigerating capacity (Btu/hr)	−10	1950	3455	2505	2375	3780
Coefficient of performance						
Actual		0.92	1.23	0.98	1.00	1.20
Indicated[b]		1.76	1.94	1.63	1.77	1.77
Theoretical		2.82	2.61	2.49	2.75	2.58
Compression ratio		7.90	7.78	7.34	7.99	7.17
Pressure (psia)						
Suction		19.2	31.3	28.4	22.5	36.9
Discharge		151	243	209	180	265

Temperature (°F)						
Suction		125	128	96	119	105
Discharge		233	272	187	231	226
Motor input (W)		625	823	750	693	925
Evaporator input (W)		572	1013	737	696	1108
Measured refrigerating capacity, (Btu/hr)	40	8660	13,660	10,040	10,155	14,090
Coefficient of performance						
Actual		2.86	3.28	2.72	3.08	3.13
Indicated[b]		4.34	4.35	3.77	4.47	4.05
Theoretical		5.91	5.64	4.83	5.80	5.42
Compression ratio		2.93	2.90	2.79	2.91	2.77
Pressure (psia)						
Suction		51.7	84.0	74.7	61.7	95.7
Discharge		151	243	209	180	265
Temperature (°F)						
Suction		86	86	75	84	79
Discharge		170	191	139	168	162
Motor input (W)		885	1220	1080	965	1320
Evaporator input (W)		2537	4003	2942	2975	4129

[a]Condensing temperature, 110°F; temperature of vapor leaving evaporator and entering compressor, 65°F.

[b]Indicated COP $= \dfrac{\text{evaporator input (W)}}{\text{(motor wattage at load conditions)} - \text{[no-load conditions (300 W)]}}$.

Source: Ref. 3–4.

TABLE 3–4 CAPACITY: CALCULATED VERSUS MEASURED[a]

		Capacity (Btu/hr)				
			Calculated			
R	Evap. temp. (°F)	Measured	65°F[b]	Percent[c]	At suction temperature[d]	Percent[c]
12	−20	1250	3042	41	2666	47
22		2360	4842	49	4177	57
115		1675	3594	47	3407	49
500		1525	3548	43	3129	49
502		2650	5042	53	4551	58
				Av. 47		Av. 52
12	−10	1950	3833	51	3407	57
22		3455	6094	57	5377	64
115		2505	4538	55	4245	59
500		2375	4469	53	4013	59
502		3780	6282	60	5778	65
				Av. 56		Av. 61
12	40	8660	10,584	82	10,053	86
22		13,660	16,771	81	15,887	86
115		10,040	12,144	83	11,785	85
500		10,155	12,377	82	11,793	86
502		14,090	16,571	85	15,935	88
				Av. 83		Av. 86

[a]Condensing temperature, 110°F.

[b]Assuming that the gas enters the compressor cylinder at 65°F.

[c]Measured capacity divided by calculated capacity × 100.

[d]Assuming that the gas enters the compressor cylinder at the measured suction temperature.

gas temperatures, the calculated discharge temperatures are higher than those actually found for the −20°F and −10°F evaporator cycles. This would normally be the case since it is likely that some heat will be lost during compression. It would be difficult to account for a gain in the heat content of the gas necessary to cause a rise in temperature greater than for truly adiabatic compression. For the 40°F evaporator test the calculated and measured temperatures are about the same (except for R-115).

Both measured and calculated compressor discharge temperatures for R-502 are substantially lower than those for R-22. This difference would probably increase at higher inlet temperatures and is the principal reason that R-502 has largely replaced R-22 in supermarket frozen-food display cases, where long suction lines and higher temperatures are common.

Heat capacity is the relationship between enthalpy and temperature—how much heat it takes to raise the temperature 1 degree or how many degrees the temperature will change for a given amount of heat. The work of McHarness and Chapman provides an opportunity to study these relationships in a compressor by measurements of temperature as a refrigerant enters the compressor shell and as it enters the compression cylinder. Some comparisons are given in Table 3–6. The operation of

TABLE 3–5. COMPRESSOR DISCHARGE TEMPERATURES[a]

R	Evap. temp. (°F)	Discharge temperatures (°F) Measured	Calculated 65°F	Calculated Suction temperature
12	−20	236	225	301
22		282	273	358
115		194	173	208
500		238	229	299
502		238	216	269
12	−10	233	210	275
22		272	252	321
115		187	163	193
500		231	214	271
502		226	202	243
12	40	170	144	165
22		191	166	187
115		131	125	151
500		168	147	165
502		162	143	156

[a]Condensing temperature, 110°F.

TABLE 3–6 HEAT CAPACITY PER POUND[a]

	Temperature (°F)			Enthalpy			Heat capacity at constant pressure (Btu/lb·°F)	
R	Evap. temp. (°F)	(1) Entering compressor cylinder	(2) Rise in shell	(3) At 65°F	(4) At (1)	(5) Rise in shell	(6) Exptl.	(7) Charts
12	−20	134	69	87.169	97.501	10.332	0.150	0.146
22		142	77	155.488	128.016	12.528	0.163	0.158
115		102	37	70.354	76.886	6.533	0.177	—
500		130	65	104.44	116.08	11.64	0.179	
502		116	51	89.030	97.630	8.601	0.169	0.166
12	−10	125	60	86.984	95.991	9.007	0.150	0.147
22		128	63	115.187	125.464	10.277	0.163	0.160
115		96	31	70.119	75.643	5.524	0.178	
500		119	54	104.19	113.878	9.688	0.179	
502		105	40	88.756	95.519	6.763	0.169	0.169
12	40	86	21	85.373	88.676	3.303	0.157	0.158
22		86	21	112.541	116.214	3.673	0.175	0.177
115		75	10	68.107	70.017	1.909	0.191	
500		84	19	101.935	105.564	3.629	0.191	
502		79	14	86.373	88.899	2.526	0.180	0.180

[a]Condensing temperature, 110°F; temperature of gas entering compressor shell, 65°F.

the calorimeter was adjusted so that the temperature of all the refrigerants studied was 65°F just outside the compressor (T4 in Figure 3–7). Column 1 in Table 3–6 shows the measured temperature as the gas enters the cylinder and column 2 the increase in temperature inside the compressor shell—from 37°F for R-115 to 77°F for R-22. The enthalpy at 65°F and at the compressor cylinder—both at the pressure of the evaporating refrigerant—are columns 3 and 4 and the change is column 5. Dividing column 5 by column 2 gives the heat capacity at constant pressure:

$$\frac{\Delta H \text{ (Btu/lb)}}{\Delta T \text{ (°F)}} = C_p \text{ (Btu/lb·°F)}$$

These measured values of C_p are listed in column 6 and can be compared with the heat capacity in column 7, obtained from charts shown elsewhere. The agreement is very good. This heat capacity gives the number of Btu needed to raise the temperature of 1 pound of refrigerant 1 degree Fahrenheit. In this property, the refrigerants are not so very far apart—varying from 0.150 for R-12 to 0.179 for R-500 at −20°F evaporating.

When refrigerant flow rate is considered, the relationship of heat and temperature is altered as outlined in Table 3–7. The circulation rate can be calculated from the net refrigeration effect and the measured capacity:

$$\text{Btu/min} \times \text{lb/Btu} = \text{lb/min}$$

The flow rate in terms of weight is higher for some refrigerants than for others partly because of differences in molecular weight and partly from differences in the net refrigeration effect.

TABLE 3–7 HEAT CAPACITY PER MINUTE[a]

			Heat capacity			Net refrigeration	
R	Evap. temp. (°F)	Flow rate (lb/min)	Per pound (Btu/lb·°F)	Per minute (Btu/min·°F)	R-12 (%)	Btu/lb	R-12 (%)
12	−20	0.828	0.150	0.124	100	53.6	100
22		0.953	0.163	0.155	125	73.0	136
115		1.684	0.177	0.298	240	33.7	63
500		0.805	0.179	0.144	116	64.8	121
502		1.567	0.169	0.265	214	48.4	90
12	−10	1.062	0.150	0.159	100	53.5	100
22		1.232	0.163	0.201	126	72.7	136
115		2.113	0.178	0.376	236	33.5	63
500		1.037	0.179	0.186	117	64.5	121
502		2.001	0.169	0.338	213	48.1	90
12	40	3.232	0.157	0.507	100	51.8	100
22		3.777	0.175	0.661	130	70.1	135
115		6.240	0.191	1.192	235	31.5	61
500		3.157	0.191	0.603	119	62.3	120
502		5.973	0.180	1.075	212	45.7	88

[a]Condensing temperature, 110°F; temperature of gas entering compressor shell, 65°F.

From Table 3–7 it can be seen that R-115 and R-502 can absorb considerably more heat than other refrigerants for each degree rise in temperature. Not only will motors in hermetic or semihermetic compressors tend to be cooler with R-115 and R-502 but gas temperatures entering the compression cylinder will be lower and compressor discharge temperatures also lower.

REFERENCES

3–1. B. F. Dodge, *Chemical Engineering Thermodynamics*. New York: McGraw-Hill Book Company (1944).

3–2. H. K. Suttle, "Fundamental Conceptions of Refrigeration Processes," *Chem. Process Eng. Bombay*, 44 (Apr. 1963), 198.

3–3. R. C. Downing, "Pressure–Enthalpy Charts and Their Use," *Service Manual*. Des Plaines, Ill.: Refrigeration Service Engineers Society (1962) (also DuPoint Bulletin RT-40).

3–4. R. C. McHarness and D. D. Chapman, "Refrigerating Capacity and Performance Data for Various Refrigerants, Azeotropes and Mixtures," *ASHRAE J.* 4 (Jan. 1962), 49.

3–5. W. A. Pennington, "Refrigerants," *Air Cond., Heat., and Vent.*, 55 (Nov. 1958), 71.

3–6. W. B. Gosney, "A Survey of the Newer Refrigerants," *J. Refrig.* 5 (Sept.–Oct. 1962), 113.

3–7. J. P. Barger, W. M. Rohsenow, and K. M. Treadwell, "A Comparison of Refrigerants When Used in Vapor Compression Cycles over an Extended Temperature Range," *Trans. ASME*, 79 (Apr. 1957), 681.

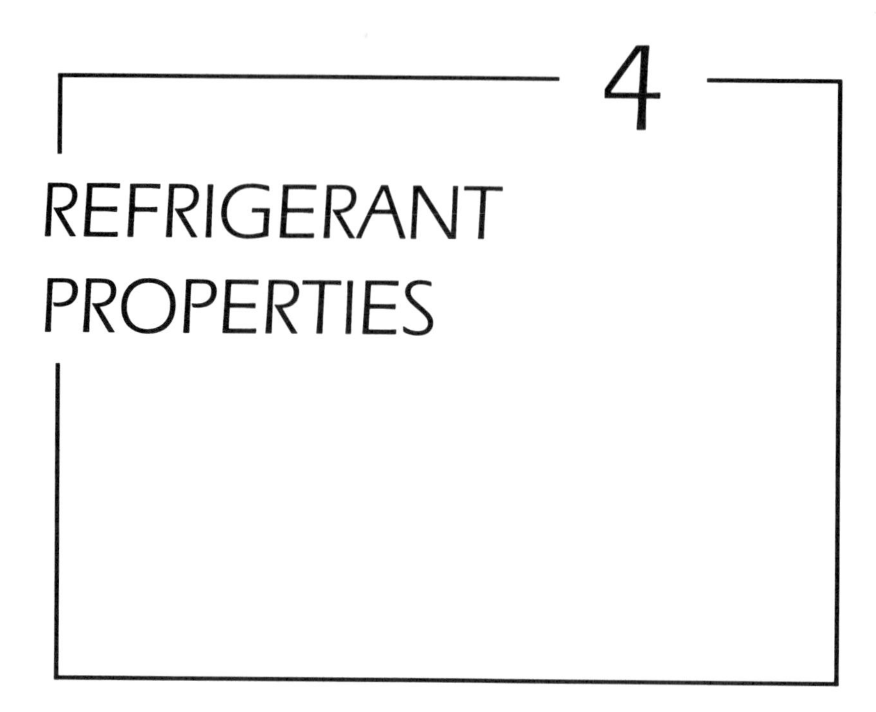

The fluorinated refrigerants are a family of compounds closely related in chemical structure and properties. However, it should be recognized that each is a separate entity with somewhat different properties and that these differences may sometimes be important in their use as refrigerants.

The duty of a refrigerant is to carry heat from some area where it is not wanted to some other place where it can be released. Small amounts of heat can be transferred by gases (such as air) or even liquids operating between different temperatures. A much larger amount of heat can be absorbed and transferred, however, when a liquid is vaporized to a gas. This method is used in nearly all cooling equipment today, whether the moving force is mechanical, as in a compressor, or heat, as in an absorption system.

In heat pump applications heat is recovered from the refrigerant as it changes from a gas to a liquid. Although most of the heat comes from condensation of the gas, some heat is also obtained in cooling the hot discharge gas to the condensing temperature. The latter source may amount to 15 to 20% of the total under some conditions.

A number of materials have physical and thermodynamic properties suitable for their use as refrigerants. Many different refrigerants have been used in the past and are in use today with good ability to transfer heat from one place to another.

From the thermodynamic or heat-carrying standpoint, differences among refrigerants with similar boiling points are small—other properties may be more important. Safety is perhaps the single foremost factor in the selection of a refrigerant.

The best way to get a good start on safety is to choose a refrigerant that is low in toxicity and is not flammable. Most of the fluorinated refrigerants meet these requirements. Refrigerants for use in homes, theaters, stores, hospitals, and other locations where people gather are universally selected from this group. Some properties of these refrigerants are discussed below and listed in Table 4–1.

CHEMICAL FORMULA AND MOLECULAR WEIGHT

The chemical formula shows the atoms making up the compound. Each element has a relative atomic weight, usually based on one of the isotopes of oxygen. Some atomic weights are

$$\begin{aligned} C &= 12.011 \\ H &= 1.008 \\ F &= 18.998 \\ Cl &= 35.453 \\ Br &= 79.904 \end{aligned}$$

Adding together the atomic weight gives the molecular weight. For example, for R-12 (CCl_2F_2), the molecular weight is

$$\begin{aligned} C &= 12.011 \\ 2 \times Cl &= 70.906 \\ 2 \times F &= \underline{37.996} \\ & \quad 120.913 \end{aligned}$$

The molecular weight is used in some gas-phase equations and gives an indication of the relative density of the liquid and vapor and of the atmospheric boiling point. Refrigerants with a low molecular weight tend to be low in density and have a low boiling point. Exceptions to this relationship are found with some molecules that tend to be loosely attracted to each other, so they act as if the molecular weight were much higher than the molecular formula would indicate. For example, water, H_2O, with a molecular weight of 18, has a much higher density and boiling point than such a low molecular weight would indicate.

LIQUID DENSITY

The density is weight or mass per unit of volume. It should not be confused with specific gravity, which is the weight relative to the weight of a reference material at a specific temperature. The fluorinated refrigerants are relatively dense, so the weight of refrigerant used in a system would be much greater than for a light refrigerant such as ammonia. The principal result for a dense refrigerant is the need for a larger liquid line to accommodate the greater flow rate without an increase

TABLE 4–1 PROPERTIES OF REFRIGERANTS[a]

	R-11	R-12	R-13	R-13B1
Chemical formula	CCl_3F	CCl_2F_2	$CClF_3$	$CBrF_3$
Molecular weight	137.37	120.92	104.46	148.92
Boiling point at 1 atm				
°C	23.82	−29.79	−81.4	−57.75
°F	74.87	−21.62	−114.6	−71.95
Freezing point				
°C	−111	−158	−181[d]	−168
°F	−168	−252	−294	−270
Critical temperature				
°C	198.0	112.0	28.9	67.0
°F	388.4	233.6	83.9	152.6
Critical pressure				
atm	43.5	40.6	38.2	39.1
psia	639.5	596.9	561	575
Critical volume				
cm^3/mol	247	217	181	200
ft^3/lb	0.0289	0.0287	0.0277	0.0215
Critical density				
g/cm^3	0.554	0.558	0.578	0.745
lb/ft^3	34.6	34.8	36.1	46.5
Density, liquid at 25°C (77°F)				
g/cm^3	1.476	1.311	1.298 } @ −30°C (−22°F)	1.538
lbs/ft^3	92.14	81.84	81.05 } @ −30°C (−22°F)	96.01
Density, sat'd vapor at boiling point				
g/liter	5.86	6.33	7.01	8.71
lbs/ft^3	0.367	0.395	0.438	0.544

Specific heat, liquid (heat capacity) at 25°C (77°F) [cal/g-°C (Btu/lb-°F)]	0.208	0.232	0.247 @ −30°C (−22°F)	0.208
Specific heat, vapor, at constant pressure (1 atm) at 25°C (77°F) [cal/g-°C (Btu/lb-°F)]	0.142 @ 38°C (100°F)	0.145	0.158	0.112
Specific heat ratio at 25°C and 1 atm, C_p/C_v	1.137 @ 38°C (100°F)	1.137	1.145	1.144
Heat of vaporization at boiling point cal/g Btu/lb	 43.10 77.58	 39.47 71.04	 35.47 63.85	 28.38 51.08
Thermal conductivity at 25°C (77°F) (Btu/hr-ft-°F) Liquid[b] Vapor (1 atm)	 0.050 0.00484	 0.041 0.00557	0.020	0.025
Viscosity at 25°C (77°F) (cp) Liquid Vapor (1 atm)	 0.42 0.0106	 0.20 0.0125	 0.195 (−50°F) 0.0145	 0.135 0.0158
Surface tension at 25°C (77°F) (dyn/cm)	18	9	14 @ −100°F	4
Refractive index of liquid at 25°C (77°F)	1.374	1.287	1.199 @ −100°F	1.238
Relative dielectric strength[l] at 1 atm and 25°C (77°F) (nitrogen = 1)	3.71	2.46	1.65	1.83
Dielectric constant Liquid Vapor (1 atm)[m]	 2.28 @ 29°C 1.0036 @ 24°C[n]	 2.13 } @ 29°C 1.0032 }	 1.0024 @ 29°C	
Solubility of Freon in water at 1 atm and 25°C (77°F) (wt %)	0.11	0.028	0.009	0.03
Solubility of water in Freon at 25°C (77°F) (wt %)	0.011	0.009		0.0095 (70°F)
Toxicity	Group 5a[p]	Group 6[p]	Probably Group 6[q]	Group 6[p]

TABLE 4–1 (Cont.)

	R-14	R-21	R-22	R-23
Chemical formula	CF_4	$CHCl_2F$	$CHClF_2$	CHF_3
Molecular weight	88.00	102.93	86.47	70.01
Boiling point at 1 atm				
°C	−127.96	8.92	−40.75	−82.03
°F	−198.32	48.06	−41.36	−115.66
Freezing point				
°C	−184[e]	−135	−160	−155.2
°F	−299	−211	−256	−247.4
Critical temperature				
°C	−45.67	178.5	96.0	25.9
°F	−50.2	353.3	204.8	78.6
Critical pressure				
atm	36.96	51.0	49.12	47.7
psia	543.2	750	721.9	701.4
Critical volume				
cm^3/mol	141	197	165	133
ft^3/lb	0.0256	0.0307	0.0305	0.0305
Critical density				
g/cm^3	0.626	0.522	0.525	0.525
lb/ft^3	39.06	32.6	32.76	32.78
Density, liquid at 25°C (77°F)				
g/cm^3	1.317 } @ −80°C (−112°F)	1.366	1.194	0.670
lbs/ft^3	82.21 } @ −80°C (−112°F)	85.28	74.53	41.82
Density, sat'd vapor at boiling point				
g/liter	7.62	4.57	4.72	4.66
lbs/ft^3	0.476	0.285	0.295	0.291
Specific heat, liquid (heat capacity) at 25°C (77°F) [cal/g-°C (Btu/lb-°F)]	0.294 @ −80°C (−112°F)	0.256	0.300	1.553

Specific heat, vapor, at constant pressure (1 atm) at 25°C (77°F) [cal/g-°C (Btu/lb-°F)]	0.169	0.140	0.157	0.176
Specific heat ratio at 25°C and 1 atm, C_p/C_v	1.159	1.175	1.184	1.191 @ 0 pressure
Heat of vaporization at boiling point				
cal/g	32.49	57.86	55.81	57.23
Btu/lb	58.48	104.15	100.45	103.02
Thermal conductivity at 25°C (77°F) (Btu/hr-ft-°F)				
Liquid[b]	0.040 (−100°F)	0.063	0.052	0.008
Vapor (1 atm)		0.00569	0.00678	
Viscosity at 25°C (77°F) (cp)				
Liquid	0.09 (−100°F)	0.32	0.18	0.22 (−50°F)
Vapor (1 atm)	0.0145	0.0114	0.013	0.0148
Surface tension at 25°C (77°F) (dyn/cm)	4 @ −100°F	18	8	15 @ −100°F
Refractive index of liquid at 25°C (77°F)	1.151 @ −100°F	1.354	1.256	1.215 @ −100°F
Relative dielectric strength[l] at 1 atm and 25°C (77°F) (nitrogen = 1)	1.06	1.85	1.27	1.04
Dielectric constant				
Liquid		5.34 @ 28°C	6.11@ 24°C	1.0073 @ 25°C[m]
Vapor (1 atm)[m]	1.0012 @ 24.5°C	1.0070 @ 30°C	1.0071 @ 25.4°C	
Solubility of Freon in water at 1 atm and 25°C (77°F) (wt %)	0.0015	0.95	0.30	0.10
Solubility of water in Freon at 25°C (77°F) (wt %)		0.13	0.13	
Toxicity	Probably Group 6[q]	Much less than Group 4, somewhat more than Group 5[p]	Group 5a[p]	Probably Group 6[q]

TABLE 4–1 (Cont.)

	R-112	R-113	R-114	R-114B2	R-115
Chemical formula	$CCl_2F\text{-}CCl_2F$	$CCl_2\text{-}CClF_2$	$CClF_2\text{-}CClF_2$	$CBrF_2\text{-}CBrF_2$	$CClF_2\text{-}CF_3$
Molecular weight	203.84	187.38	170.93	259.85	154.47
Boiling point at 1 atm					
°C	92.8	47.57	3.77	47.26	−39.1
°F	199.0	117.63	38.78	117.06	−38.4
Freezing point					
°C	26	−35	−94	−110.5	−106[f]
°F	79	−31	−137	−166.8	−159
Critical temperature					
°C	278	214.1	145.7	214.5	80.0
°F	532	417.4	294.3	418.1	175.9
Critical pressure					
atm	34[h]	33.7	32.2	34.4	30.8
psia	500	495	473.2	506.1	453
Critical volume					
cm^3/mol	370[h]	325	293	329	259
ft^3/lb	0.029	0.0278	0.0275	0.0203	0.0269
Critical density					
g/cm^3	0.55[h]	0.576	0.582	0.790	0.596
lb/ft^3	34	36.0	36.32	49.32	37.2
Density, liquid at 25°C (77°F)					
g/cm^3	1.634 } [i] @ 30°C	1.565	1.456	2.163	1.291
lbs/ft^3	102.1 } @ 30°C	97.69	90.91	135.0	80.60
Density, sat'd vapor at boiling point					
g/liter	7.02[j]	7.38	7.83		8.37
lbs/ft^3	0.438	0.461	0.489		0.522
Specific heat, liquid (heat capacity) at 25°C (77°F) [cal/g-°C (Btu/lb-°F)]		0.218	0.243	0.166	0.285

Specific heat, vapor, at constant pressure (1 atm) at 25°C (77°F) [cal/g-°C (Btu/lb-°F)]		0.161 @ 60°C (140°F)	0.170		0.164
Specific heat ratio at 25°C and 1 atm, C_p/C_v		1.080 @ 60°C (140°F)	1.084		1.091
Heat of vaporization at boiling point cal/g Btu/lb	 37 (est.) 67	 35.07 63.12	 32.51 58.53	 25 (est.) 45 (est.)	 30.11 54.20
Thermal conductivity at 25°C (77°F) (Btu/hr-ft-°F) Liquid[b] Vapor (1 atm)	0.040	0.038 0.00450 (0.5 atm)	0.034 0.00646	0.027	0.026 0.00803
Viscosity at 25°C (77°F) (cp) Liquid Vapor (1 atm)	1.21[i]	0.68 0.010 (0.1 atm)	0.36 0.0112	0.72	0.16 0.0127
Surface tension at 25°C (77°F) (dyn/cm)	23 @ 30°C 86°F	17.3	12	18	5
Refractive index of liquid at 25°C (77°F)	1.413	1.354	1.288	1.367	1.214
Relative dielectric strength[l] at 1 atm and 25°C (77°F) (nitrogen = 1)	5 (est.)	3.9 (0.44 atm)	3.34	4.02 (0.44 atm)	2.54
Dielectric constant Liquid Vapor (1 atm)[m]	2.54 @ 25°C	2.41 @ 25°C	2.26 @ 25°C 1.0043 @ 26.8°C	2.34 @ 25°C	1.0035 @ 27.4°C
Solubility of Freon in water at 1 atm and 25°C (77°F) (wt %)	0.012 (Sat'n press.)	0.017 (Sat'n press.)	0.013		0.006
Solubility of water in Freon at 25°C (77°F) (wt %)		0.011	0.009		
Toxicity	Probably less than Group 4, more than Group 5[q]	Much less than Group 4, somewhat more than Group 5[p]	Group 6[p]	Group 5a[p]	Group 6[p]

TABLE 4–1 (Cont.)

	R-116	R-C318	R-502	R-500	R-503
Chemical formula	CF_3-CF_3	*c*-C_4F_8			
Molecular weight	138.01	200.03	111.63	99.31	87.5
Boiling point at 1 atm					
°C	−78.2	−5.85	−45.42	−33.5	−88.7
°F	−108.8	21.47	−49.76	−28.3	−127.6
Freezing point					
°C	−100.6	−41.4		−159	
°F	−149.1	−42.5		−254	
Critical temperature					
°C	19.7[g]	115.3	82.2	105.5	19.5
°F	67.5	239.6	179.9	221.9	67.1
Critical pressure					
atm	29.4[g]	27.5	40.2	43.7	41.3
psia	432	403.6	591.0	641.9	607
Critical volume					
cm^3/mol	225	323	199	200	178
ft^3/lb	0.0262	0.0258	0.0286	0.0323	0.0326
Critical density					
g/cm^3	0.612	0.620	0.561	0.496	0.491
lb/ft^3	38.21	38.7	35.00	30.96	30.67
Density, liquid at 25°C (77°F)					
g/cm^3	1.587 }[g] @ −100°F	1.500	1.217	1.168	1.326
lbs/ft^3	99.08 }	93.64	75.95	72.16	81.93 (−50°F)
Density, sat'd vapor at boiling point					
g/liter	9.01[g]	9.62	6.22	5.32	5.78
lbs/ft^3	0.562	0.601	0.388	0.332	0.361

Specific heat, liquid (heat capacity) at 25°C (77°F) [cal/g-°C (Btu/lb-°F)]	0.232 @ −100°F	0.266	0.293	0.284	0.252 (−50°F)
Specific heat, vapor, at constant pressure (1 atm) at 25°C (77°F) [cal/g-°C (Btu/lb-°F)]	0.182[k] @ 0 pressure	0.190	0.164	0.176	0.162
Specific heat ratio at 25°C and 1 atm, C_p/C_v	1.085 (est.) @ 0 pressure	1.067	1.132		
Heat of vaporization at boiling point cal/g Btu/lb	 27.97 50.35	 27.77 49.99	 41.21 74.18	 48.03 86.45	 42.99 77.38
Thermal conductivity at 25°C (77°F) (Btu/hr-ft-°F) Liquid[b] Vapor (1 atm)	 0.045 (−100°F) 0.0098 (est.)	 0.025 0.0059 (est.)	 0.038	 0.0432[c]	 0.0532 (−50°F)[c]
Viscosity at 25°C (77°F) (cp) Liquid Vapor (1 atm)	 0.30 0.0148	 0.38 0.0119	 0.16 0.013	 0.19 0.030	 0.193 (−50°F)
Surface tension at 25°C (77°F) (dyn/cm)	16 @ −100°F	7	8		
Refractive index of liquid at 25°C (77°F)	1.206 @ −100°F	1.217	1.235		
Relative dielectric strength[l] at 1 atm and 25°C (77°F) (nitrogen = 1)	2.02	2.86	2.34		
Dielectric constant Liquid Vapor (1 atm)[m]	 1.0021 @ 23°C	 1.0032 @ 25°C[o]			
Solubility of Freon in water at 1 atm and 25°C (77°F) (wt %)					
Solubility of water in Freon at 25°C (77°F) (wt %)			0.056		
Toxicity	Probably Group 6[q]	Probably Group 6[q]	Group 5a[p]		

TABLE 4–1 (Cont.)

[a]Unless otherwise noted, data were developed by the DuPont Company in its laboratories or under contract with outside laboratories.

[b]W. Tauscher, "Thermal Conductivity of Liquid Refrigerants Measured by an Unsteady-State, Hot-Wire Method," *Kaltetechnik*, 19 (1967), 288; data for R-114B2, R-C318, and R-502 from P. Grassman, Zurick, Switzerland, private communication.

[c]American Society of Heating, Refrigerating, and Air-Conditioning Engineers, *ASHRAE Handbook of Fundamentals*, Atlanta, Ga.:ASHRAE (1977), Chap. 15.

[d]O. Ruff and R. Keim, "The Fluorination of Carbon Compounds . . . with Iodine Pentafluoride or Fluorine," *Z. Allg. Chem.*, 201 (1931), 245.

[e]O. Ruff and O. Bretschneider, "Preparation of C_2F_4 and C_2F_6 from CF_4," *Z. Allg. Chem.*, 210 (1933), 173.

[f]J. D. Calfee, N. Fukuhara, D. S. Young, and L. Bigelow, "Action of Elementary Fluorine upon Organic Compounds," *J. Am. Chem. Soc.*, 62 (1940), 267.

[g]E. L. Pace and J. G. Aston, "The Thermodynamics of Hexafluoroethane from Calorimetric and Spectroscopic Data," *J. Am. Chem. Soc.*, 70 (1948), 566.

[h]H. P. Meissner and E. M. Redding, "Prediction of Critical Constants," *Ind. Eng. Chem.*, 34 (1942), 521.

[i]F. Havorka and F. E. Geiger, "Thermodynamic Properties of Trifluorotrichloroethane and Difluorotetrachloroethane," *J. Am. Chem. Soc.*, 55 (1933), 4759.

[j]Calculated value based on normal deviation from the ideal gas law.

[k]R. A. Carney, E. A. Piotowski, A. G. Meister, J. H. Braun, and F. F. Cleveland. "Substituted Ethanes: V. Raman and Infrared Spectra Assignments, Potential Constants, and Calculated Thermodynamic Properties for C_2F_6, C_2Cl_6, and C_2Br_6," *J. Mol. Spectros.*, 7 (1961), 209.

[l]Data from the DuPont Company at 1 atm., 0.1-in. gap and $\frac{3}{4}$-in. sphere-to-plane gap.

[m]Vapor dielectric constants from Ref. 4–44.

[n]A. A. Maryott and F. Buckley, "Table of Dielectric Constants and Electric Dipole Moments of Substances in the Gaseous State," National Bureau of Standards Circular 537 (June 1953), p. 149.

[o]A. DiGiacomo and C. P. Smyth, "The Dipole Moments and Molecular Structure of Some Highly Fluorinated Hydrocarbons and Ethers," *J. Am. Chem. Soc.*, 77 (1955), 774.

[p]Underwriters' Laboratories' classification.

[q]Based on preliminary toxicological data.

Source: Ref. 4–1.

in the pressure drop. The expansion valve must have a larger opening to regulate the liquid flow rate properly. For example, for a ton of refrigeration at an evaporating temperature of 5°F and a condensing temperature of 86°F, the liquid flow rate for R-12 is about 86 in^3/min and for ammonia about 20 in^3. The respective densities are 80.7 lb/ft^3 for R-12 and 37.2 lb/ft^3 for ammonia. Liquid densities at several temperatures are given in Table 4–2. The specific volume is the reciprocal of the specific density:

$$\text{volume (ft}^3\text{/lb)} = \frac{1}{\text{density (lb/ft}^3\text{)}}$$

TABLE 4–2 LIQUID DENSITY OF REFRIGERANTS

	°F	lb/ft³	lb/gal	g/cm³	°F	lb/ft³	lb/gal	g/cm³
R-11	−50	102.0	13.64	1.634	50	94.35	12.61	1.511
	−40	101.3	13.54	1.622	60	93.54	12.50	1.498
	−30	100.5	13.44	1.610	70	92.72	12.40	1.485
	−20	99.79	13.34	1.598	80	91.90	12.29	1.472
	−10	99.03	13.24	1.586	90	91.06	12.17	1.459
	0	98.27	13.14	1.574	100	90.21	12.06	1.445
	10	97.50	13.03	1.562	110	89.35	11.94	1.431
	20	96.72	12.93	1.549	120	88.48	11.83	1.417
	30	95.94	12.83	1.537	130	87.59	11.71	1.403
	40	95.15	12.72	1.524				
R-12	−50	95.62	12.78	1.532	50	85.14	11.38	1.364
	−40	94.66	12.65	1.516	60	83.94	11.22	1.345
	−30	93.69	12.52	1.501	70	82.72	11.06	1.325
	−20	92.70	12.39	1.485	80	81.45	10.89	1.305
	−10	91.69	12.26	1.469	90	80.14	10.71	1.284
	0	90.66	12.12	1.452	100	78.79	10.53	1.262
	10	89.61	11.98	1.435	110	77.38	10.34	1.239
	20	88.53	11.84	1.418	120	75.91	10.15	1.216
	30	87.43	11.69	1.400	130	74.37	9.94	1.191
	40	86.30	11.54	1.382				
R-22	−50	89.00	11.90	1.426	50	78.03	10.43	1.250
	−40	88.01	11.77	1.410	60	76.77	10.26	1.230
	−30	86.99	11.63	1.393	70	75.47	10.09	1.209
	−20	85.96	11.49	1.377	80	74.12	9.91	1.187
	−10	85.01	11.36	1.362	90	72.71	9.72	1.165
	0	83.83	11.21	1.343	100	71.24	9.52	1.141
	10	82.72	11.06	1.325	110	69.69	9.32	1.116
	20	81.60	10.91	1.307	120	68.05	9.10	1.090
	30	80.44	10.75	1.289	130	66.31	8.86	1.062
	40	79.26	10.60	1.270				
R-113	−30	105.6	14.12	1.692	60	99.05	13.24	1.587
	−20	105.0	14.03	1.681	70	98.26	13.14	1.574
	−10	104.3	13.94	1.670	80	97.45	13.03	1.561
	0	103.6	13.84	1.659	90	96.63	12.92	1.548
	10	102.8	13.75	1.647	100	95.79	12.81	1.534
	20	102.1	13.65	1.635	110	94.95	12.69	1.521
	30	101.4	13.55	1.624	120	94.09	12.58	1.507
	40	100.6	13.45	1.611	130	93.04	12.44	1.490
	50	99.83	13.35	1.599				

TABLE 4–2 (Continued)

	°F	lb/ft^3	lb/gal	g/cm^3	°F	lb/ft^3	lb/gal	g/cm^3
R-114	−50	102.9	13.76	1.649	50	93.70	12.53	1.501
	−40	102.1	13.65	1.635	60	92.69	12.39	1.485
	−30	101.2	13.53	1.621	70	91.65	12.25	1.468
	−20	100.3	13.41	1.607	80	90.59	12.11	1.451
	−10	99.42	13.29	1.592	90	89.51	11.97	1.434
	0	98.50	13.17	1.578	100	88.41	11.82	1.416
	10	97.57	13.04	1.563	110	87.27	11.67	1.398
	20	96.63	12.92	1.548	120	86.10	11.51	1.379
	30	95.67	12.79	1.532	130	84.90	11.35	1.360
	40	94.70	12.66	1.517				
R-500	−50	85.20	11.39	1.365	50	75.26	10.06	1.206
	−40	84.28	11.27	1.350	60	74.14	9.91	1.188
	−30	83.35	11.14	1.335	70	72.98	9.76	1.169
	−20	82.40	11.02	1.320	80	71.80	9.60	1.150
	−10	81.44	10.89	1.305	90	70.56	9.43	1.130
	0	80.46	10.76	1.289	100	69.28	9.26	1.110
	10	79.46	10.62	1.273	110	67.95	9.08	1.088
	20	78.45	10.49	1.257	120	66.55	8.90	1.066
	30	77.41	10.35	1.240	130	65.08	8.70	1.042
	40	76.34	10.21	1.223				
R-502	−50	92.51	12.37	1.482	50	80.06	10.70	1.282
	−40	91.39	12.21	1.464	60	78.59	10.51	1.259
	−30	90.25	12.07	1.446	70	77.06	10.30	1.234
	−20	89.09	11.91	1.427	80	75.46	10.09	1.209
	−10	87.90	11.75	1.408	90	73.76	9.86	1.182
	0	86.68	11.59	1.388	100	71.97	9.62	1.153
	10	85.43	11.42	1.368	110	70.04	9.36	1.122
	20	84.15	11.25	1.348	120	67.96	9.08	1.089
	30	82.83	11.07	1.327	130	65.66	8.78	1.052
	40	81.47	10.89	1.305				
R-13B1	−50	121.0	16.17	1.938	50	102.5	13.70	1.642
	−40	119.4	15.96	1.912	60	100.2	13.40	1.605
	−30	117.7	15.73	1.885	70	97.79	13.07	1.566
	−20	116.0	15.51	1.858	80	95.22	12.73	1.525
	−10	114.3	15.28	1.830	90	92.45	12.36	1.481
	0	112.5	15.04	1.802	100	89.42	11.95	1.432
	10	110.6	14.79	1.772	110	86.06	11.51	1.379
	20	108.7	14.53	1.741	120	82.22	10.99	1.317
	30	106.7	14.27	1.710	130	77.61	10.37	1.243
	40	104.7	13.99	1.676				
R-717	−50	43.49	5.81	0.697	40	39.49	5.28	0.633
	−40	43.08	5.76	0.690	50	39.00	5.21	0.625
	−30	42.65	5.70	0.683	60	38.50	5.15	0.617
	−20	42.22	5.64	0.676	70	38.00	5.08	0.609
	−10	41.78	5.59	0.669	80	37.48	5.01	0.600
	0	41.34	5.53	0.662	90	36.95	4.94	0.592
	10	40.89	5.47	0.655	100	36.40	4.87	0.583
	20	40.43	5.40	0.648	110	35.84	4.79	0.574
	30	39.96	5.34	0.640	120	35.26	4.71	0.565

Ammonia is included in Table 4–2 in order to compare with the denser fluorocarbon refrigerants. One aspect of this difference is found in the crankcase of reciprocating compressors. Ammonia is not very soluble in lubricating oils. When conditions permit an excess of ammonia in the crankcase, it will form a separate liquid layer on top of the oil. Although the fluorinated refrigerants are more soluble in oils, an excess of refrigerant may at times be present in the crankcase. When this occurs, the separate oil layer will be on top of the liquid refrigerant.

A number of years ago, a serviceman needed a charging cylinder for ammonia. One suitable for that use was not available, so he used a small R-12 cylinder—replacing the brass valve with a steel one. The label on the cylinder showed a net weight of 25 lb of R-12. The serviceman was not familiar with the properties of the two refrigerants, so he put 25 lb of ammonia in the cylinder. The cylinder was liquid-full and when the temperature increased, the cylinder ruptured, with tragic results.

In another instance, a system required nine 25-lb cylinders of R-114 for the proper charge. For some reason the charge had to be withdrawn. The operator cooled the shipping cylinders in carbon-ice and felt very pleased when he was able to fit the entire refrigerant charge in five cylinders. When the overfilled cylinders reached ambient temperature, hydrostatic pressure developed and all five cylinders ruptured. Again, a knowledge of liquid density change with temperature would have prevented this mishap. Furthermore, carbon-ice should not be used to cool steel cylinders. At temperatures much below about 0°F, ordinary carbon steel undergoes a change in structure and becomes more brittle and more likely to rupture.

Regulations of the Department of Transportation require that shipping containers for liquefied compressed gases must not be full of liquid at 130°F. Good data on the change of liquid density with temperature are essential to satisfy this requirement and to handle the liquid safely under other circumstances.

VAPOR DENSITY

The gas phase of the fluorocarbon refrigerants is also relatively heavy, as illustrated in Table 4–3. Whether a gas is heavy or light does not matter very much in a

TABLE 4–3 VAPOR DENSITY AT 77°F (25°C)

R	Density at 1 atm of pressure (14.7 psia) (lb/ft^3)	Density relative to air
Air	0.0739	1
11	0.364	4.93
12	0.315	4.26
13	0.269	3.63
13B1	0.385	5.21
14	0.225	3.05
22	0.224	3.03
113[a]	0.214	2.90
114	0.450	6.09
500	0.256	3.47
502	0.289	3.91
503	0.223	3.02

[a]At 6.46 psia.

refrigeration system. The power needed in a reciprocating or rotary compressor depends on the number of molecules rather than the weight of each molecule. The density of the gas is a factor in centrifugal compressors, where the pressure developed is a function of the velocity and density of the refrigerant.

Gas density is more important outside a system. If a fluorinated refrigerant is intentionally or accidentally released from a system, it tends initially to sink to the floor or ground and collect in low places. If there is a possibility of refrigerant discharge, good ventilation should be provided near the floor to hasten the dispersion of the gas. All gases present in a room will eventually become thoroughly mixed and the composition will be the same throughout the room. With heavy gases such as the fluorocarbons, however, the mixing will be slow unless they are agitated with a good fan or some other means of stirring things up.

BOILING POINT

The boiling point is any temperature at which the liquid and vapor are in equilibrium. The pressure would be the vapor pressure at that temperature. If the pressure is not specified, it is usually assumed to be 1 atm (0 psig). The atmospheric boiling point indicates the direction of gas flow in case of leakage. If the boiling point is high, the vapor pressure will be low and air and water vapor will tend to enter the system through a leak. If the boiling point is low, refrigerant will tend to leak outward. This distinction, however, really depends on the pressures inside and outside the system.

For example, the atmospheric boiling point of R-11 is 75°F and the vapor pressure is 14.7 psia. If the evaporating temperature is below 75°F and the ambient pressure is 14.7 psia, air and water vapor will come in if there is a leak in the low-pressure side of the system. At an elevation of 5000 ft, the normal atmospheric pressure is about 12.2 psia and the evaporating temperature would need to be below about 66°F for inward leakage.

Accurate measurement of the boiling point is not easy since the liquid and vapor must be in equilibrium when the temperature and pressure are measured. More often, the vapor pressure is measured at a number of different temperatures and an equation is developed to show the change with temperature. An example is

$$P = A + \frac{B}{T} + CT + DT^2$$

P is set at 14.7 psia and the equation is solved by iteration to find the corresponding boiling point.

PRESSURE

The relationships between temperature and pressure are important in refrigeration and air conditioning. Three types of pressure may be of some concern: hydrostatic, vapor, and gas. Each is affected by temperature in a different way.

Hydrostatic pressure develops when a container becomes completely filled with liquid. The rate of increase in pressure with increasing temperature depends on how close to the critical temperature the liquid is when the container becomes full. The pressure increase can be as much as 40 to 50 psi for each rise in temperature of 1°F. At this rate, pressures high enough to rupture cylinders, receivers, piping, and so on, can be reached. As a safety regulation, the Department of Transportation requires that shipping containers for liquefied compressed gasses not be liquid-full at 130°F. Similar concern should be observed by anyone using these products.

Vapor pressure is the pressure when both liquid and vapor are present and in equilibrium. The liquid is said to have one degree of freedom. This means that when the temperature is given, the vapor pressure will have a definite and unchangeable value. The relative amounts of liquid and vapor will not make any difference. A cylinder or other container can be nearly full or nearly empty, but the pressure will be the same as long as a little bit of liquid is present. The vapor pressures of some refrigerants are given in Figure 4–1. The change in pressure with temperature is similar with all the refrigerants—with some minor differences.

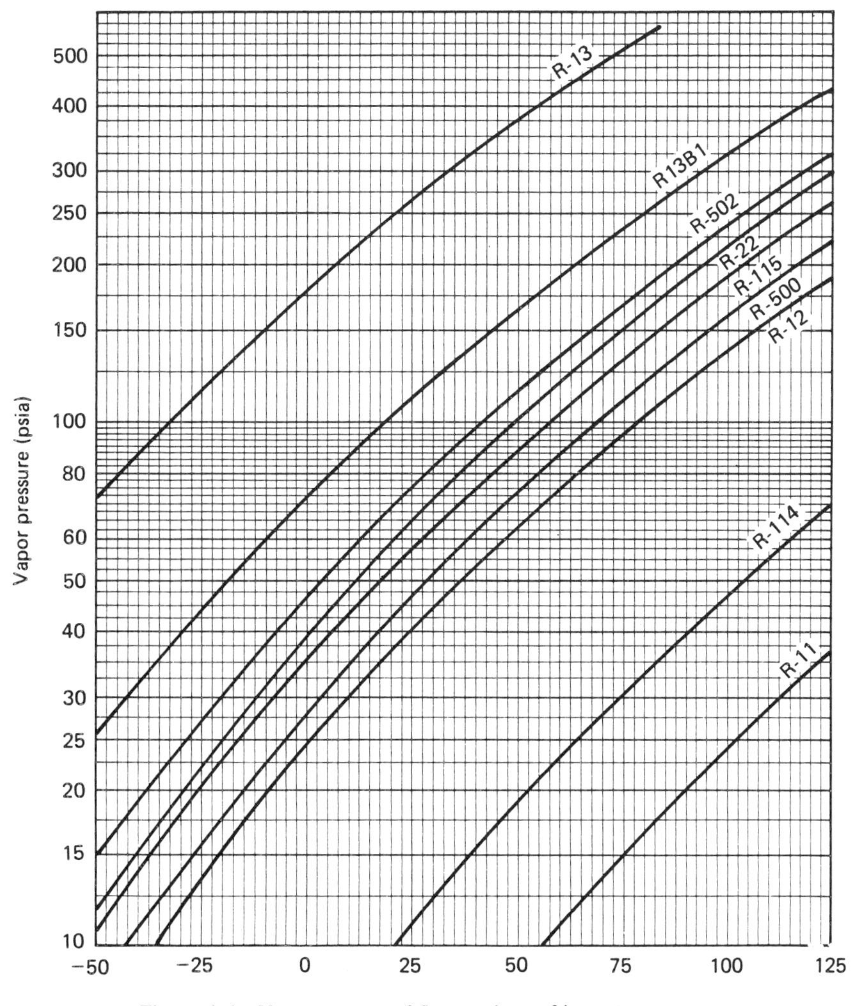

Figure 4–1 Vapor pressure of fluorocarbon refrigerants.

The *compression ratio* is an indication of the amount of work required from the compressor in a refrigeration system. It is obtained by dividing the high-side pressure by the low-side pressure—both in absolute terms generally indicated by the letter "a" or abbreviation "abs" in the units, such as "psia" or "psi, abs." A low value for the compression ratio correlates fairly well with a high coefficient of performance (COP), as illustrated in Table 4–4. The comparisons shown there were obtained using different refrigerants in the same laboratory equipment with a small reciprocating compressor [4–2]. The heat absorbed in the evaporator was measured electrically. The motor wattage was measured directly and the actual COP computed from these measurements. The actual COP is much lower than the theoretical COP because in the small compressor used, the no-load wattage was a considerable fraction of the total wattage. However, the comparison of the various refrigerants seems valid.

TABLE 4–4 COMPRESSION RATIO VERSUS COP[a]

R	Compression Ratio	COP Actual	COP Theoretical
13B1	7.30	1.10	2.02
502	8.88	0.95	2.28
22	9.74	0.84	2.30
12	9.88	0.58	2.50
500	10.07	0.72	2.46

[a]Nominal conditions: −20/65/110; temperatures in degrees Fahrenheit, evaporating/vapor entering compressor/condensing.

Source: Ref. 4–2.

The performance of R-13B1 is especially noteworthy—with the lowest compression ratio and the highest COP. The favorable compression ratio is even more apparent at more extreme conditions. For example, at an evaporating temperature of −50°F and a condensing temperature of 120°F, the ratio for R-13B1 is 15.8, within the range of most compressors for single-stage compression. Under the same conditions, the ratio for R-502 is 20.4 and for R-22, 23.5. In spite of good refrigeration properties, R-13B1 has found little use in this application, perhaps because the inherent cost of manufacture is greater than for most other fluorocarbon refrigerants. The principal use of R-13B1 at pressent is as a fire extinguishing agent for specialty applications such as libraries, museums, computer rooms, on board ships, and other places where the use of other extinguishants would cause significant damage to property or life.

It is important to remember that the vapor pressure means the pressure of the vapor above the liquid. Some liquid must be present somewhere. In a static condition, the pressure will be the vapor pressure even though the point of measurement is removed from the location of the liquid. If the gas is moving through a pipe, the measurement may be a little lower than the true vapor pressure if there is resistance to flow through the pipe. The pressure drop in the pipe depends on the flow rate of the gas, the diameter and length of the pipe, the temperature, and the nature of the gas.

In a refrigeration system, the high-side and low-side pressures are usually measured and vapor-pressure tables or curves used to find the corresponding condensing and evaporating temperatures. If the pressure drop in the piping is not large, measuring the pressure is generally easier and more reliable than measuring the temperature, unless special care and technique is used.

The vapor pressure depends only on the temperature and is the same at any elevation. Ordinary pressure gages used in refrigeration work, however, measure the difference between the system pressure and the outside atmospheric pressure. The scale on the gage is usually set for an outside pressure of 1 atm (14.7 psia) or zero gage. The gage reading will be high if used at higher altitudes where the atmospheric pressure is low. This error amounts to about 0.5 psi for every 1000 ft of elevation (see Table 4–5). The scale on the gage can be adjusted for the average local pressure or a correction can be made for each reading as follows:

PRESSURE READING ON GAGE AT 70°F (PSI)

Refrigerant	Sea level	1000 ft	3000 ft	5000 ft
12	70.192	70.715	71.717	72.660
22	121.43	121.95	122.96	123.90
502	137.62	138.14	139.15	140.09

TABLE 4–5 EFFECT OF ALTITUDE ON PRESSURE

Elevation (ft)	Decrease in pressure for each 1000-ft rise in elevation (psi)	Total change in pressure from sea level (psi)
1,000	0.523	0.52
2,000	0.509	1.03
3,000	0.493	1.53
4,000	0.479	2.00
5,000	0.464	2.47
6,000	0.451	2.92
7,000	0.437	3.36
8,000	0.424	3.78
9,000	0.411	4.19
10,000	0.399	4.59

VELOCITY OF SOUND

Vapor

The *velocity of sound* is the maximum velocity that can be reached in a pipe or channel of constant cross section and so is of some interest in refrigeration.

The sonic velocity values in Figures 4–2 to 4–5, pages 62, 63 and 66, were calculated using the following equation [4–3]:

$$V_a = \sqrt{-gV^2 \left[144 \left(\frac{dP}{dV} \right)_T - \frac{144JT(dP/dT)_V^2}{C_v} \right]}$$

or

$$V_a = V \sqrt{\frac{657.36091T(dP/dT)_V^2}{C_v} - 4633.056 \left(\frac{dP}{dV} \right)_T}$$

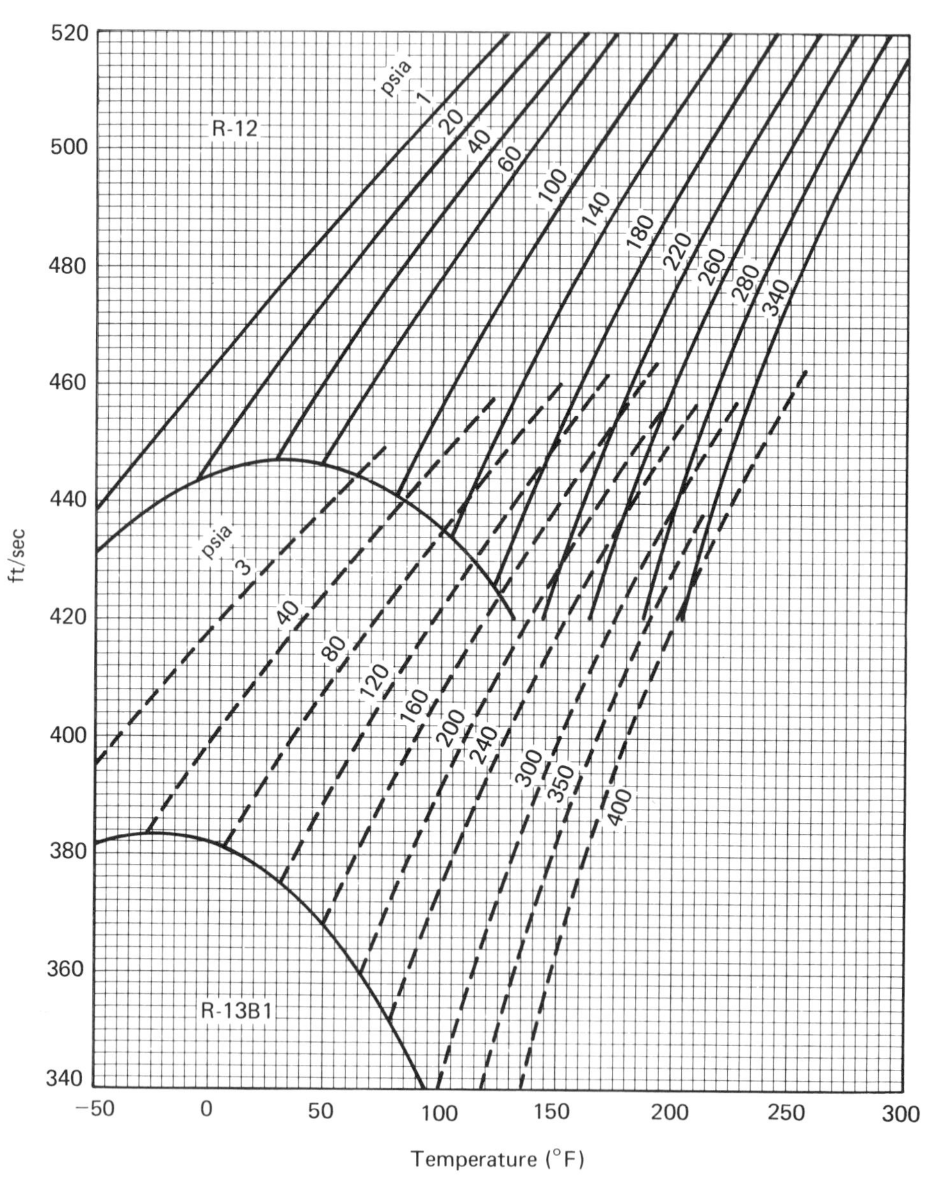

Figure 4–2 Velocity of sound in vapor.

where V_a = velocity of sound, ft/sec
g = gravitational constant, 32.1740 lb_m-ft/lb_f·sec
P = pressure, psia
T = temperature, °R
V = volume, ft^3/lb
C_v = specific heat at constant volume, Btu/lb·°F
J = factor for converting work units to heat units = 0.185053

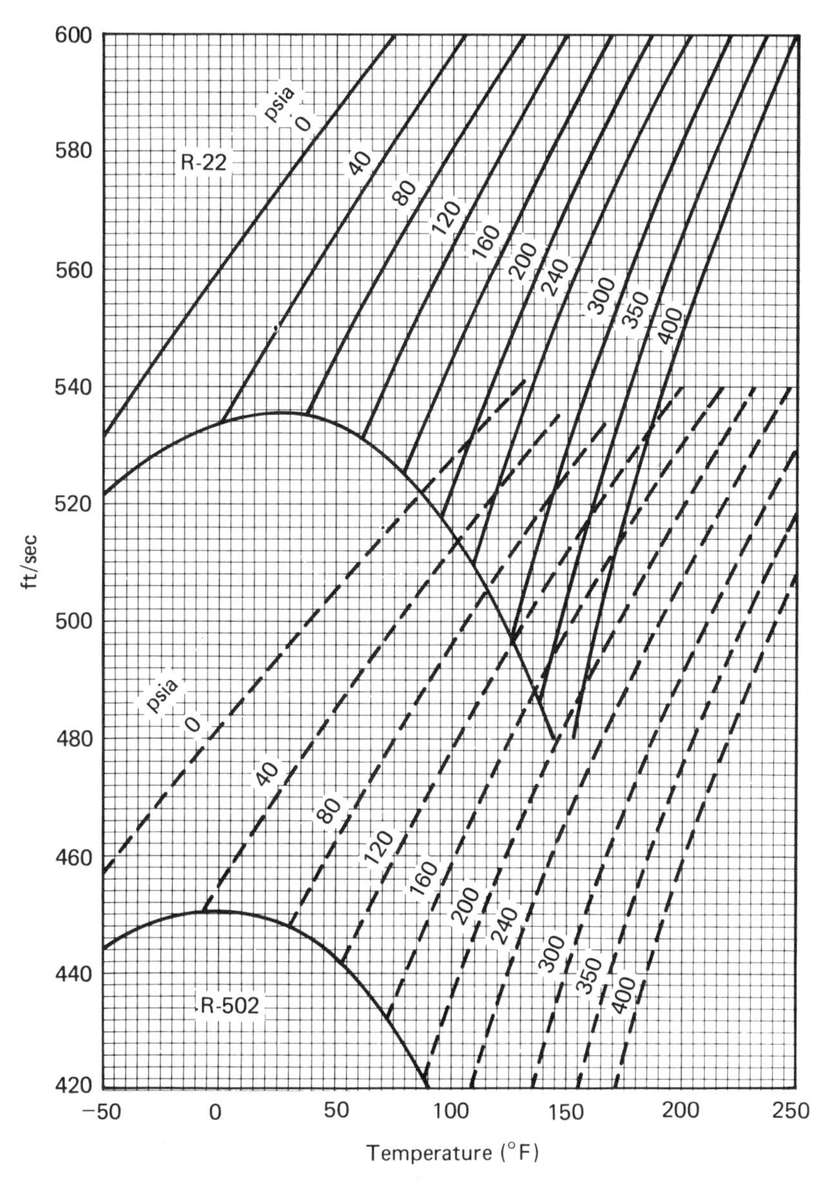

Figure 4–3 Velocity of sound in vapor.

The derivatives are based on the Martin–Hou equation of state (see Chapter 16) for most of the products. In a few cases a simpler equation of state was used with good results in the region of general interest.

In the vapor, the velocity of sound increases with increasing temperature at constant pressure and decreases with pressure at constant temperature. In other words, the sonic velocity increases directly with an increase in the specific volume of the gas. This change is also illustrated in Figure 4–6. At the same temperature and pressure, the molecular weight reflects the density. Products with low molecular weight and therefore high specific volume have greater sonic velocity than those with high molecular weight.

A comparison of some measured velocity of sound data and calculations as outlined above are shown in Table 4–6. In general, the agreement is good.

TABLE 4–6 COMPARISON OF MEASURED AND CALCULATED SONIC VELOCITIES

R	Pressure (psia)	Temperature (°F)	Sonic velocity (ft/sec)		Difference (%)	Reference
			Measured	Calculated		
11	14.7	117	472	474		4–4
	14.7	212	513			4–4
22	14.7	117	615	615.5		4–4
	14.7	211	663	663.3		4–4
12	14.7	77	494	498		4–5
	14.7	117	511	515		4–4
	14.7	212	551	555		4–4
	60	78	468.8	466.2	0.55	4–6
	60	81.5	469.1	468.4	0.13	4–6
	60	84	470.9	469.9	0.20	4–6
	60	122.5	494.6	492.6	0.41	4–6
	60	140.0	503.2	502.1	0.21	4–6
	60	192.5	530.3	528.6	0.32	4–6
	70	100.0	477.8	474.3	0.74	4–6
	70	120.0	489.3	486.4	0.59	4–6
	70	123.0	491.0	488.2	0.58	4–6
	70	142.0	503.1	498.9	0.82	4–6
	70	191.0	527.1	524.6	0.47	4–6
	80	75.0	453.6	451.3	0.49	4–6
	80	81.5	456.2	455.8	0.10	4–6
	80	120.0	485.1	481.5	0.75	4–6
	80	131.0	492.6	488.1	0.92	4–6
	80	152.5	503.8	500.6	0.64	4–6
	100	131.5	481.7	478.8	0.55	4–6
	100	147.0	492.5	488.8	0.76	4–6
	100	171.0	507.1	503.2	0.78	4–6
	100	187.5	515.9	512.6	0.64	4–6

Liquid

The velocity of sound in the liquid phase of refrigerants is not as well known as in the vapor phase. A few measurements have been made and are summarized in Figure 4–7.

The circled point in Figure 4–7, page 67, is a measurement by Kokernak and Feldman [4–7] of 1640 ft/sec for R-12 at 43°F. The square points are DuPont measurements with R-113 of 2050 ft/sec at 86°F and 2450 ft/sec at 32°F. The solid points are calculated values based on the R-12 measurement. The linear curves are measured data by Meyers [4–8] at low temperatures. The Meyer data are a little higher than the other measurements but the agreement is not bad.

Rao [4–9] proposed the following equation for estimating sonic velocities in liquids:

$$V_a = \left(\frac{\beta d}{M}\right)^3$$

where V_a = sonic velocity, cm/sec
d = liquid density, g/cm^3
M = molecular weight of liquid
β = additive coefficient obtained from molecular structural considerations

Additive coefficients have been suggested by Sakiades and Coates [4–10] and others, and have been reasonably successful in correlating various properties. Reported values are as follows:

	Coefficient
—C— (with single bonds above and below)	1850
—C—C— (with single bonds above and below each C)	2722
Cl	610
Br	692
F	80

Hydrogen has a value of zero when connected to carbon in the basic methane or ethane structure. For example, the coefficient for R-12 is

C	1850
2Cl	1220
2F	160
	3230

The value of 80 for each fluorine atom was suggested by Kokernak and Feldman based on their measured value for sonic velocity at 43°F. The solid points in Figure 4–7, page 67, were calculated using 80 for the fluorine coefficient for the other products. The measured value for R-113 at 86°F agrees very well with the calculation. However, the measurement at 32°F is somewhat higher than calculated, indicating perhaps that the additive coefficient should reflect some change in temperature.

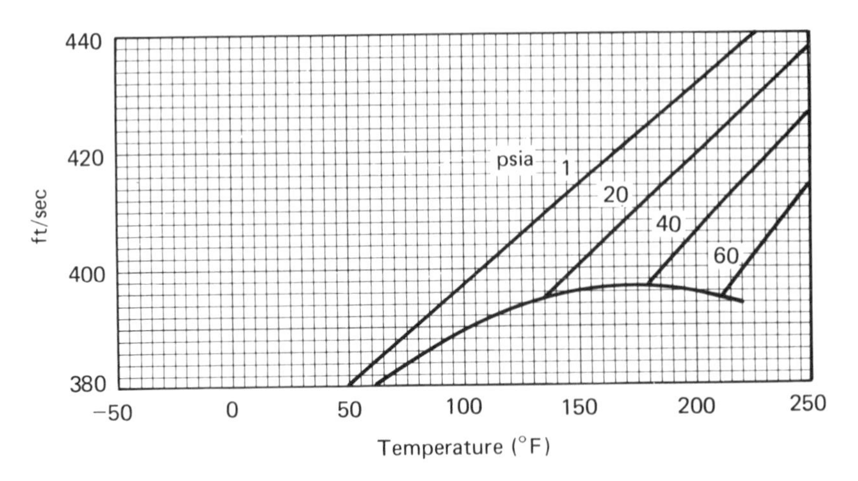

Figure 4–4 Velocity of sound in R-113 vapor.

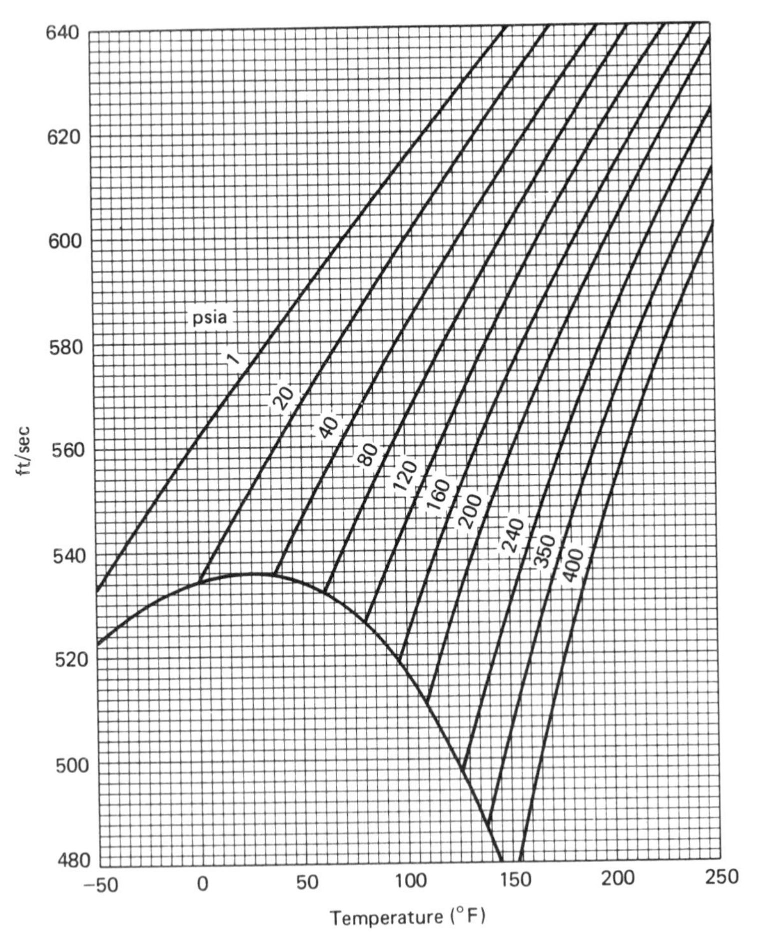

Figure 4–5 Velocity of sound in R-114 vapor.

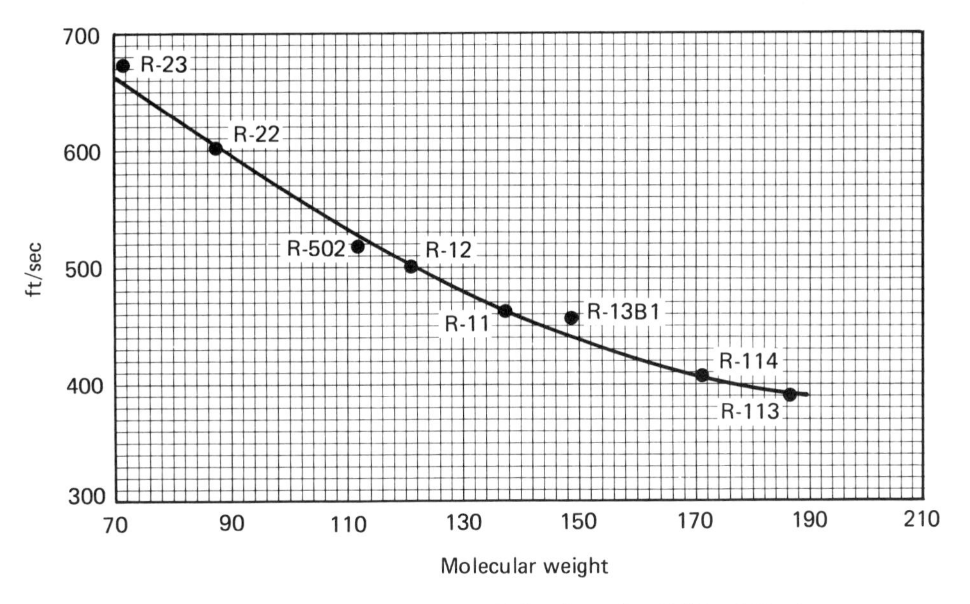

Figure 4–6 Molecular weight versus velocity of sound at 100°F and 10 psia.

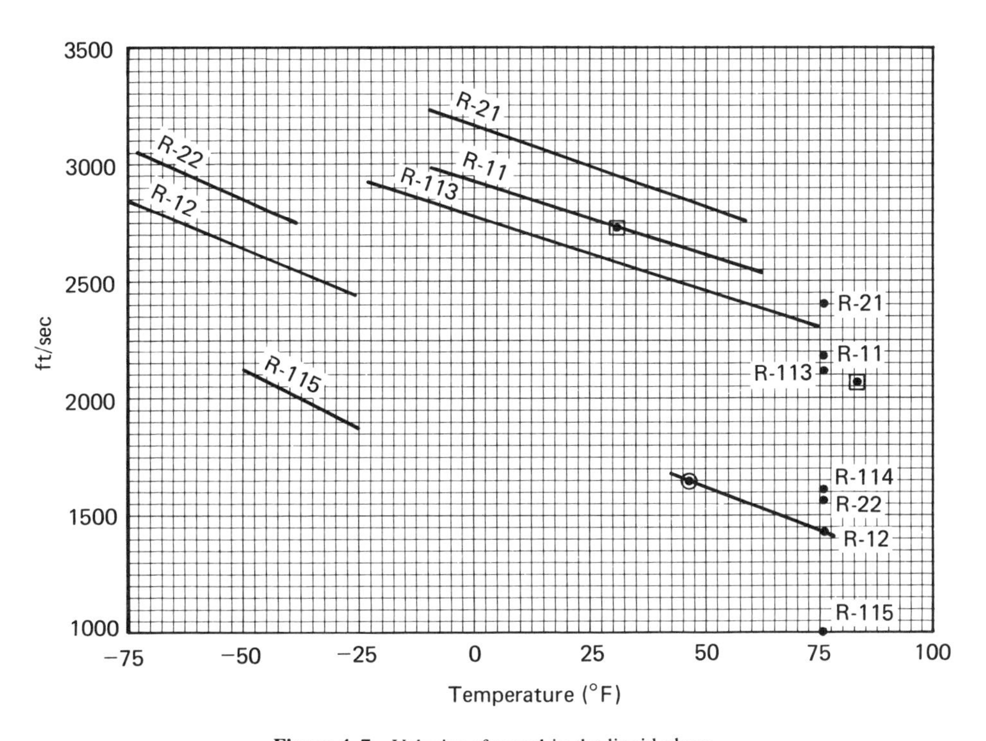

Figure 4–7 Velocity of sound in the liquid phase.

PRESSURE DROP AND VELOCITY

In designing and installing refrigeration systems, choosing the proper pipe size is important. The pipe or tubing should be as small as possible to reduce the original cost and provide adequate refrigerant velocity. On the other hand, it should be large enough to avoid excessive pressure drop and friction losses. Such losses reduce capacity and therefore increase operation costs. These considerations are different for lines carrying liquid or vapor refrigerant. Charts for estimating pressure drop and velocity for R-12, R-22, and R-502 are given in Figures 4–8 to 4–13, pages 71–76.

TABLE 4–7 RELATIONSHIP OF PRESSURE DROP AND SATURATION TEMPERATURE FOR THE LIQUID

R	Change in temperature (°F)	Pressure drop (psi)			
		120°F	100°F	80°F	60°F
11	1	0.5	0.4	0.3	0.2
	2	1.1	0.8	0.6	0.4
	3	1.6	1.2	0.9	0.7
	4	2.1	1.6	1.2	0.9
	5	2.6	2.0	1.5	1.1
12	1	2.2	1.8	1.5	1.2
	2	4.4	3.6	2.9	2.3
	3	6.6	5.4	4.4	3.5
	4	8.7	7.2	5.8	4.6
	5	10.9	8.9	7.2	5.7
22	1	3.5	2.9	2.3	1.9
	2	7.0	5.7	4.6	3.7
	3	10.4	8.6	6.9	5.5
	4	13.8	11.3	9.2	7.3
	5	17.2	14.1	11.4	9.1
113	1	0.3	0.2	0.2	0.1
	2	0.6	0.4	0.3	0.2
	3	0.9	0.6	0.5	0.3
	4	1.1	0.8	0.6	0.4
	5	1.4	1.0	0.7	0.5
114	1	0.9	0.7	0.6	0.4
	2	1.9	1.5	1.1	0.9
	3	2.8	2.2	1.7	1.3
	4	3.7	2.9	2.2	1.7
	5	4.6	3.6	2.8	2.1
502	1	3.6	3.0	2.5	2.0
	2	7.2	6.0	4.9	4.0
	3	10.8	9.0	7.3	5.9
	4	14.3	11.9	9.7	7.9
	5	17.8	14.8	12.1	9.8
500	1	2.6	2.2	1.7	1.4
	2	5.2	4.3	3.5	2.7
	3	7.8	6.4	5.2	4.1
	4	10.4	8.5	6.9	5.4
	5	12.9	10.6	8.5	6.7

Liquid

The effect of pressure drop in a liquid line is relatively small but may be significant. When pumping liquid into or out of a storage tank, power requirements will be directly affected by pressure drop in the liquid lines. In a refrigeration system, pressure drop in the liquid line will be reflected in higher condensing temperatures, higher compressor discharge temperatures, and more power to operate the compressor. If the pressure drop in the liquid line is greater than it should be for a given pipe size and liquid flow rate, the line may have a restriction such as a sharp bend or some solid material lodged in the pipe, or the liquid drier may be partially plugged.

The relationship between pressure drop and temperature change is shown in Table 4–7 for several refrigerants at four different temperature levels. At higher pressures the effect on temperature is less than at lower pressures. For example, with R-12 a 1°F change in temperature requires a change of 2.2 psi at 120°F but only 1.2 psi at 60°F. Comparing refrigerants, those with high vapor pressure require a greater pressure drop for a given change in temperature. At 120°F the pressure drop corresponding to a 1°F change in temperature is 3.5 psi for R-22 (vapor pressure is 275 psia), 2.2 psi for R-12 (vapor pressure 172 psi), and 0.5 psi for R-11 (vapor pressure 33 psia). The effect of temperature on pressure drop for liquid refrigerants is not great, as illustrated in Table 4–8.

TABLE 4–8 PRESSURE DROP WITH LIQUID REFRIGERANTS[a]

		Flow rate (lb/min)					
	Change in pressure	0.545 in. ID		0.785 in. ID		1.025 in. ID	
R	(psi)	−40°F	120°F	−40°F	120°F	−40°F	120°F
11	30	77	80				
12	30	81	76				
22	30	79	72				
11	25	70	73	186	193	379	388
12	25	74	69	194	180	394	367
22	25	72	65	190	170	383	351
11	20	61	65	164	170	334	346
12	20	65	61	172	160	350	324
22	20	63	58	168	152	340	308
11	15	52	55	140	146	285	295
12	15	55	52	147	137	300	278
22	15	54	49	143	130	292	263
11	10	41	44	111	117	228	238
12	10	44	42	118	110	241	225
22	10	43	40	115	105	233	214

[a]100 ft of type L copper pipe.

Pressure drop in a liquid line can be overcome to some extent by elevating the condenser with respect to the evaporator in a refrigeration or air-conditioning system. The weight of a liquid refrigerant is significant as shown in Table 4–9.

TABLE 4–9 WEIGHT OF LIQUID REFRIGERANT[a]

R	120°F	100°F	80°F	60° F
11	0.61	0.63	0.64	0.65
12	0.53	0.55	0.57	0.58
22	0.47	0.50	0.52	0.53
113	0.65	0.67	0.68	0.69
114	0.60	0.61	0.63	0.64
502	0.47	0.50	0.52	0.55

[a]Weight of 1-ft column, psi.

For example, a 1-ft column of liquid R-22 at 100°F weighs 0.5 psi. If the condenser were located about 4 ft above the evaporator, a pressure drop of 2 psia would be overcome.

Comparisons of pressure drops with different refrigerants are shown in Table 4–8 on a basis of lb/min. This comparison is somewhat misleading since the capacity of a system does not depend on the density of the refrigerant. In Table 4–10, page 77, refrigerant flow rates are compared at the same capacity. For example, the flow rate for R-12 is about 1.4 times that for R-22 for 1 ton of refrigeration. From Table 4–8 at 120°F and 0.785-in. ID pipe, the pressure drop is 20 psi per 100 ft at a flow rate of 152 lb/min for R-22 and 160 lb/min for R-12. However, at the same capacity the R-12 flow rate would be $1.4 \times 152 = 213$ lb/min and the pressure drop would be considerably higher.

Vapor

Pressure drop in the suction line is directly related to loss in capacity. The compressor must work harder to produce the desired pressure at the evaporator. As a rule, a temperature change of 2°F in the evaporator is considered the limit for good operating practice. The corresponding pressure change is different for different refrigerants and for different conditions. High suction line pressure drop is especially serious at low evaporating temperatures.

It is also important to maintain gas velocity in the suction line at a level that will ensure good oil travel. Velocities in the range 1200 to 4000 ft/min have been suggested. A frequent recommendation is not less than 1500 ft/min for vertical risers and not less than 750 ft/min for horizontal lines. A compromise must be reached between large pipe diameter for low pressure drop and small diameter for high velocity.

Pressure drop in the compressor discharge line is not as critical as in the suction line but still should be considered. The equivalent of a temperature change of 3°F and velocity in the range 2000 to 3500 ft/min is suggested.

Figure 4–8 Velocity in lines for Freon 12 refrigerant. (Reprinted by permission of the DuPont Company.)

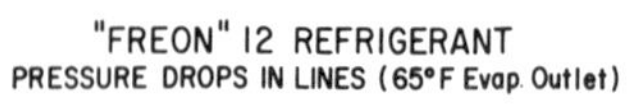

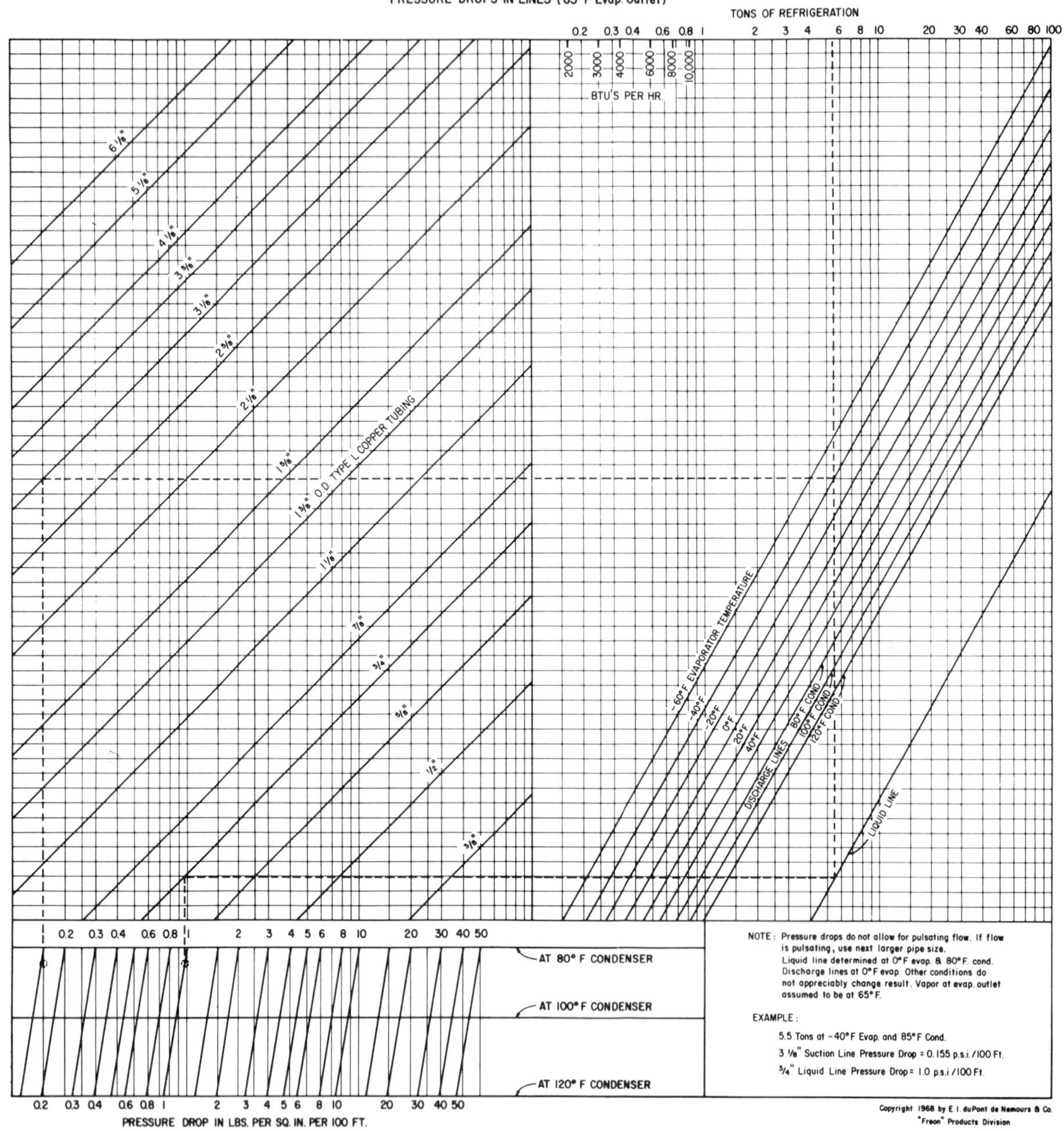

Figure 4–9 Pressure drops in lines for Freon 12 refrigerant. (Reprinted by permission of the DuPont Company.)

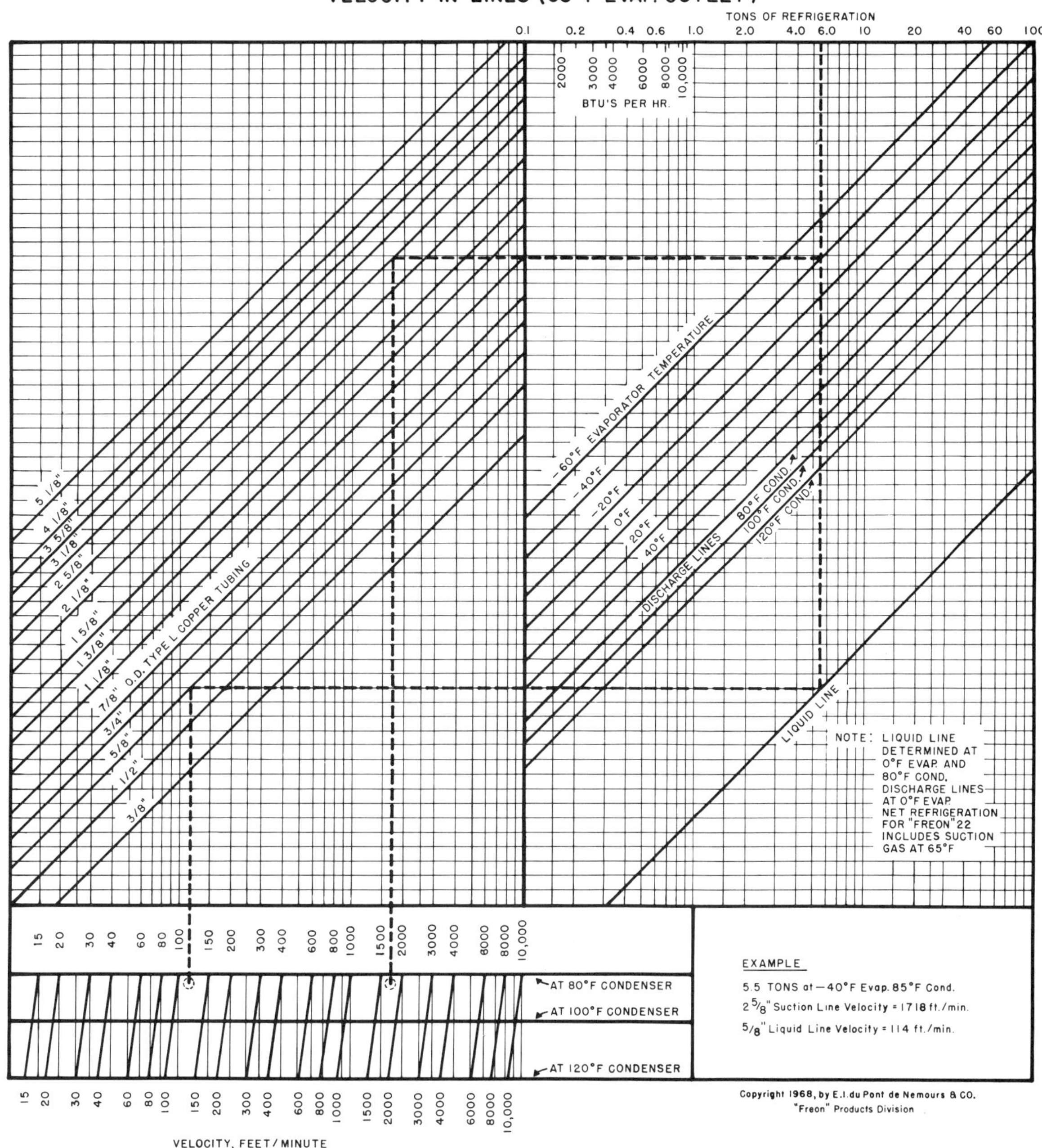

Figure 4–10 Velocity in lines for Freon 22 refrigerant. (Reprinted by permission of the DuPont Company.)

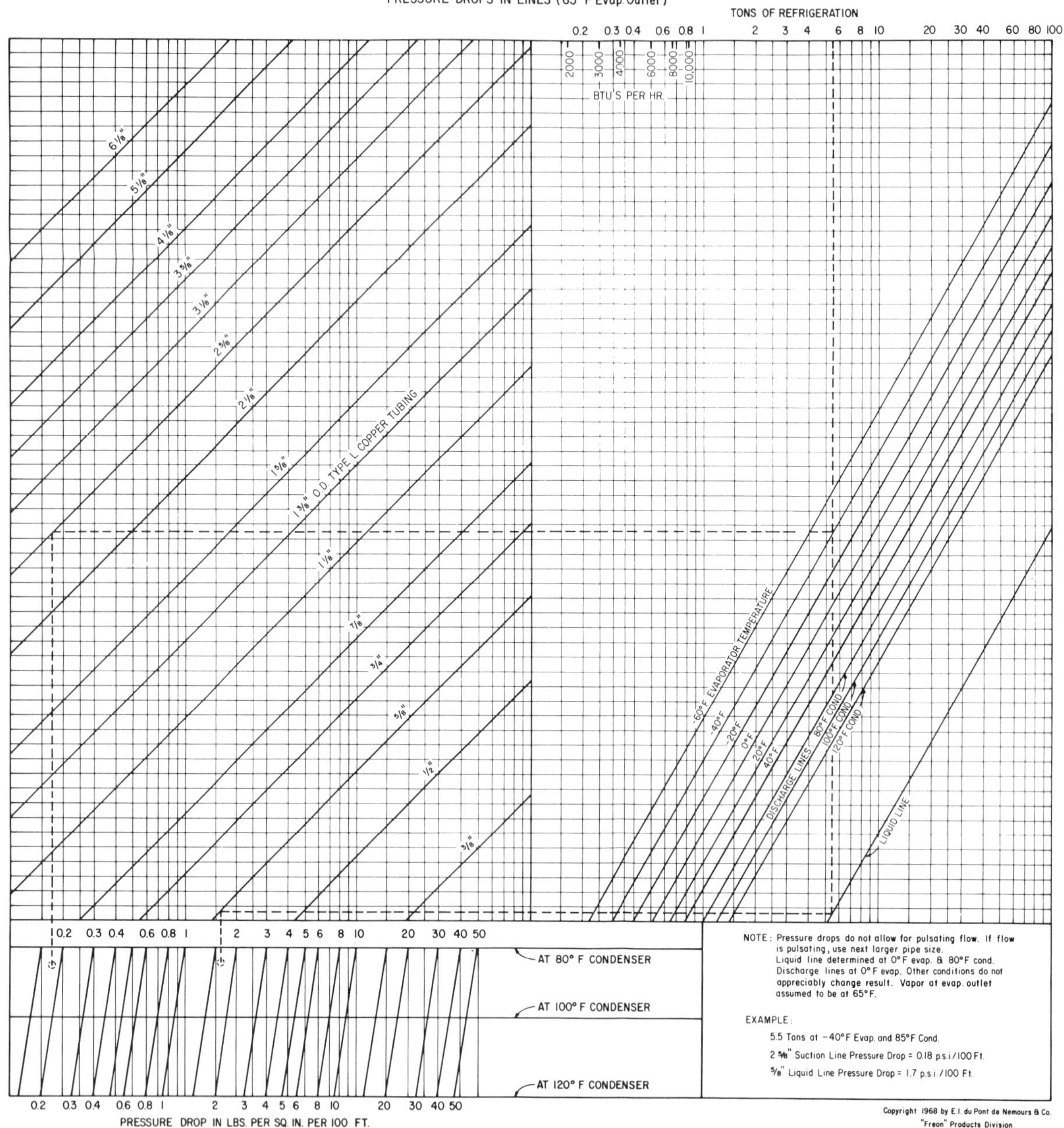

Figure 4–11 Pressure drops in lines for Freon 22 refrigerant. (Reprinted by permission of the DuPont Company.)

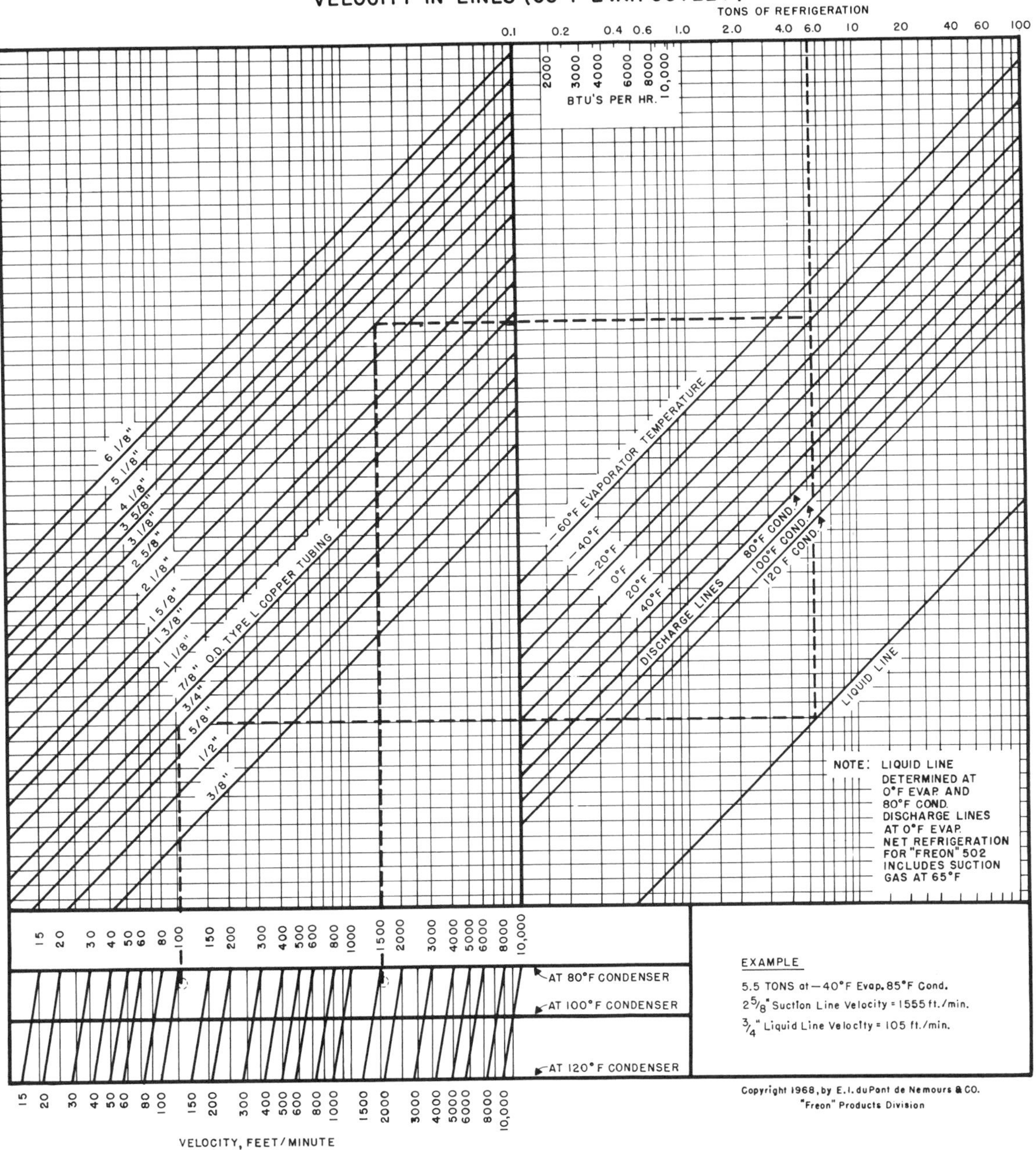

Figure 4–12 Velocity in lines for Freon 502 refrigerant. (Reprinted by permission of the DuPont Company.)

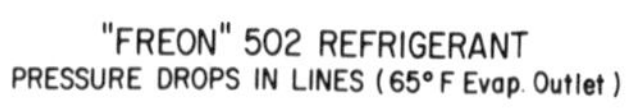

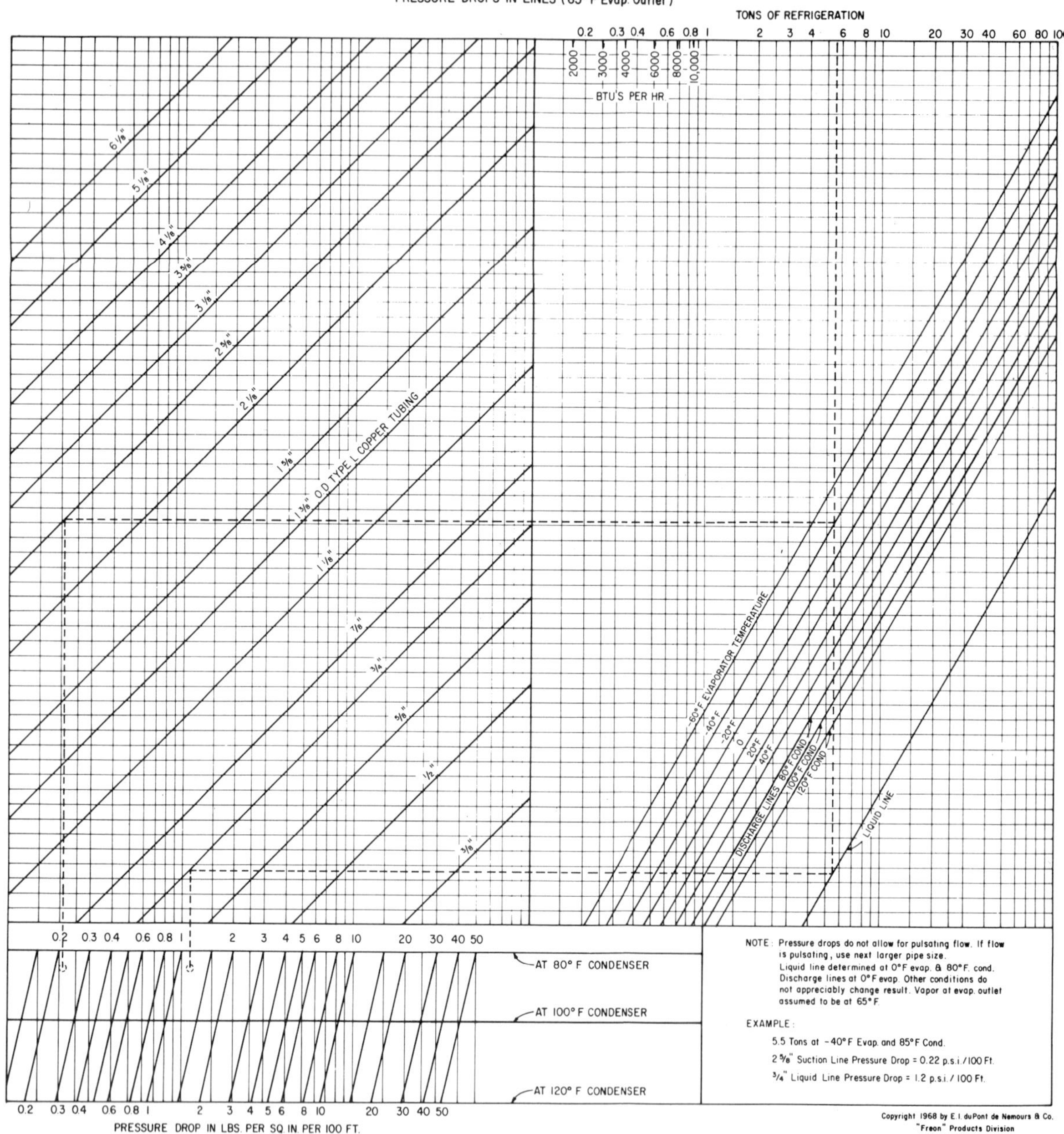

Figure 4–13 Pressure drops in lines for Freon 502 refrigerant. (Reprinted by permission of the DuPont Company.)

TABLE 4–10
REFRIGERANT FLOW RATE (LB/MIN)[a]

R	lb/min per ton
11	2.8
12	3.6
22	2.6
114	4.1
115	5.5
502	3.9

[a]Evaporating temperature, 10°F; return gas temperature, 65°F; condensing temperature, 100°F.

Source: Ref. 4–2.

Methods of Determining Pressure Drop and Velocity

Charts. The pressure drop and velocity charts in Figures 4–8 to 4–13 are based on the following conditions:

Liquid line:	0°F evaporating 80°F condensing
Discharge line:	0°F evaporating

Net refrigeration includes suction gas at 65°F.

The pressure drop and velocity charts [C-39(65) and C-39(65)V] for Freon 502 (Figures 4–13 and 4–14 respectively) were developed from estimated properties. Calculations with newer data (copyright 1969) varied from 5% higher to 10% lower for the pressure drop, as shown below. The calculations are based on a temperature of 65°F at the evaporator outlet.

Evaporator temperature (°F)	−40	0	+40
Condenser temperature (°F)	85	100	100
Capacity (tons)	5.5	1	1
Velocity (ft/min)			
Calculated	1545	4064	4182
C-39(65)V	1555	4000	4100
Pressure drop (psi/100 ft)			
Calculated	0.20	22.1	6.1
C-39(65)	0.22	21.0	6.5
Suction line (in.)			
OD	2.625	0.500	0.75
ID	2.465	0.430	0.666

If flow is pulsating, use the next-larger pipe size. Common pipe and tubing sizes are shown in Table 4–11.

TABLE 4–11 DIMENSIONS OF COPPER TUBING AND PIPE

	Soft tubing			Type L hard pipe		
Nominal size	Outside diameter (in.)	Inside diameter (in.)	Inside cross section (in.^{2})	Outside diameter (in.)	Inside diameter (in.)	Inside cross section (in.^{2})
$\frac{1}{8}$	0.125	0.065	0.0033			
$\frac{3}{16}$	0.1875	0.1275	0.0128			
$\frac{1}{4}$	0.250	0.190	0.0284	0.375	0.315	0.0779
$\frac{5}{16}$	0.3125	0.2425	0.0462			
$\frac{3}{8}$	0.375	0.311	0.0760	0.500	0.430	0.1452
$\frac{1}{2}$	0.500	0.436	0.1493	0.625	0.545	0.2333
$\frac{5}{8}$	0.625	0.555	0.2419	0.750	0.666	0.3484
$\frac{3}{4}$	0.750	0.666	0.3484	0.875	0.785	0.4840
$\frac{7}{8}$	0.875	0.785	0.4840			
1	1.00	0.888		1.125	1.025	0.8252
$1\frac{1}{8}$	1.125	1.025	0.8252	1.375	1.265	1.2568
$1\frac{3}{8}$	1.375	1.265	1.2568	1.625	1.505	1.7789

Source: Ref. 4–12.

Equations. An alternative to the charts is the use of the equations on which the charts are based. With the help of a computer, using the equations may be faster and give better answers for unusual conditions.

Additional Information. The *ASHRAE Handbook of Fundamentals* [4–11, Chap. 32] has a complete discussion of pipe sizing and many charts and tables relating capacity to pressure drop.

Example

Refrigerant:	R-12
Capacity:	5 tons = 60,000 Btu/hr = 1000 Btu/min
Evaporator:	−20°F
Evaporator outlet:	65°F
Condenser:	100°F

Charts

Velocity (Figure 4–8)

Suction line:	2600 ft/min, $2\frac{1}{8}$-in. OD pipe
Discharge line:	1300 ft/min, $1\frac{1}{8}$-in. OD pipe
Liquid line:	140 ft/min, $\frac{5}{8}$-in. OD pipe

Pressure Drop (Figure 4–9)

Suction line:	0.62 psi per 100 ft
Discharge line:	2.0 psi per 100 ft
Liquid line:	2.3 psi per 100 ft

Equations

Pressure Drop in Pipe

$$P = \frac{4fL}{D}\left(\frac{dV^2}{2g_c}\right) \tag{4–1}$$

where $\frac{P}{144}$ = psi
L = length, ft
f = Fanning friction factor
D = inside diameter of pipe, ft
d = density, lb/ft^3 (liquid or vapor)
V = velocity, ft/sec
g_c = 32.17

Reynold's Number

$$N_{\text{Re}} = \frac{VDd}{\mu} = \frac{4W}{\pi D \mu} \tag{4–2}$$

where V = velocity, ft/sec
D = inside diameter of pipe, ft
d = density in lb/ft^3
μ = viscosity, lb/ft·sec
W = velocity, lb/sec
π = 3.1416

Friction Factor

$$\frac{1}{\sqrt{f}} = -4 \log\left[\frac{0.27E}{D} + \left(\frac{7}{N_{\text{Re}}}\right)^{0.9}\right] \tag{4–3}$$

where f = Fanning friction factor
E = roughness factor = 0.00006 for drawn tubing
= 0.0018 for commercial steel
D = inside diameter of pipe, in.

Velocity

$$V = \frac{(W)(v)}{A} \tag{4–4}$$

where V = velocity, ft/sec
W = flow rate, lb/sec
v = specific volume, ft^3/lb
A = cross section of pipe, ft^2

Use of equations. Pressure drop and velocity can be calculated using the equations listed below. Computer programs can easily be developed to permit rapid determination of these properties.

R-12 vapor is at a suction pressure of 15.2 psia (−20°F saturation temperature) and 65°F.

Specific volume:	2.980 ft^3/lb
Heat content:	87.169 Btu/lb
Heat content of R-12 liquid at 100°F:	31.100 Btu/lb
Net refrigeration:	56.069 Btu/lb
Circulation rate:	

$$\frac{1000 \text{ Btu/min}}{56.069 \text{ Btu/lb}} = 17.835 \text{ lb/min} = 0.297 \text{ lb/sec}$$

***Velocity* [*equation (4–4)*]**

Suction line

2.125 in. OD
1.985 in. ID
0.02149 ft^2, inside cross section

$$V = \frac{(0.297)(2.98)}{0.02149} = 41.18 \text{ ft/sec} = 2471 \text{ ft/min}$$

Discharge line

1.125 in. OD
1.025 in. ID
0.00573 ft^2, inside cross section
0.4144 ft^3/lb, specific volume at 215°F and 132 psia

$$V = \frac{(0.297)(0.4144)}{0.00573} = 21.48 \text{ ft/sec} = 1289 \text{ ft/min}$$

Liquid line

0.625 in. OD
0.545 in. ID
0.00162 ft^2, inside cross section
0.01269 ft^3/lb, specific volume at 100°F

$$V = \frac{(0.297)(0.01269)}{0.00162} = 2.326 \text{ ft/sec} = 140 \text{ ft/min}$$

***Reynold's Number* [*equation (4–2)*]**

Suction line

$$N_{Re} = \frac{(41.18 \text{ ft/sec})(0.165 \text{ ft})(0.3356 \text{ lb/ft}^3)}{0.00000822 \text{ lb/ft-sec}} = 278{,}080$$

Discharge line

$$N_{Re} = \frac{(21.48 \text{ ft/sec})(0.08542 \text{ ft})(2.413 \text{ lb/ft}^3)}{0.00001121 \text{ lb/ft-sec}} = 394{,}939$$

Liquid line

$$N_{Re} = \frac{(2.326 \text{ ft/sec})(0.04542 \text{ ft})(78.78 \text{ lb/ft}^3)}{0.000009 \text{ lb/ft-sec}} = 924{,}763$$

***Friction Factor* [*equation (4–3)*]**

Suction line

$$\frac{1}{\sqrt{f}} = -4 \log \left[\frac{(0.27)(0.00006)}{1.985} + \left(\frac{7}{278{,}080}\right)^{0.9}\right] = 0.00373$$

Discharge line

$$\frac{1}{\sqrt{f}} = -4 \log \left[\frac{(0.27)(0.00006)}{1.025} + \left(\frac{7}{394{,}939}\right)^{0.9}\right] = 0.003607$$

Liquid line

$$\frac{1}{\sqrt{f}} = -4 \log \left[\frac{(0.27)(0.00006)}{0.545} + \left(\frac{7}{924{,}763}\right)^{0.9}\right] = 0.003436$$

***Pressure Drop* [*equation (4–1)*]**

Suction line

$$P = \left[\frac{(4)(0.00373)(100 \text{ ft})}{0.1654 \text{ ft}}\right]\left[\frac{(0.3356 \text{ lb/ft}^3)(41.18 \text{ ft/sec})}{(2)(32.17)}\right]$$

$$= 79.81 \text{ lb/ft}^2 = 0.554 \text{ psi per 100 ft}$$

Discharge line

$$P = \left[\frac{(4)(0.003607)(100 \text{ ft})}{0.08542 \text{ ft}}\right]\left[\frac{(2.413 \text{ lb/ft}^3)(21.48 \text{ ft/sec})}{(2)(32.17)}\right]$$

$$= 31.06 \text{ lb/ft}^2 = 2.03 \text{ psi per 100 ft}$$

Liquid line

$$P = \left[\frac{(4)(0.003436)(100 \text{ ft})}{0.04542 \text{ ft}}\right]\left[\frac{(78.78 \text{ lb/ft}^3)(2.326 \text{ ft/sec})}{(2)(32.17)}\right]$$

$$= 200 \text{ lb/ft}^2 = 1.4 \text{ psi per 100 ft}$$

HEAT OF FUSION

The heat of fusion is the amount of heat needed to change solid to liquid. It is usually measured at the atmospheric melting point or at the triple point (solid, liquid, and vapor in equilibrium). This property has not been widely studied, but some data are given in Table 4–12.

TABLE 4–12 HEAT OF FUSION

R	Chemical formula	Freezing point °F	Freezing point °R	Heat of fusion, H_f (cal/mol)	Ratio H_f/m.p. (°R)	Reference
			Chlorofluorocarbons			
11	CCl_3F	−166	293	1648	5.6	4–13
12	CCl_2F_2	−253	207	990	4.8	4–14
14	CF_4	−300	160	167	1.0	4–14
22	$CHClF_2$	−256	203	985	4.9	4–15
23	CHF_3	−247	212	970	4.6	4–16
112	CCl_2FCCl_2F	78	538	897	1.7	4–17
112a	CCl_3CClF_2	105	565	1040	1.8	4–17
			Chlorocarbons			
40	CH_3Cl	−132	328	1570	4.9	4–18
20	$CHCl_3$	−82	378	2270	6.0	4–14
10	CCl_4	−10	450	590	1.3	4–14
160	C_2H_5Cl	−220	240	1100	4.6	4–14
			Hydrocarbons			
50	CH_4	−296	164	225	1.4	4–19
170	C_2H_6	−298	162	683	4.2	4–19
290	C_3H_8	−305	155	842	5.4	4–19
	C_4H_{10}	−217	243	1114	4.6	4–19
	n—C_5H_{12}	−200	259	2006	7.7	4–19

Attempts to correlate the heat of fusion with other physical properties have not been very successful. The ratio of the molar latent heat to the absolute melting point has been suggested, but no useful pattern could be detected.

HEAT OF FORMATION

The heat of formation data given in Table 4–13 are for standard conditions of gas at 1 atm and 77°F as reported in the various references. In some cases, the heat associated with direct combination of the elements was measured. In others, the heat of two or more related reactions was measured and the desired heat of formation determined by differences. For example, several investigators measured the reaction of carbon and fluorine to form carbon tetrafluoride:

$$C + 2F_2 \longrightarrow CF_4$$

Other reactions that have been used include:

$$CF_2{=}CF_2 \longrightarrow C + CF_4$$

$$CF_2{=}CF_2 + 2H_2 \longrightarrow 2C + 4HF$$

$$CF_2{=}CF_2 + O_2 \longrightarrow CO_2 + CF_4$$

$$CO_2 + CF_4 \longrightarrow 2COF_2$$

TABLE 4–13 HEAT OF FORMATION

R	Chemical formula	Heat of formation (kcal/mol)	Reference
10	CCl_4	−22	4–20
11	CCl_3F	−72	4–21
11	CCl_3F	−68.1	4–22
12	CCl_2F_2	−111.3	4–23
12	CCl_2F_2	−120	4–21
12	CCl_2F_2	−117.9	4–22
13	$CClF_3$	−172.0	4–23
13	$CClF_3$	−167.5	4–24
13	$CClF_3$	−169.2	4–22
14	CF_4	−212.7	4–25
14	CF_4	−217.1	4–26
14	CF_4	−222.5	4–27
14	CF_4	−218	4–20
20	$CHCl_3$	−25	4–20
21	$CHCl_2F$	−72	
21	$CHCl_2F$	−68	4–22
22	$CHClF_2$	−118	Est'd
22	$CHClF_2$	−115.6	4–22
23	CHF_3	−163	4–28
23	CHF_3	−165.7	4–23
23	CHF_3	−169	4–29
31	CH_2ClF	−63.2	4–22
32	CH_2F_2	−106	4–28
41	CH_3F	−58	4–28
41	CH_3F	−56.8	4–22
114	$CClF_2CClF_2$	−215	4–30
125	CHF_2CF_3	−264	4–23
134a	CH_2FCF_3	−214	4–22
143a	CH_3CF_3	−178.2	4–22
152a	CH_3CHF_2	−113.6	4–31
152a	CH_3CHF_2	−119.7	4–22
C318	C_4F_8	−352	4–25
	$CBrCl_2F$	−64.4	4–22
	$CBrClF_2$	−112.7	4–22
13B1	$CBrF_3$	−155.1	4–22
	CBr_2F_2	−102.7	4–22
	$CHBrClF$	−70.5	4–22
	$CHBrF_2$	−110.8	4–22
	CH_2BrF	−60.4	4–22

REFRIGERANT HYDRATES

Gas hydrates are solid, ice-like materials formed by the physical entrapment of gas molecules by a group of water molecules. They may be stable at temperatures above the freezing point of water at higher pressures.

Water has unusual properties in that it has a relatively high boiling point but a low molecular weight. For example, nitrogen has a molecular weight of 28 and an atmospheric boiling point of −320°F. The critical temperature is −232°F. Water with a lower molecular weight, 18, has a much higher boiling point, 212°F, and a much higher critical temperature, 706°F. This great difference in properties is due to an attraction between the water molecules called *hydrogen bonding*. A number of molecules are strongly associated with each other, so they tend to act as a single molecule with a much higher molecular weight. These associated molecules form a cage that can entrap foreign molecules. The water is known as the *host* and the entrapped product as the *guest* or *hydrate former*.

Hydrate formation is not usually a problem in refrigeration since sufficiently large amounts of water are not present. In some special or unusual circumstances, however, hydrates may form.

Purification of Seawater [4–32,4–33]

The use of hydrates has been proposed [4–34 through 4–36] for the purification of seawater, and a number of chlorofluorocarbon and hydrocarbon hydrates have been evaluated for this application. The process is similar to the direct freezing of water. In either case, the ice or hydrate is separated from water containing dissolved impurities by filtration. Pure water is recovered by melting the ice or decomposing the hydrate. The principal advantage of hydrates is that they can be handled at higher temperatures (under pressure). The heat of formation of hydrates from liquids varies from 130 to 180 Btu/lb of water solidified [4–36], in the neighborhood of the heat of formation of ice (144 Btu/lb).

Hydrate properties important in a desalting process include stability in water, solubility in water and salt solutions, nature of the hydrate (easy to filter, etc.), and cost of the agent [4–35]. The decomposition temperature and pressure are also important. A minimum decomposition temperature of 41°F is considered reasonable so that the process will have a significant advantage over freezing. An upper pressure limit of 70 to 90 psia will tend to keep equipment cost within reasonable limits. The decomposition properties of several hydrates are shown in Table 4–14.

Some studies with R-11 and other hydrate formers [4–34,4–37] indicate that stirring speed affects the rate of hydrate formation but not the total amount of hydrate formed. Gradual addition of the hydrate former rather than a single addition did not change the rate or amount of hydrate formed. It is possible to form hydrates in aqueous solutions containing sizable amounts of carbohydrates, proteins, lipids, and so on. Water-soluble gases such as ethylene oxide or sulfur dioxide formed abundant hydrate crystals, in contrast to low-solubility materials such as R-11.

TABLE 4–14 DECOMPOSITION PROPERTIES OF HYDRATES

Agent	Formula	Temperature (°F)	Pressure (psia)
Methyl chloride	CH_3Cl	70	75 (est.)
R-31	CH_2ClF	42	42
R-152a	CH_3CHF_2	60	66
R-142b	CH_3CClF_2	56	35
R-12	CCl_2F_2	54	67
R-12B1	$CClBrF_2$	50	25
R-22B1	$CHBrF_2$	50	39
R-21	$CHCl_2F$	48	15
R-11	CCl_3F	46	8
Propane	C_3H_8	42	80
R-22	$CHClF_2$	61[a]	
R-32	CH_2F_2	69[a]	
R-31	CH_2ClF	64[a]	
R-41	CH_3F	66[a]	

[a]Reference [4–40].
Sources: Refs. 4–33 and 4–34.

Composition

The difficulties of directly measuring the composition of hydrates and some methods of indirect determination are discussed by Barduhn [4–36]. A few examples of hydrate composition are given in Table 4–15.

TABLE 4–15 HYDRATE COMPOSITION

Agent	Moles of water per mole of agent	Reference
R-21	16.8	4–36
R-21	17.0	4–39
R-31	8.0	4–36

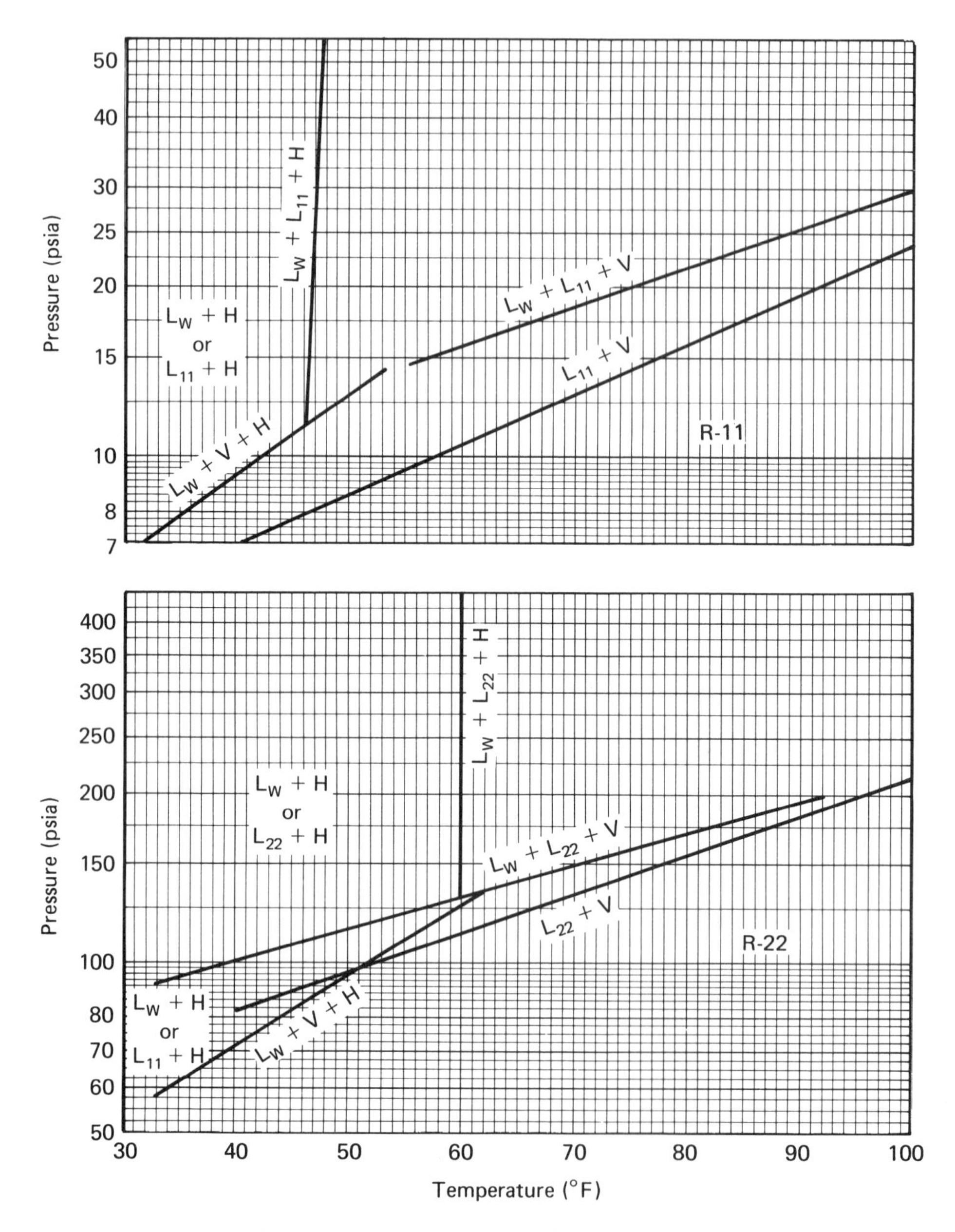

Figure 4–14 Phase diagram for water and refrigerant. (From Ref. 4–38.)

Phase Diagrams

Phase relationships of water and several refrigerants are illustrated in Figures 4–14 to 4–17. In these diagrams:

$$H = \text{hydrate}$$

$$V = \text{vapor (refrigerant and/or water)}$$

$$L_W = \text{liquid water}$$

$$L_R = \text{liquid refrigerant } (L_{11}, L_{12}, \text{etc.})$$

The vapor pressure of the liquid refrigerant is shown as $L_R + V$. In some cases the pressure when liquid water is also present ($L_R + L_W + V$) is not much higher than with refrigerant alone (or was not reported). With other refrigerants (R-11, R-22) the pressure is higher when water is present and is shown on the chart. The hydrate decomposition point is the intersection of the $L_W + H + V$ curve and the vapor-pressure curve.

Phase diagrams for water and several refrigerants follow.

Figure 4–14. R-11 and R-22	Figure 4–16. R-21
Figure 4–15. R-12	Figure 4–17. R-12Bl and R-142b

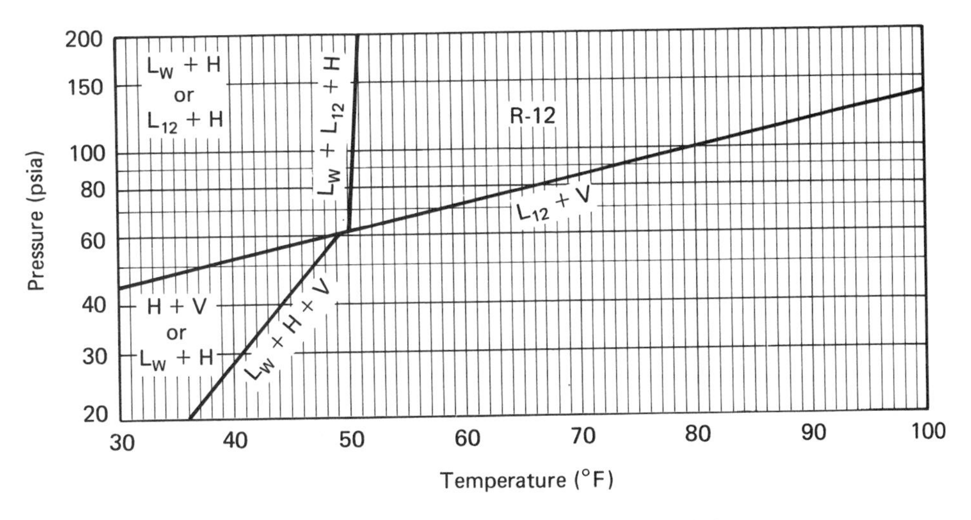

Figure 4–15 Phase diagram for water and R-12. (From Ref. 4–38.)

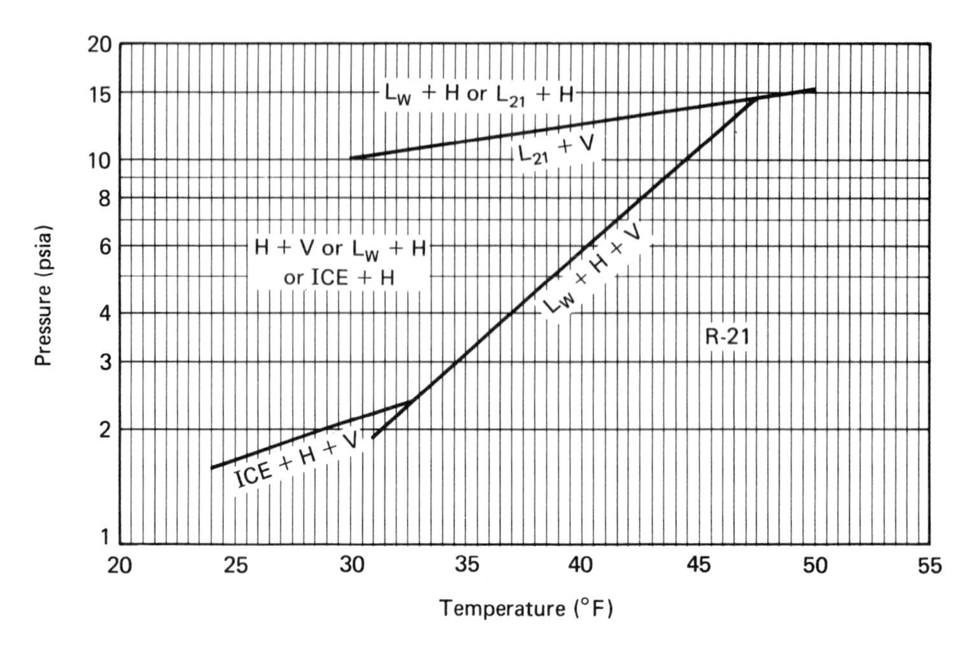

Figure 4–16 Phase diagram for water and R-21. (From Ref. 4–39.)

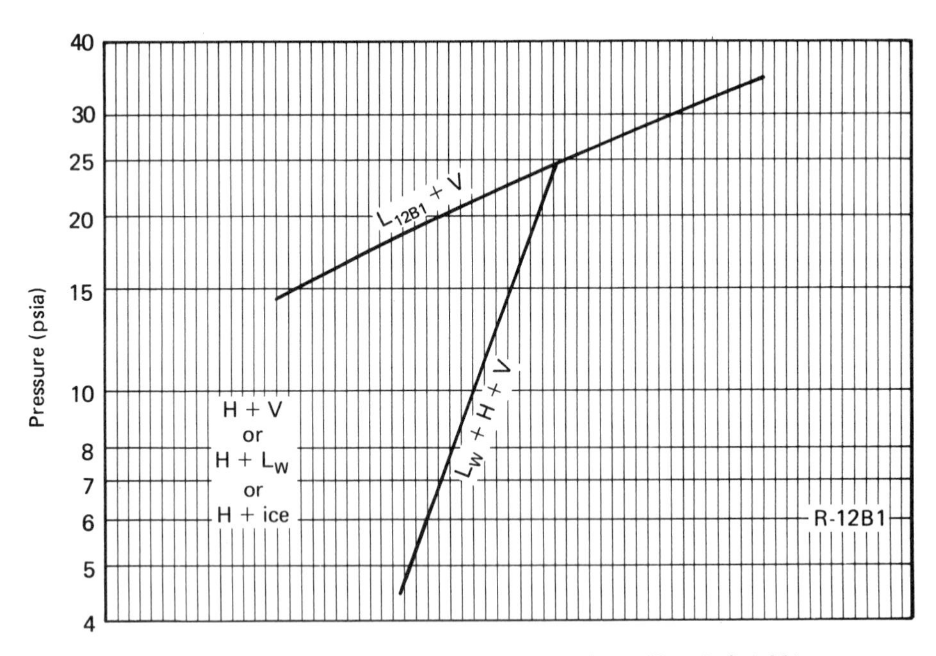

Figure 4–17 Phase diagram for water and fluorocarbons. (From Ref. 4–35.)

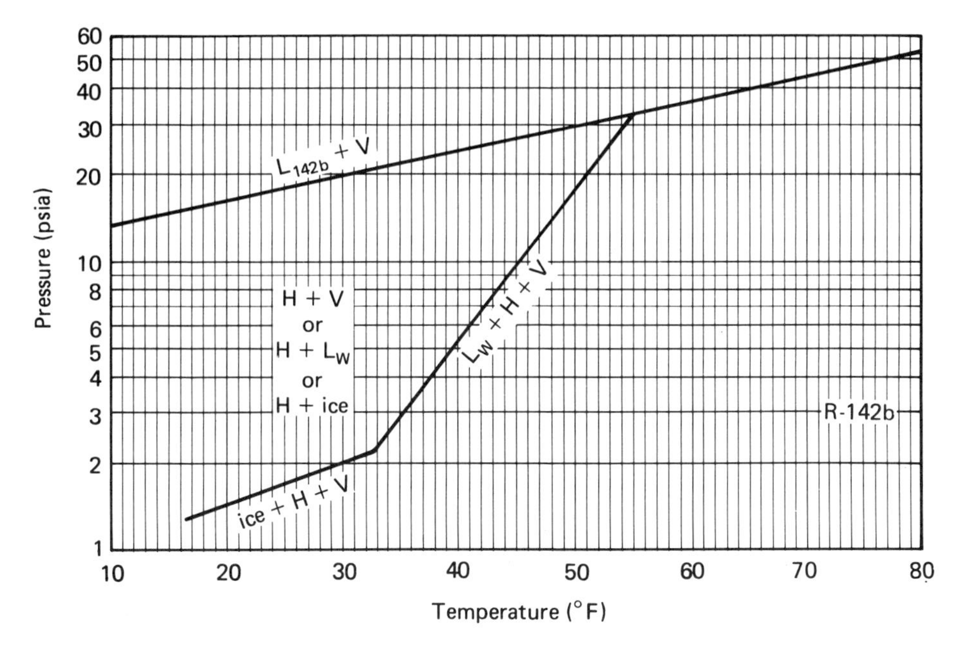

Figure 4–17 (Continued)

ABSORPTION SPECTRA

Most fluorinated compounds do not show any absorbance at wavelengths below 2 μm and very little, if any, below 3 or 4 μm. The liquids are colorless in the ultraviolet and visible ranges.

Ultraviolet	Visible	Infrared

0.38 0.75

Wavelength (μm)

REFRACTIVE INDEX

The refractive indices of fluorocarbon compounds can be calculated using the Lorentz–Lorenz equation.

$$n = \left[\frac{1 + 2\left(\dfrac{d \times \mathrm{Rm}}{\mathrm{MW}}\right)}{1 - \left(\dfrac{d \times \mathrm{Rm}}{\mathrm{MW}}\right)} \right]^{1/2}$$

where n = refractive index
d = density, g/cm^3 (liquid or vapor)
MW = molecular weight
Rm = molar refraction

The molar refraction can be calculated by adding the atomic refractions for the molecule.

The following atomic refractions have been reported by Miller [4–43] for a temperature of 68°F and using the D line of sodium.

Atom or group	Atomic refraction
C	2.584
Br	8.57
Cl (terminal)	5.85
Cl (internal)	5.69
F	1.165
4C ring	0.10

Sample calculation for R-12:

$$\begin{aligned} C &= 2.584 \\ 2Cl &= 11.70 \\ 2F &= \underline{2.330} \\ & 16.614 \text{ molar refraction at } 68°F \end{aligned}$$

Some calculated values of the refractive index for the liquid are listed in Table 4–16 together with a few experimentally measured points. Similar calculations for the vapor are given in Table 4–17.

TABLE 4–16 REFRACTIVE INDEX OF THE LIQUID

		Refractive index	
R	°F	Calculated	Experimental
11	−40	1.412	
	0	1.391	
	79.7	1.374	1.384 [4–4]
	130	1.356	
12	−40	1.334	
	0	1.319	
	79.7	1.287	1.285 [4–4]
	130	1.261	
13	−40	1.241	
	0	1.220	
	77	1.144	
13B1	−40	1.299	
	0	1.281	
	130	1.192	
14	−200	1.207	
	−150	1.185	
	−100	1.158	

TABLE 4–16 (Continued)

R	°F	Refractive index	
		Calculated	Experimental
22	−40	1.306	
	0	1.291	
	79.7	1.257	1.252 [4–4]
23	−100	1.222	
	−40	1.200	
113	−30	1.386	
	0	1.378	
	79.7	1.356	1.355 [4–4]
	130	1.341	
114	−40	1.330	
	0	1.324	
	79.7	1.290	1.290 [4–4]
	130	1.274	
115	−40	1.267	
	0	1.253	
	77	1.222	
	130	1.190	
116	−40	1.199	
	0	1.187	
	77	1.164	

Source: Ref. 4–42.

TABLE 4–17 REFRACTIVE INDEX OF THE VAPOR AT 70°F

R	MW	Rm	psia	Density (g/cm^3)	Refractive index
11	137.37	21.299	10	0.003979	1.0009255
12	120.92	16.614	14.7	0.005117	1.001055
			80	0.03034	1.006260
13	104.46	11.929	3.7	0.004329	1.0007417
			100	0.03159	1.005416
14	88.00	7.244	14.7	0.003625	1.0004477
			100	0.02546	1.003145
22	86.47	10.764	14.7	0.003638	1.0006793
			100	0.02753	1.005146
113	187.38	26.213	5.523	0.002964	1.0006221
114	170.93	21.528	14.7	0.007315	1.001302
			26	0.01332	1.002518
115	154.47	16.843	14.7	0.006483	1.001061
			100	0.05254	1.008605

ELECTRICAL PROPERTIES

Dielectric Constant

The dielectric constant is the ratio of the capacitance of a condenser filled with a gas or liquid to its capacitance when evacuated. Vapor dielectric constants for a number of halogenated compounds were measured by Fuoss [4–44] and the results for those of possible interest in refrigeration are summarized in Table 4–18. Pressures were measured in millimeters and corrected to 0°C. All pressures were below 1 atm (760 mm). The measurements were made at 60 Hz with 100 V across the cell. The temperature varied from 22°C to 30°C depending on room temperature.

Makita et al. [4–45] measured the liquid and vapor dielectric constants of four fluorocarbon refrigerants at the following ranges of temperature and pressure:

R-12: 20 to 120°C, 1 to 108 bar (14.5 to 1566 psia)
R-13: 10 to 75°C, 1 to 155 bar (14.5 to 2248 psi)
R-22: 20 to 100°C, 1 to 49 bar (14.5 to 711 psi), vapor only
R-23: 10 to 50°C, 2 to 168 bar (29 to 2437 psi)

TABLE 4–18 VAPOR DIELECTRIC CONSTANTS

R-11 (CCl_3F) 26.0°C (78.8°F)		R-12 (CCl_2F_2) 29.0°C (84.2°F)		R-13 ($CClF_3$) 29.0°C (84.2°F)		R-14 (CF_4) 24.8°C (76.6°F)	
P[a]	Constant	P[a]	Constant	P[a]	Constant	P[a]	Constant
498	1.00251	719	1.00305	619	1.00196	723	1.00113
448	1.00224	504	1.00214	607	1.00193	629	1.00098
339	1.00167	417	1.00176	484	1.00153	515	1.00081
270	1.00132	331	1.00141	330	1.00100	429	1.00067
189	1.00092	188	1.00079	211	1.00066	332	1.00052
129	1.00060	157	1.00065	99	1.00031	223	1.00036
73	1.00073	83	1.00032			131	1.00021

R-21 ($CHCl_2F$) 30.0°C (86.0°F)		R-22 ($CHClF_2$) 25.4°C (77.7°F)		R-114 ($CClF_2CClF_2$) 26.8°C (80.2°F)		R-115 (CF_3CClF_2) 27.4°C (81.3°F)	
P[a]	Constant	P[a]	Constant	P[a]	Constant	P[a]	Constant
719	1.00659	711	1.00658	715	1.00404	722	1.00334
709	1.00651	611	1.00564	605	1.00340	584	1.00271
419	1.00379	509	1.00469	512	1.00287	506	1.00236
312	1.00280	410	1.00374	416	1.00230	333	1.00158
222	1.00199	311	1.00282	328	1.00180	260	1.00123
134	1.00120	205	1.00187	225	1.00125	195	1.00093
		154	1.00137	150	1.00083	121	1.00058
		105	1.00096	93	1.00052	59	1.00028

TABLE 4–18 (Continued)

R-116 (CF_3CF_3) 23.0°C (73.4°F)		R-142b (CH_3CClF_2) 27.4°C (81.3°F)		R-133 ($C_2H_2ClF_3$) 26.6°C (79.9°F)		R-143b (CH_3CF_3) 25.2°C (77.4°F)	
P[a]	Constant	P[a]	Constant	P[a]	Constant	P[a]	Constant
711	1.00197	706	1.01338	715	1.01025	644	1.01323
608	1.00169	600	1.01131	612	1.00874	554	1.01134
509	1.00140	511	1.00959	513	1.00730	436	1.00891
411	1.00114	379	1.00709	414	1.00586	337	1.00683
307	1.00086	279	1.00521	309	1.00437	219	1.00443
201	1.00056	204	1.00379	207	1.00295	120	1.00242
110	1.00031	155	1.00289	153	1.00219	83	1.00167
61	1.00017	123	1.00229	107	1.00153	61	1.00123
		95	1.00177	65	1.00091	40	1.00082
		71	1.00133	43	1.00060	24	1.00047

[a]Pressure in millimeters adjusted to 0°C.
Source: Ref. 4–44.

The average error in the measurements is reported to be less than 10^{-4}. The results are given in Tables 4–19 to 4–22. Smoothed data are reported. For both liquids and gases, the dielectric constant increases as the pressure increases and decreases as the temperature increases.

Hess et al. [4–49] measured the refractive index and dielectric constant of R-13 in the vapor phase as functions of density, with the results shown in Figure 4–18. R-13 was slowly introduced into an evacuated chamber and periodic measure-

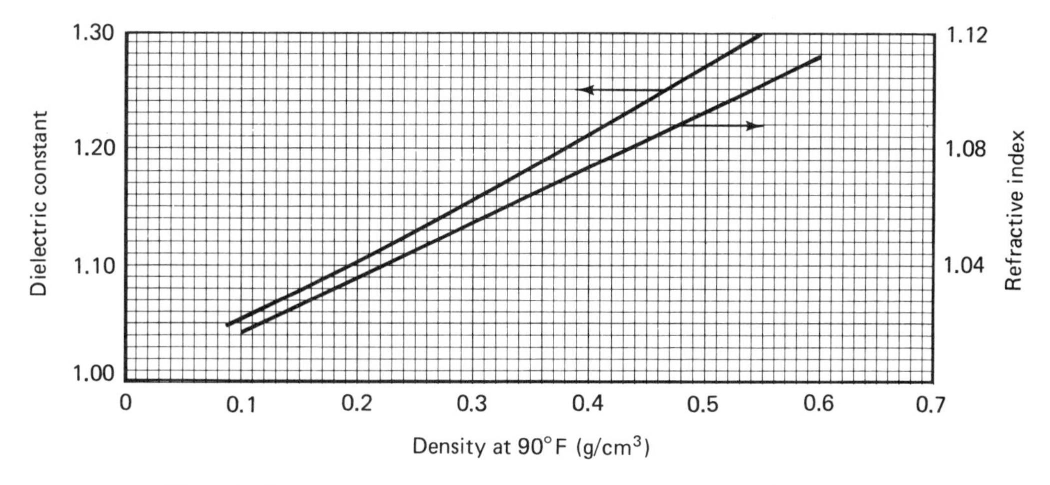

Figure 4–18 Dielectric constant and refractive index as a function of density for R-13.

TABLE 4–19 VAPOR DIELECTRIC CONSTANTS FOR R-22

	Temperature [°C(°F)]				
	20 (68)	30 (86)	50 (122)	75 (167)	100 (212)
Saturation pressure [bar (psia)]	9.10 (132.0)	11.9 (172.6)	19.4 (281.4)	33.8 (490.2)	
Pressure [bar (psia)]	1.074	1.103	1.174	1.383	
Saturated vapor					
1 (14.50)	1.005	1.005	1.004	1.004	1.003
5 (72.52)	1.033	1.031	1.026	1.023	1.022
10 (145.0)		1.076	1.065	1.056	1.049
15 (217.6)			1.114	1.093	1.079
20 (290.1)				1.139	1.112
25 (36.26)				1.193	1.155
30 (435.1)				1.277	1.199
35 (507.6)					1.261
40 (580.2)					1.344

Source: Ref. 4–45.

ments of capacitance and container weight were made. Similar measurements were made during withdrawal of the gas. The temperature of the chamber was held near 90°F. Care had to be taken to make the procedure slow enough since the R-13 was very close to its critical temperature and the density at a given pressure is a very steep function of the temperature. The pressure corresponding to various densities at a temperature of 90°F for R-13 is as follows:

Density		
g/cm^3	lb/ft^3	Pressure (psia)
0.1	6.243	286
0.2	12.49	457
0.3	18.73	544
0.4	24.97	580
0.5	31.21	595

Makita et al. [4–45] also noted the direct relationship between dielectric constant and density.

TABLE 4–20 DIELECTRIC CONSTANTS FOR R-12

A. At saturation

	Temperature [°C(°F)]								
	20 (68)	25 (77)	30 (86)	40 (104)	50 (122)	60 (140)	80 (176)	100 (212)	110 (230)
Pressure [bar (psia)]	5.7 (83)	6.5 (94)	7.4 (107)	9.6 (139)	12.2 (177)	15.3 (222)	23.1 (335)	33.4 (484)	39.8 (577)
Vapor	1.022	1.025	1.029	1.036	1.048	1.060	1.096	1.172	1.245
Liquid	2.125	2.100	2.075	2.026	1.974	1.922	1.816	1.663	1.500

B. Liquid and vapor

	Temperature [°C(°F)]										
Pressure bar (psia)	20 (68)	25 (77)	30 (86)	40 (104)	50 (122)	60 (140)	80 (176)	100 (212)	110 (230)	115 (239)	120 (248)
1 (14.5)	1.005	1.004	1.004	1.004	1.004	1.004	1.004	1.004	1.004	1.004	1.004
5 (72.5)	1.019	1.018	1.018	1.017	1.016	1.016	1.015	1.015	1.014	1.014	1.014
10 (145)	2.129	2.103	2.077	2.027	1.035	1.034	1.031	1.028	1.027	1.027	1.026
15 (218)	2.132	2.106	2.081	2.032	1.979	1.059	1.049	1.044	1.042	1.041	1.040
20 (290)	2.135	2.109	2.084	2.036	1.983	1.928	1.076	1.063	1.059	1.057	1.056
25 (363)	2.138	2.113	2.088	2.040	1.987	1.935	1.820	1.087	1.079	1.075	1.072

TABLE 4–20 (Continued)

B. Liquid and vapor

Pressure bar (psia)	Temperature [°C(°F)]										
	20 (68)	25 (77)	30 (86)	40 (104)	50 (122)	60 (140)	80 (176)	100 (212)	110 (230)	115 (239)	120 (248)
30 (435)	2.140	2.116	2.092	2.044	1.992	1.941	1.828	1.122	1.105	1.099	1.095
35 (508)	2.143	2.119	2.095	2.048	1.997	1.946	1.837	1.672	1.146	1.136	1.127
40 (580)	2.146	2.122	2.098	2.052	2.001	1.952	1.845	1.701	1.507	1.193	1.168
50 (725)	2.152	2.127	2.104								
60 (870)	2.158	2.133	2.110	2.060	2.009	1.962	1.860	1.735	1.648	1.590	1.480
70 (1015)	2.163	2.139	2.116	2.065	2.018	1.970	1.873	1.762	1.697	1.653	1.602
80 (1160)	2.168	2.144	2.122	2.071	2.026	1.980	1.885	1.781	1.724	1.692	1.651
90 (1305)	2.172	2.149	2.127	2.079	2.034	1.988	1.897	1.799	1.743	1.717	1.682
100 (1450)	2.177	2.154	2.132	2.090	2.047	2.003	1.917	1.827	1.779	1.757	1.728

Source: Ref. 4–45.

TABLE 4–21 DIELECTRIC CONSTANTS FOR R-13

	Temperature [°C(°F)]						
	10 (50)	20 (68)	25 (77)	30 (86)	40 (104)	50 (122)	75 (167
Saturation pressure [bar (psia)]	25.2 (365)	31.9 (463)	35.5 (515)				
Saturated vapor	1.111	1.142	1.165				
Saturated liquid	1.740	1.641	1.592				
Pressure [bar (psia)]							
1 (14.50)	1.003	1.003	1.003	1.003	1.002	1.002	1.002
5 (72.52)	1.014	1.013	1.012	1.011	1.011	1.010	1.008
10 (145.0)	1.029	1.027	1.026	1.025	1.023	1.021	1.018
15 (217.6)	1.046	1.042	1.040	1.038	1.035	1.033	1.028
20 (290.0)	1.068	1.062	1.059	1.056	1.051	1.046	1.039
25 (362.6)	1.108	1.092	1.086	1.080	1.069	1.059	1.050
30 (435.1)	1.751	1.129	1.120	1.110	1.093	1.079	1.065
35 (507.6)	1.762	1.652	1.157	1.144	1.122	1.102	1.080
40 (580.2)	1.773	1.678	1.619	1.357	1.176	1.127	1.097
50 (725.2)	1.791	1.712	1.669	1.610	1.410	1.217	1.140
60 (870.2)	1.804	1.736	1.700	1.652	1.551	1.370	1.194
70 (1015)	1.818	1.758	1.727	1.686	1.606	1.495	1.261
80 (1160)	1.831	1.774	1.746	1.712	1.642	1.552	1.332
90 (1305)	1.844	1.788	1.762	1.730	1.669	1.590	1.395
100 (1450)	1.857	1.802	1.776	1.745	1.687	1.617	1.443
110 (1595)	1.868	1.814	1.790	1.760	1.704	1.638	1.479
120 (1740)	1.878	1.826	1.802	1.773	1.719	1.657	1.512
130 (1885)	1.888	1.836	1.811	1.785	1.733	1.672	1.540
140 (2031)	1.898	1.846	1.821	1.796	1.744	1.684	1.562
150 (2176)	1.907	1.855	1.830	1.805	1.753	1.693	1.574

Source: Ref. 4–45.

TABLE 4–22 DIELECTRIC CONSTANTS FOR R-23

	Temperature [°C(°F)]					
	10 (50)	20 (68)	20 (77)	30 (86)	40 (104)	50 (122)
Saturated pressure [bar (psia)]	32.5 (471.4)	42.1 (610.6)	47.8 (693.3)			
Saturated vapor	1.597	1.95	2.30			
Saturated liquid	6.83	5.51	4.70			
Pressure [bar (psia)]						
1 (14.50)	1.008	1.007	1.007	1.007	1.007	1.006
5 (72.52)	1.044	1.042	1.041	1.039	1.037	1.035
10 (145.0)	1.096	1.091	1.087	1.084	1.079	1.073
15 (217.6)	1.161	1.147	1.140	1.135	1.124	1.115
20 (290.1)	1.238	1.212	1.201	1.191	1.175	1.162
25 (362.6)	1.339	1.293	1.274	1.260	1.235	1.215
30 (435.1)	1.481	1.391	1.363	1.339	1.303	1.273
35 (507.6)	6.900	1.541	1.483	1.444	1.386	1.341
40 (580.2)	7.020	1.783	1.660	1.580	1.486	1.418
50 (725.2)	7.248	5.978	5.040	2.200	1.790	1.628
60 (870.2)	7.463	6.345	5.694	4.870	2.480	1.944
70 (1015)	7.658	6.603	6.040	5.443	3.870	2.503
80 (1160)	7.810	6.818	6.317	5.780	4.619	3.328
90 (1305)	7.945	6.980	6.520	6.042	5.040	3.965
100 (1450)	8.065	7.110	6.680	6.243	5.338	4.419
110 (1595)	8.180	7.274	6.837	6.418	5.577	4.741
120 (1740)	8.280	7.398	6.980	6.578	5.766	4.990
130 (1885)	8.379	7.509	7.105	6.707	5.938	5.190
140 (2031)	8.481	7.612	7.220	6.837	6.082	5.372
150 (2176)	8.577	7.717	7.322	6.949	6.222	5.540
160 (2321)	8.650	7.816	7.420	7.052	6.345	5.679

Source: Ref. 4–45.

In the course of his work, Fuoss [4–44] measured the vapor density of the materials with which he worked. The results are listed in Table 4–23.

TABLE 4–23 VAPOR DENSITIES AND POLARIZATIONS OF FLUORINE COMPOUNDS[a]

Compound	p	t_1	d	M/M_0	t_2	P
CF_4	722.9	23.6	3.451	1.0039	24.8	9.7
$CClF_3$	613.8	28.3	3.430	1.0058	29.0	19.8
CCl_2F_2	719.3	28.0	4.709	1.0169	29.0	26.2
CCl_3F	551.2	27.2	4.138	1.0238	26.0	30.7
$CHClF_2$	711.2	25.4	3.377	1.0223	25.4	56.1
$CHCl_2F$	712.8	29.2	3.958	1.0172	30.0	56.5
CH_3CF_3	644.2	26.0	2.948	1.0161	25.2	124.2
CH_3CClF_2	706.8	26.6	3.895	1.0252	27.4	115.5
CF_3CF_3	710.5	22.6	5.335	1.0036	22.6	17.2
$CClF_2CF_3$	721.8	27.0	6.007	1.0092	26.2	28.7
$CClF_2CClF_2$	715.2	26.2	6.569	1.0047	26.2	34.5

[a] p, Pressure for density measurement, mm; t_1, temperature (°C) for density measurement; d, density, mg/cm^3; M/M_0, ratio of molecular weight based on density to that calculated from the chemical formula; t_2, temperature (°C) at which the polarization was calculated; P, polarization—average based on direct measurements and extrapolations.

Polarization and Dipole Moments

The relationship between total polarization and dielectric constant is given by the Clausius–Mosotti equation [4–48]:

$$P_{\text{total}} = \frac{E-1}{E+2}\left(\frac{\text{MW}}{d}\right) \tag{4–5}$$

where P = total polarization
MW = molecular weight
d = density, g/cm^3
E = dielectric constant

Debye and others have modified this equation to allow for more precise definitions of various types of polarization in a molecule. As reported by Fuoss [4–44], the total polarization consists of three parts:

$$P_{\text{total}} = P_{\text{molar}} + P_{\text{electrical}} + P_{\text{atomic}} \tag{4–6}$$

P_{molar} can be obtained from the permanent dipole moment from the following equation:

$$u^2 = \left(\frac{9kT}{4\pi N}\right) P_M \qquad \text{or} \qquad u = 0.01281PT \tag{4–7}$$

where u = permanent dipole moment (vapor)
k = Boltzmann's constant = 1.37×10^{-16}
T = temperature, K
N = Avogadro's number = 6.06×10^{23}
π = 3.1416
P_M = polarization, molar

Molar Polarization [4–50]

1. The molar polarization can be found from equation (4–6) if the total polarization, the electrical polarization, and the atomic polarization are known. The total polarization can be calculated with equation (4–5). The electrical polarization can be considered equal to the molar refraction which can be estimated from atomic refractions as illustrated on page 000.
2. Another method of finding the molar polarization could be called the *temperature variation procedure*. The total polarization is found by direct measurement or by calculation from equation (4–5) at several temperatures and then plotted versus $1/T$ to give a straight line.

$$P_{\text{total}} = A + \frac{B}{T}$$

The atomic and electrical polarizations are not temperature dependent and are included in A. The molar polarization equals B/T, where B is the slope of the line. The dipole moment can then be calculated from equation (4–7).

3. The molar refraction can be used as an estimate for the electrical polarization. The index of refraction, n, is related to the molar refraction in equation (4–8).

$$R_{\text{molar}} = \frac{(n^2 - 1)(\text{MW})}{(n^2 + 2)d} \tag{4–8}$$

where R_{molar} = molar refraction
n = refractive index
d = liquid density, g/cm^3
MW = molecular weight

If there are not sufficient values of dielectric constant or measured polarization for the temperature plot, a reasonable approximation may be found using a single value.

$$\begin{aligned} u &= 0.01281\sqrt{P_M T} \\ &= 0.01281\sqrt{(P_T - P_E - P_A)T} \end{aligned} \tag{4–9}$$

where P_T = total polarization and is measured or is calculated with equation (4–5)
P_E = molar refraction = R_M
P_A = estimated as follows:

The dipole moment is calculated by the temperature variation method described above and equation (4–9) rearranged.

$$P_A = P_T - P_E - \frac{u^2}{(0.01281)^2 T}$$

The average P_A was calculated for compounds based on methane and those based on ethane as shown in Table 4–24.

TABLE 4–24 CALCULATED VALUES FOR P_A

R	Dipole moment	P_T	P_M	P_E	P_A (calc'd)
		Methane Derivatives[a]			
11	0.45	30.7	4.17	21.30	5.27
12	0.51	26.2	5.25	16.61	4.33
13	0.46	19.8	4.27	11.93	3.60
14	0.00	9.7	0.00	7.24	2.46
21	1.29	56.5	33.47	16.48	6.55
22	1.40	56.1	40.02	11.79	4.28
		Ethane Derivatives[b]			
114	0.56	34.5	6.39	21.53	6.58
115	0.52	28.7	5.51	16.84	6.34
116	0.00	17.28	0.00	12.16	5.12

[a]Average P, 4.41; standard deviation, 0.57.
[b]Average P, 6.01; standard deviation, 0.45.

If no other data are available, the dipole moment can be calculated from the Onsager equation:

$$u^2 = \frac{9kT}{4\pi N}\left[\frac{(2E + E_\infty)(E + 2)}{3E(E_\infty + 2)}\right]\left[\frac{M}{d}\left(\frac{E - 1}{E + 2} - \frac{E_\infty - 1}{E_\infty + 2}\right)\right]$$

where u = vapor dipole moment
E = liquid dielectric constant
E_∞ = dielectric constant measured at a frequency so high that the permanent dipole cannot contribute; the square of the index of refraction serves as a good estimate for E_∞ for nonpolar or only slightly polar molecules
d = liquid density, g/cm^3
T = temperature, K

The term

$$\frac{(2E + E_\infty)(E + 2)}{3E(E_\infty + 2)}$$

is added by Onsager to compensate for the internal field of the molecule. The term

$$\frac{\text{MW}}{d}\left(\frac{E - 1}{E + 2} - \frac{E_\infty - 1}{E_\infty + 2}\right) \quad \text{or} \quad \frac{\text{MW}}{d}\left(\frac{E - 1}{E + 2} - \frac{n^2 - 1}{n^2 + 2}\right)$$

is analogous to $P_T - P_E$ in the vapor relationship, where

$$P_T = \frac{\text{MW}}{d}\left(\frac{E - 1}{E + 2}\right) \quad \text{and} \quad P_E = R_M = \frac{\text{MW}}{d}\left(\frac{n^2 - 1}{n^2 + 2}\right)$$

Somewhat better values for the dipole moment are obtained if the estimated values for P_A (4.41 for methane derivatives and 6.01 for ethane derivatives) are added to the last term of the equation. The three methods of calculating dipole moments are compared below:

R	Temperature variation	Onsager	Onsager corrected
11	0.45	0.58	0.34
12	0.51	0.70	0.49
21	1.29	1.43	1.32
22	1.40	1.66	1.55
114	0.56	0.83	0.62

The dipole moments listed in Table 4–25 are from National Bureau of Standards Circular 537 [4–51]; the original references are included here.

Change with Temperature

The change of the liquid dielectric constant with temperature can be estimated from charts like Figure 4–19, page 104, if one or more values for the dielectric constant are known [4–48]. The polarization is calculated by the Clausius–Mosotti equation:

$$P = \frac{E - 1}{E + 2}\left(\frac{MW}{d}\right)$$

The polarization is multiplied by the absolute temperature and the result plotted against temperature. A straight line is drawn between the point or points and the origin. The procedure is reversed to find the dielectric constant at a new temperature. If the dielectric constant is known at several different temperatures and extrapolation of a straight line connecting them passes through the origin, the compound is thought not to have a permanent dipole moment and the equation above is suitable for calculating the polarization. Most of the fluorocarbon refrigerants are in this category or at least close enough for reasonable estimation. Compounds containing hydrogen, such as R-22 and R-21, have fairly high dielectric constants and the PT versus T curve (straight line) probably does not extrapolate to the origin. However, estimations using the procedure above will still give reasonable results. The curves in Figure 4–19 are based on the data in Table 4–26.

Review

In 1955, B. J. Eiseman, Jr. published a review of the electrical properties of fluorocarbon refrigerants and their significance in hermetic refrigeration systems [4–46]. The article contains so much information that is still relevant that parts of it are included here.*

Probably the most important electrical property is the dielectric strength, measured by the maximum potential difference that can be applied to an insulating material without causing electrical breakdown. The dielectric strengths of the refrigerants in hermetic systems must be high since the motor operates in an atmosphere of the refrigerant. In the course of measuring resistance, the capacitance was also

*Reprinted by permission of the American Society of Heating, Refrigerating and Air-Conditioning Engineers, Inc., Atlanta, from *Refrig. Eng.*, 63 (Apr. 1955), 61.

TABLE 4–25 ELECTRIC DIPOLE MOMENTS

R	Formula	Dipole moment [u × 10^{18} (esu)]	Temperature (K)	Reference
10	CCl_4	0	296, 368	4–52
20	$CHCl_3$	1.02	298, 368	4–52
		1.013	301–427	4–53
30	CH_3Cl	1.87	290–456	4–54
		1.87	298–418	4–55
		1.89	296, 368	4–52
		1.87		4–56
		1.869		4–57
11	CCl_3F	0.45	299–367	4–58
		0.68	299	4–44
12	CCl_2F_2	0.505	305–470	4–58
		0.70	302	4–44
		0.55	300–410	4–59
13	$CClF_3$	0.65	302	4–44
		0.39	273–420	4–59
14	CF_4	0	193–368	4–60
		0	298–368	4–61
21	$CHCl_2F$	1.293	305–424	4–58
		1.41	303	4–44
22	$CHClF_2$	1.409	304–479	4–58
		1.48	298	4–44
41	CH_3F	1.81	224–498	4–62
		1.85	193–368	4–52
32	CH_2F_2	1.93		4–63
23	CHF_3	1.60	193–368	4–60
		1.645		4–64
12B2	CBr_2F_2	1.02	302	4–44
22B1	$CHBrF_2$	1.50	300	4–44
30B2	CH_2Br_2	1.43	338–427	4–53
115	$CClF_2CF_3$	0.80	300	4–44
114	$CClF_2CClF_2$	0.80	299	4–44
116	CF_3CF_3	0	296	4–44
132	$CHCl_2CHF_2$	1.34	334	4–65
		1.40	395	
		1.47	474	
131	$CHCl_2CHClF$	1.38	379	4–65
		1.44	512	
142b	CH_3CClF_2	2.21	300	4–44
		2.14	357–507	4–65
143	CH_3CF_3	2.35	298	4–44
		2.321		4–57
151	CH_2ClCH_2F	1.84	309	4–65
		1.97	506	
161	CH_3CH_2F	1.92	236–535	4–62

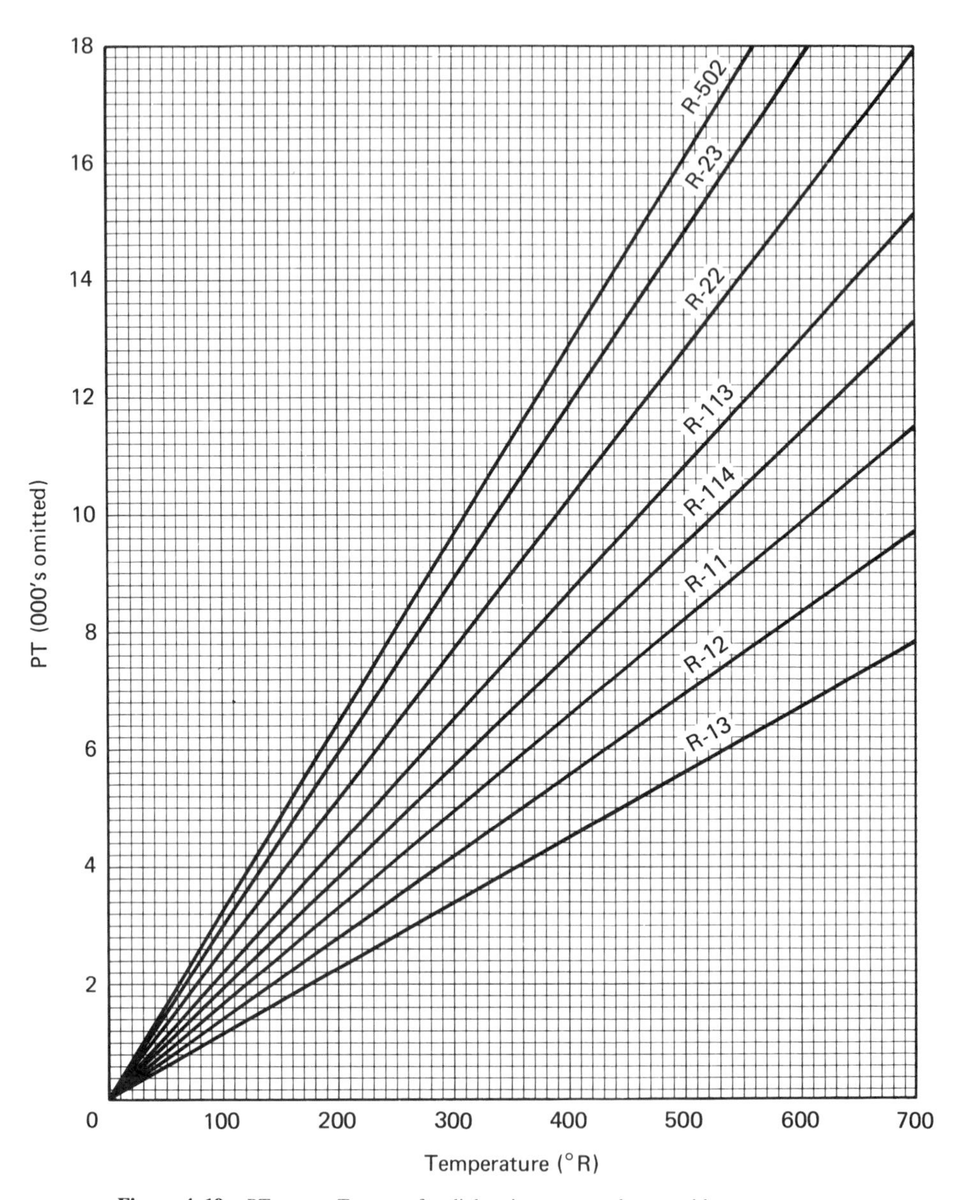

Figure 4–19 *PT* versus *T* curves for dielectric constant change with temperature.

balanced, providing data for estimating the dielectric constant. The values in Table 4–27 are believed to be accurate within ±10 to 15%.

The dielectric constant is the ratio of the capacitance of a condenser filled with some material to its capacitance when evacuated. The dielectric constant of a vacuum is 1 and increases with the density of the material. The dielectric constant of a liquid is much higher than that of a gas. The dielectric constant of gases at moderate pressures is only slightly higher than 1. These small differences have little effect on the operation of the motor in a hermetic compressor.

TABLE 4–26 LIQUID DIELECTRIC CONSTANTS

R	Temperature °F	Temperature °K	Density lb/ft^3	Density g/cm^3	Dielectric constant	Polarization	*PT*	Reference
11	77	298	92.144	1.476	2.5	31.02	9,250	4–46
11	84.2	302	91.477	1.465	2.28	28.04	8,471	4–1
12	77	298	81.835	1.311	2.1	24.75	7,379	4–46
12	84.2	302	80.906	1.296	2.13	25.53	7,714	4–1
12	77	298	81.835	1.311	2.100	24.75	7,379	4–45
13	−22	243	81.053	1.298	2.3	24.32	5,914	4–46
13	68	293	57.528	0.9215	1.641	19.96	5,850	4–45
23	−22	243	77.774	1.246	6.3	35.88	8,725	4–46
23	68	293	50.344	0.8064	5.51	52.14	15,284	4–45
113	77	298	97.69	1.565	2.6	41.65	12,418	4–46
113	77	298	97.69	1.565	2.41	38.29	11,415	4–1
114	77	298	90.913	1.456	2.2	33.54	9,999	4–46
114	77	298	90.913	1.456	2.26	34.72	10,351	4–1
22	77	298	74.528	1.194	6.6	47.17	14,062	4–46
22	75.2	297	74.772	1.198	6.11	45.49	13,517	4–1
502	77	298	75.947	1.217	6.11	57.82	17,240	4–1

TABLE 4–27 ELECTRICAL PROPERTIES OF LIQUID FLUOROCARBONS[a]

Number	Formula	Resistivity (MΩ-cm)	Dielectric constant
10	CCl_4	12,000	2.4
11	CCl_3F	9,000	2.5
12	CCl_2F_2	12,000	2.1
13	$CClF_3$	12,000 (−22°F)	2.3 (−22°F)
12B1	$CBrClF_2$	9,000	2.7
12B2	CBr_2F_2	12,000	3.0
13B1	$CBrF_3$	12,000	2.7
113	CCl_2FCClF_2	12,000	2.6
114	$CClF_2CClF_2$	7,000	2.2
113B2	$CBrClFCBrF_2$	12,000	3.0
114B2	$CBrF_2CBrF_2$	12,000	2.7
124a	CHF_2CClF_2	5,000	4.0
20	$CHCl_3$	3,000	5.6
21	$CHCl_2F$	3,000	6.2
22	$CHClF_2$	7,500	6.6
23	CHF_3	1,600 (−22°F)	6.3 (−22°F)
22B1	$CHBrF_2$	200	6.6
501	$CHClF_2/CCl_2F_2$ (75/25)	4,000	5.0
30	CH_3Cl	100	9.1
32	CH_2F_2	250 (32°F)	8.2 (32°F)

[a]Frequency, 1000 Hz; temperature, 77°F except where indicated.

When there is no hydrogen in the molecule the dielectric constant is low and the resistivity high. Conversely, when hydrogen is present the dielectric constant is higher and the resistivity lower. A high dielectric constant is favorable to ionization and may help explain the difference in resistivity and the greater sensistivity of the hydrogen-containing compounds to contaminants.

Dissolved impurities will tend to decrease the resistivity if they ionize as illustrated in Table 4–28. Anhydrous hydrogen chloride was added to R-22 and R-12 and in each case the resistance was lowered.

TABLE 4–28 EFFECT OF HYDROGEN CHLORIDE ON RESISTANCE[a]

	HCl (wt %)	Resistivity (MΩ-cm)
R-22	0	16
	0.1	4
R-12	0	16×10^6
	0.1	4×10^6

[a]Temperature, 25°C (77°F).

The addition of lubricating oil also lowered the resistance even though the resistivity of oil itself is very high, as illustrated in Table 4–29. The addition of a small amount of lubricating grease to liquid R-22 lowered the resistivity by a factor of about 10 and the addition of phenol by a factor of about 20.

TABLE 4–29 EFFECT OF REFRIGERATION OIL ON ELECTRICAL PROPERTIES OF R-22[a]

Weight percent		Resistivity (MΩ-cm)	Dielectric constant
R-22	Oil		
0	100	200,000,000	2.0
41.5	58.5	490	3.3
75.3	24.7	120	4.9
79.1	20.9	45	6.3
84.8	15.2	10–50	7.2–5.2
91.8	8.2	15–35	7.1–5.1
95.1	4.9	50	5.1
98.0	2.0	40	6.0
99.8	0.2	80	7.2
100	0	5,200	6.6

[a]Temperature, 25°C (77°F).

There was no correlation between resistivity and water content of the oil/R-22 mixtures at a level of 30 to 60 ppm water. However, when a mixture of 92% R-22 and 8% oil was saturated with water the resistivity fell from 20 to 6 MΩ-cm. Ionization of dissolved impurities in the water may account for the change but the level is still high, and in fact even the addition of large amounts of water does not lead to electrical failure unless the motor insulation is attacked.

Eiseman demonstrated that magnet wire insulation suitable for use with R-12 may not be adequate for R-22 since the latter tends to penetrate and permeate organic coatings more readily than does R-12. The result is some reduction in resistivity but more important, a softening of the coating that permits mechanical wear and abrasion, leading to electrical failure. A "thumbnail" scraping test was developed and used to distinguish coatings that could be used with R-22.

REFERENCES

4–1. DuPont Company, Freon Products Division, "Freon Fluorocarbons—Properties and Applications," Bulletin B-2 (1969).

4–2. R. C. McHarness and D. D. Chapman, "Refrigerating Capacity and Performance Data for Various Refrigerants, Azeotropes, and Mixtures," *ASHRAE J.*, 4 (1962), 49.

4–3. R. C. Downing and L. J. Long, Jr., "Velocity of Sound in Some Fluorocarbon Refrigerants," *ASHRAE J.*, 5 (1963), 41.

4–4. DuPont Company, Freon Products Division, unpublished information.

4–5. R. M. Buffington and J. Fleischer, "Thermodynamic Properties of Dichlorodifluoromethane, a New Refrigerant," *Ind. Eng. Chem.*, 23 (1931), 1290.

4–6. J. Woodburn, M. T. Mettrey, and B. L. Hoa, "Velocity of Sound Measurements in Refrigerant 12," *ASHRAE J.*, 8 (1966), 74.

4–7. R. P. Kokernak and C. L. Feldman, "The Velocity of Sound in Liquid R-12," *ASHRAE J.*, 13 (July 1971), 59.

4–8. K. J. Meyer, "The Connection between Sound Velocity and Heat Conductivity of Liquid Fluorochloro Derivatives of Methane and Ethane," *Kaltetechnik*, 21 (1969), 270.

4–9. R. Rao, "Velocity of Sound in Liquids and Chemical Constitution," *J. Chem. Phys.*, 9 (1941), 682.

4–10. B. C. Sakiades and J. Coates, "Thermal Conductivity of Liquids," *AIChE J.*, 1 (1955), 275.

4–11. American Society of Heating, Refrigerating and Air-Conditioning Engineers, *ASHRAE Handbook of Fundamentals*. Atlanta: Ga.: ASHRAE (1977), Chap. 32.

4–12. R. Miller, *Refrigeration and Air-Conditioning Technology*. Peoria, Ill.: Bennett Publishing Company (1983).

4–13. D. W. Osborne, C. S. Garney, R. N. Doescher, and D. M. Yost, "The Heat Capacity, Entropy, Heats of Fusion and Vaporization, and Vapor Pressure of Fluorotrichloromethane," *J. Am. Chem. Soc.*, 63 (1941), 3496.

4–14. *American Institute of Physics Handbook*, 2nd ed. McGraw-Hill, New York (1963).

4–15. E. F. Neilson and D. White, "Heat Capacity, Heat of Fusion, Heat of Transition, and Heat of Vaporization of Chlorodifluoromethane," *J. Am. Chem. Soc.*, 79 (1957), 5618.

4–16. R. H. Valentine, G. E. Brodie, and W. F. Giauque, "Trifluoromethane: Entropy, Low Temperature Heat Capacity, Heats of Fusion and Vaporization, and Vapor Pressure," *J. Phys. Chem.*, 66 (1962), 392.

4–17. DuPont Company, Freon Products Division, unpublished information.

4–18. J. H. Awbery, "The Latent Heats of Fusion of Some Organic Refrigerants," Chap. 26, *A Text Book on Heat,* New York, Longmans, Green, and Co. (1949), p. 247.

4–19. R. C. Reid and T. K. Sherwood, *The Properties of Gases and Liquids*, 2nd ed. New York: McGraw-Hill Book Company (1966).

4–20. F. W. Kirkbride and F. G. Davidson, "Heats of Formation of Gaseous Fluoro- and Fluorochlorocarbons," *Nature*, 174 (1954), 79.

4–21. W. H. Mears and R. F. Stahl, Thermochemical Bulletin No. 2, International Union of Pure and Applied Chemistry (IUPAC) (Mar. 1956).

4–22. *Lange's Handbook of Chemistry*, 12th ed. Sandusky, Ohio: Handbook Publishers, Inc. (1978), pp. 9–65.

4–23. V. P. Kolesov, I. D. Zenkov, and S. M. Skuratov, "Standard Enthalpies of Formation of Chlorotrifluoromethane and Dichlorodifluoromethane," *Zh. Fiz. Khim.*, 36 (Sept. 1962), 2082.

4–24. A. Lord, C. A. Goy, and H. O. Pritchard, "The Heats of Formation of Trifluoromethyl Chloride and Bromide," *J. Phys. Chem.*, 71 (July 1967), 2705.

4–25. H. C. Duus, "Thermochemical Studies on Fluorocarbons," *Ind. Eng. Chem.*, 47 (1955), 1445.

4–26. C. A. Neugebauer and J. L. Margrave, "The Heats of Formation of Tetrafluoroethylene, Tetrafluoromethane, and 1,1-Difluoroethylene," *J. Phys. Chem.*, 60 (1956), 1318.

4–27. E. Greenberg and W. N. Hubbard, "Fluorine Bomb Calorimetry: XX111. The Enthalpy of Formation of Carbon Tetrafluoride," *J. Phys. Chem.*, 72 (1968), 222.

4–28. C. A. Neugebauer and J. L. Margrave, "The Heats of Formation of CHF_3 and CH_2F_2," *J. Phys. Chem.*, 62 (1958), 1043.

4–29. G. O. Pritchard, et al, "The Reactions of Trifluoromethyl Radicals," *Chem. Ind.*, (1954), 1046; *Trans. Faraday Soc.*, 52 (1956), 849.

4–30. J. D. Cox and G. Pilcher, *Thermochemistry of Organic Compounds*. New York: Academic Press, Inc. (1970), p. 402.

4–31. V. P. Kolesov, A. M. Martynov, and S. M. Skoratov, "Standard Enthalpy Formation of 1,1,1-Trifluoroethane," *Russ. J. Phys. Chem.*, 39 (Feb. 1965), 223.

4–32. Office of Saline Water Research and Development, "The Properties of Gas Hydrates and Their Use in Demineralizing Sea Water," Progress Report No. 44, PB 171,031.

4–33. W. Carey, N. A. Klausutis, and A. J. Barduhn, "Solubility of Four Gas Hydrate Formers in Water and Aqueous Sodium Chloride Solutions," *Desalination*, 1 (1966), 342.

4–34. C. P. Huang, O. Fennema, and W. D. Powrie, "Gas Hydrates in Aqueous-Organic Systems," *Cryobiology*, 2 (1965), 109.

4–35. F. A. Briggs and A. J. Barduhn, *Properties of the Hydrates of Fluorocarbons 142b and 12B1*, Advances in Chemistry Series 38, Saline in Chemistry Series 38, Saline Water Conversion II, Washington, D.C.: American Chemical Society (1963).

4–36. A. J. Barduhn, H. E. Towlson, and Y. Chien Hu, "The Properties of Some New Gas Hydrates and Their Use in Demineralizing Sea Water, *AICHE J.*, 8 (1962), 176.

4–37. G. Van Hulle, O. Fennema, and W. D. Powrie, "Gas Hydrates in Aqueous-Organic Systems," *Cryobiology*, 2 (1966), 246.

4–38. H. E. Chinworth and D. L. Katz, "Refrigerant Hydrates," *Refrig. Eng.*, 54 (Oct. 1947), 359.

4–39. W. P. Banks, B. O. Heston, and F. F. Blankenship, "Formula and Pressure-Temperature Relationships of the Hydrate of Dichlorofluoromethane," *J. Phys. Chem.*, 58 (1954), 962.

4–40. S. R. Orfeo and K. P. Murphy, "Absorption Refrigeration Systems," Canadian Patent 811,956, May 6, 1969.

4–41. DuPont Company, Freon Products Division, Bulletin B-42 (1965).

4–42. DuPont Company, Freon Products Division, Bulletin B-32 (1966).

4–43. A. H. Fainberg and W. T. Miller, "Molar Refractivity in Fluorine-Containing Perhalo Compounds," *J. Org. Chem.*, 30 (1965), 864.

4–44. R. M. Fuoss, "Dielectric Constants of Some Fluorine Compounds," *J. Am. Chem. Soc.*, 60 (1938), 1633.

4–45. T. Makita, H. Kubota, Y. Tanaka, and H. Kashiwagi, "Dielectric Constants of Refrigerants R-12, R-13, R-22, and R-23," Gakujutsu Koenkai Koen Rombunshu, *Refrigeration Tokyo Nippon Reito Kyokui*, 52 (1976), 543.

4–46. B. J. Eiseman, Jr., "How Electrical Properties of 'Freon' Compounds Affect Hermetic System's Insulation," *Refrig. Eng.*, 63 (Apr. 1955), 61.

4–47. DuPont Company, Freon Products Division, Bulletin B-2 (1971).

4–48. C. P. Smyth, *Dielectric Behavior and Structure*. New York: McGraw-Hill Book Company (1955), p. 271.

4–49. W. N. Hess, R. L. Mather, and R. A. Nobles, "Refraction Index and Dielectric Constant for Freon 13," *J. Chem. Eng. Data*, 7 (1962), 317.

4–50. B. P. A. Smith, DuPont Company, Freon Products Division, Bulletin X-154 (1966).

4–51. A. A. Maryott and F. Buckley, National Bureau of Standards, Circular 537 (June 25, 1953).

4–52. K. L. Ramaswamy, *Proc. Indian Acad. Sci.*, A4 (1936), 108.

4–53. A. A. Maryott, M. E. Hobbs, and P. M. Gross, "Bond-Moment Additivity and the Electric Moments of Some Halogenated Hydrocarbons," *J. Am. Chem. Soc.*, 63 (1941), 659.

4–54. O. Fuchs, "Temperature and Pressure Dependence of the Dielectric Constant of Some Organic Vapors," *Z. Phys.*, 63 (1930), 824.

4–55. R. Sanger, O. Steiger, and K. Gachter, *Helv. Phys. Acta*, 5 (1932), 200.

4–56. R. Karplus and A. H. Sharbaugh, "Second Order Stark Effect of Methyl Chloride," *Phys. Rev.*, 75 (1949), 1449.

4–57. R. G. Shulman, B. P. Dailey, and C. H. Townes, "Molecular Dipole Moments and Stark Effects: III. Dipole Moment Determination," *Phys. Rev.*, 78 (1950), 145.

4–58. C. P. Smyth and K. B. McAlpine, "Induction between Bond Moments in Some Halogenated Methanes," *J. Chem. Phys.*, 1 (1933), 190.

4–59. G. W. Epprecht, *Z. Angew. Math. Phys.*, 1 (1950), 138.

4–60. K. L. Ramaswamy, *Proc. Indian Acad. Sci.*, A2 (1935), 364, 630.

4–61. H. E. Watson, G. P. Kane, and K. L. Ramaswamy, *Proc. R. Soc. London*, A156 (1936), 130, 144.

4–62. C. P. Smyth, and K. B. McAlpine, "Dipole Moments of the Methyl and Ethyl Halides," *J. Chem. Phys.*, 2 (1934), 499.

4–63. D. R. Lide, *Phys. Rev.*, 87, 227.

4–64. J. N. Shoolery, and A. H. Sharbaugh, *Phys. Rev.*, 82 (1951), 95.

4–65. C. P. Smyth, private communication to National Bureau of Standards.

5

TRANSPORT PROPERTIES

The transport properties of liquids and gases include thermal conductivity, viscosity, and heat capacity. These properties as a function of temperature are given here for some fluorinated refrigerants. The charts are based on tabular data from *Thermophysical Properties of Refrigerants* published by ASHRAE [5–1], also included in the *ASHRAE Handbook of Fundamentals* [5–2]. Similar data for more than 30 other refrigerants are available in the ASHRAE publications.

THERMAL CONDUCTIVITY AND VISCOSITY

Thermal conductivity and viscosity for the vapor are shown at 1 atm of pressure and for saturated vapor (vapor in equilibrium with liquid). The pressure at saturation is the vapor pressure. The effect of pressure on these properties is slight at lower pressures but becomes significant at higher pressures. For values between atmospheric and saturation pressures, a linear interpolation is probably sufficient for most needs. For a better value, more complex estimation procedures can be used.

HEAT CAPACITY

The charts for vapor heat capacity are from equations based on experimental data.

Liquids

Figure 5–1 Thermal conductivity

Figure 5–2 Viscosity

Figure 5–3 Heat capacity

Vapor

Figure 5–4 Thermal conductivity at 1 atm

Figure 5–5 Thermal conductivity of saturated vapor

Figure 5–6 Viscosity

Figure 5–7 Viscosity

Heat capacity at constant pressure, C_p

Figure 5–8 R-11 and R-22

Figure 5–9 R-12

Figure 5–10 R-113 and R-114

Figure 5–11 R-502

Heat capacity ratio, C_p/C_v

Figure 5–12 R-11

Figure 5–13 R-12

Figure 5–14 R-22

Figure 5–15 R-113 and R-114

Figure 5–16 R-502

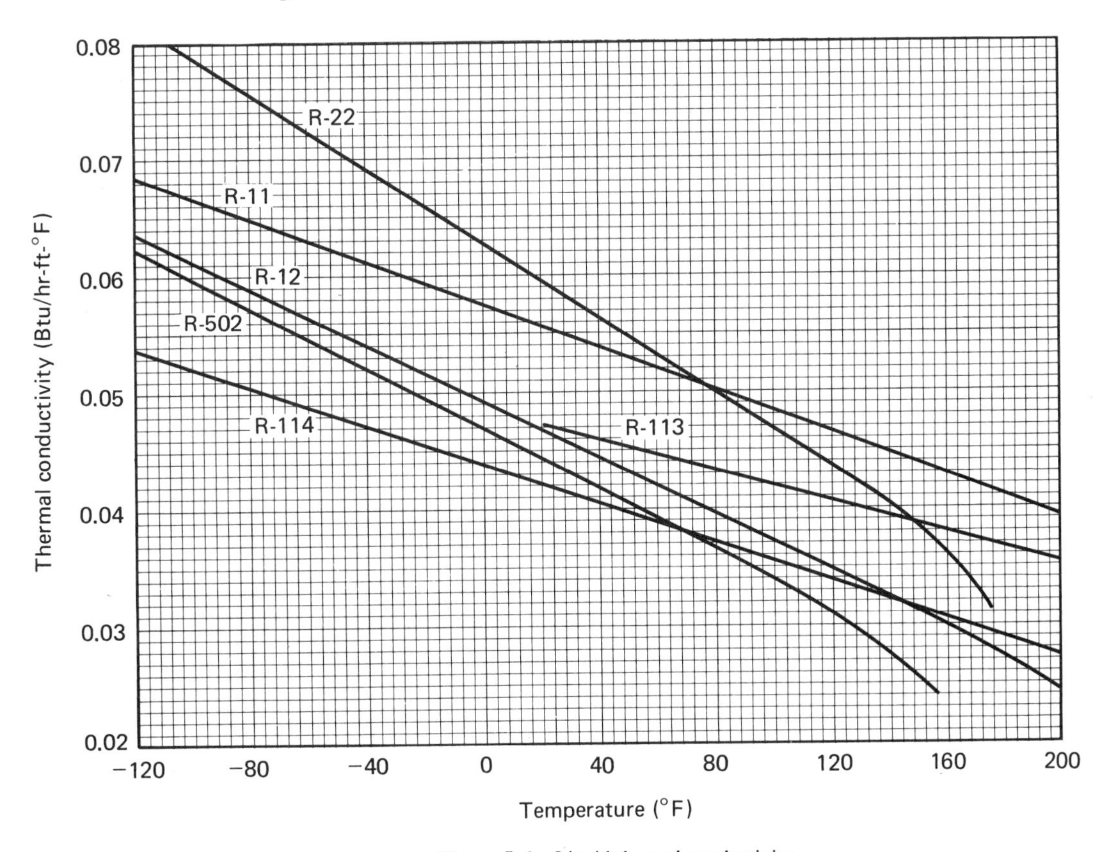

Figure 5–1 Liquid thermal conductivity.

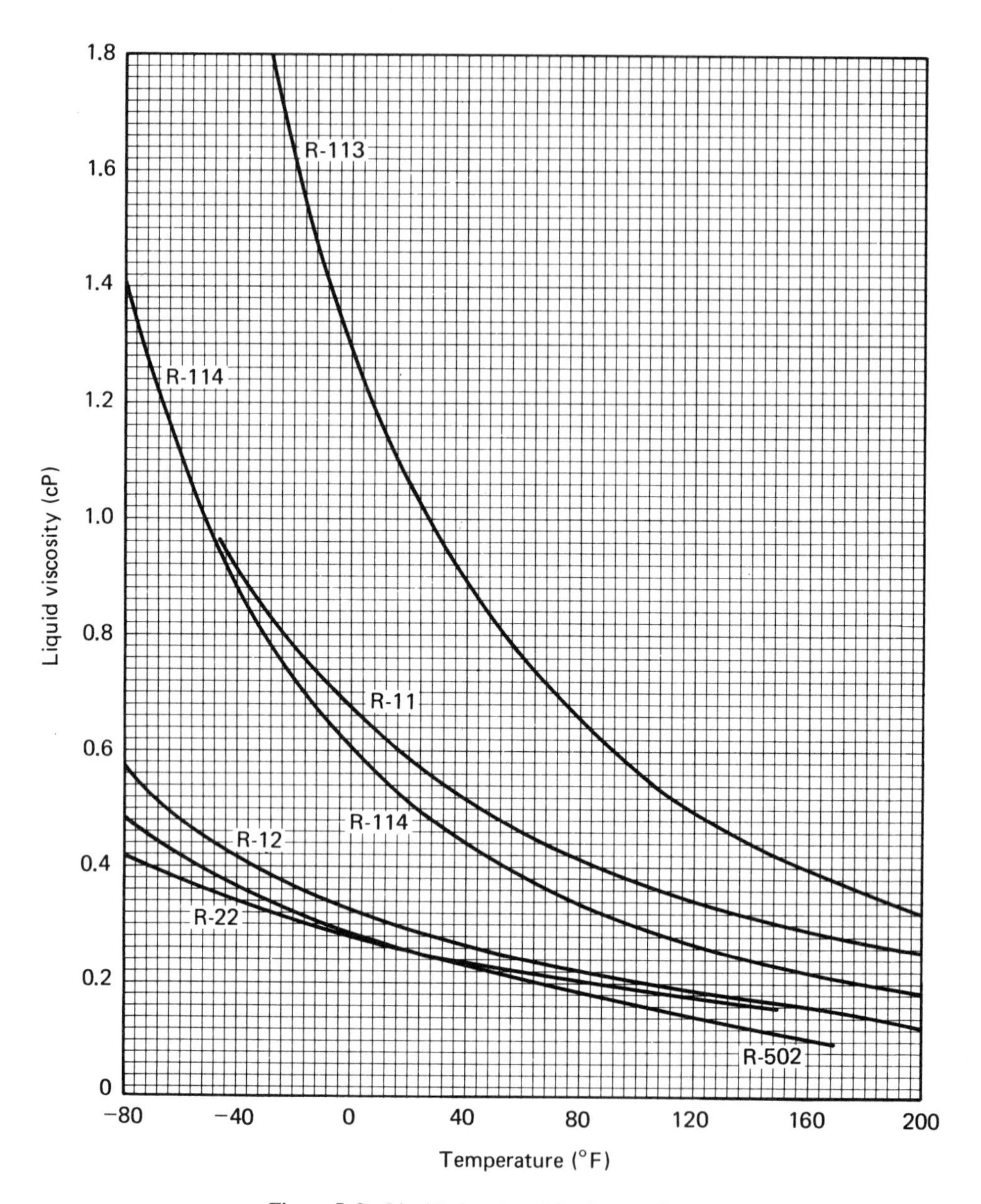

Figure 5–2 Liquid viscosity of the fluorocarbons.

Liquid Heat Capacity

The ASHRAE values for heat capacity as well as thermal conductivity and viscosity are based on review and evaluation of all available information. The liquid heat capacity can also be obtained from tables of thermodynamic properties. Such tables generally list the enthalpy or heat content at various temperatures (based on an arbitrary value at some reference temperature). The difference in enthalpy for a 1-

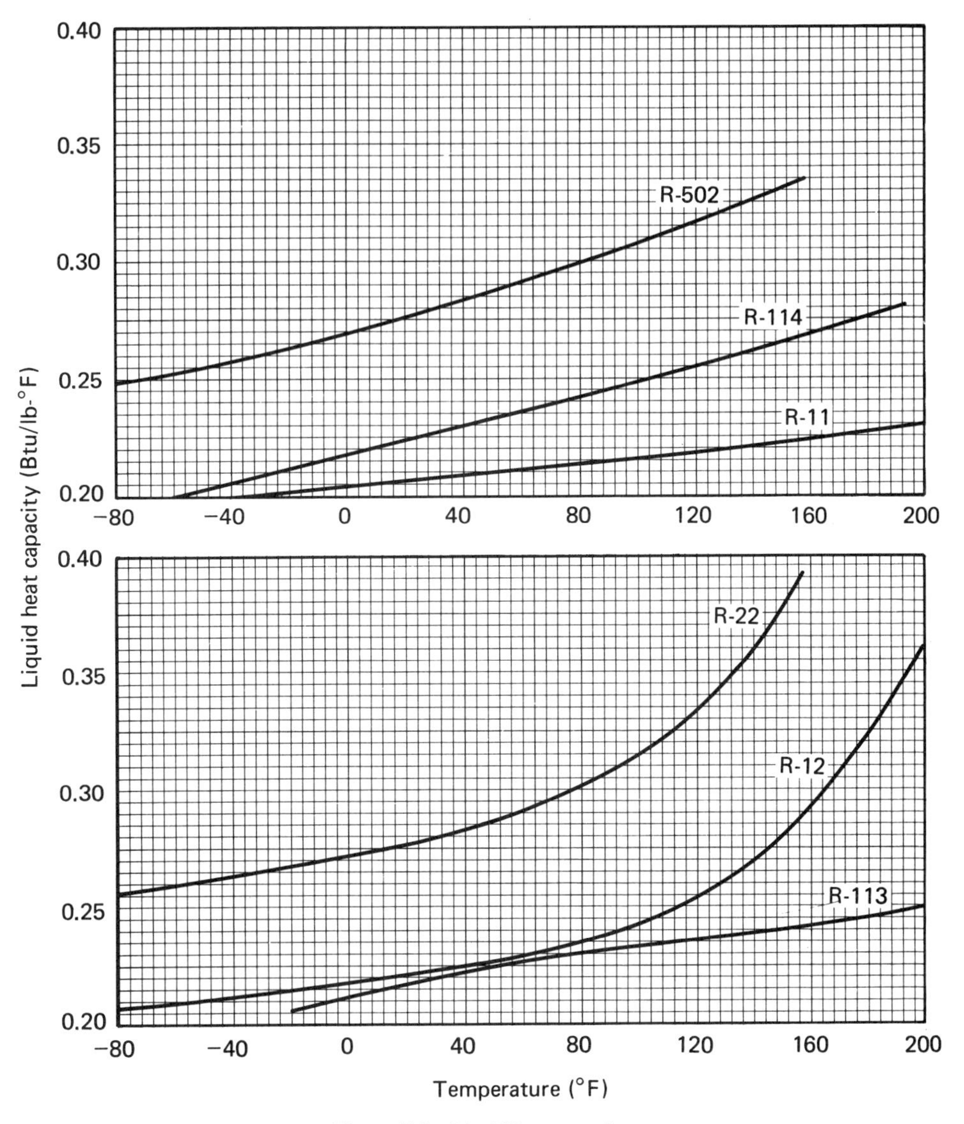

Figure 5–3 Liquid heat capacity.

degree change in temperature gives the specific heat or heat capacity. For example, with R-12:

°F	Tabulated Enthalpy (Btu/lb)
21	13.081
−19	− 12.644
2	0.437

$$\frac{0.437}{2} = 0.2185 \text{ Btu/lb-°F}$$

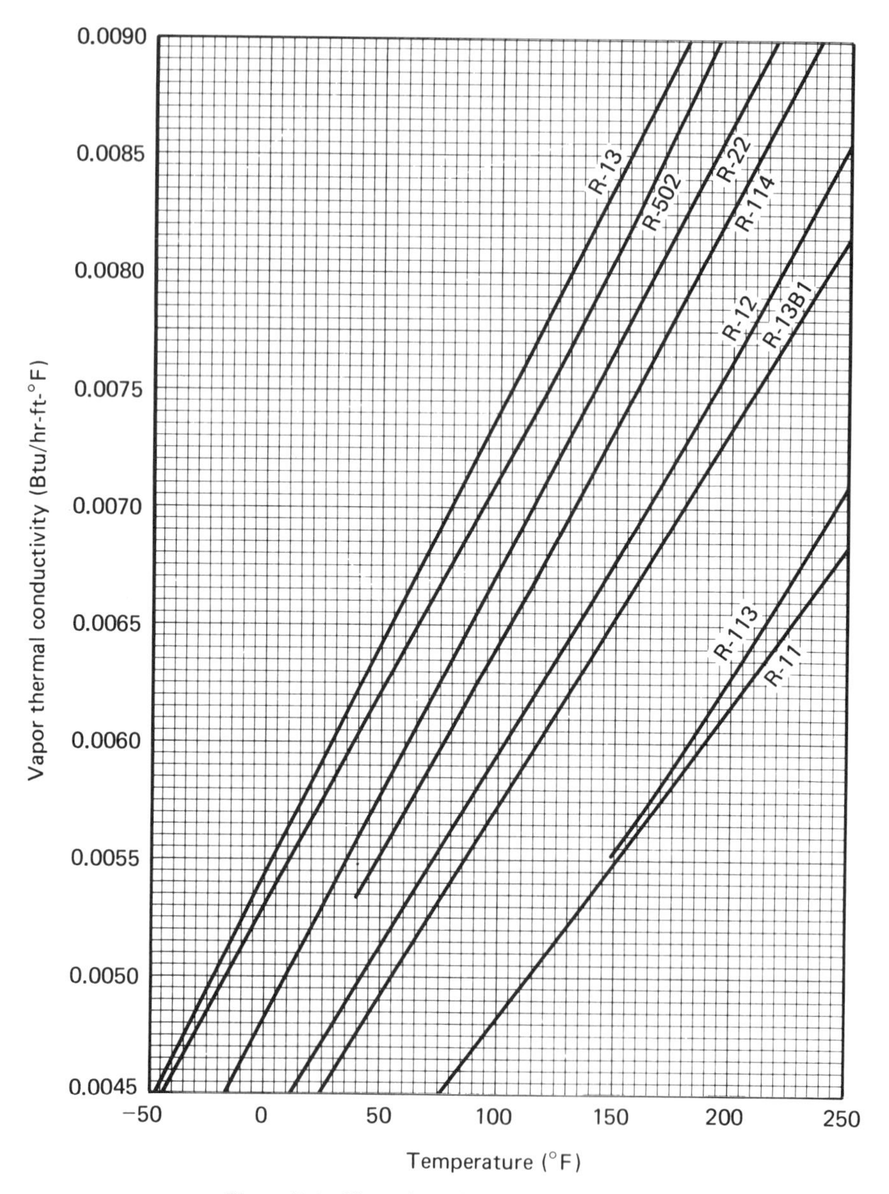

Figure 5–4 Vapor thermal conductivity at 1 atm.

The tabular values for enthalpy are calculated. The specific heat derived in this way depends on the quality of the data and equations used in calculating the tables. In most cases the calculated heat capacity agrees quite well with available measurements.

The liquid heat capacity data in Figure 5–3 are for saturated liquid when liquid and vapor are in equilibrium. Measured data or equations are not readily available for the subcooled liquid of the refrigerants, but changes in the heat content

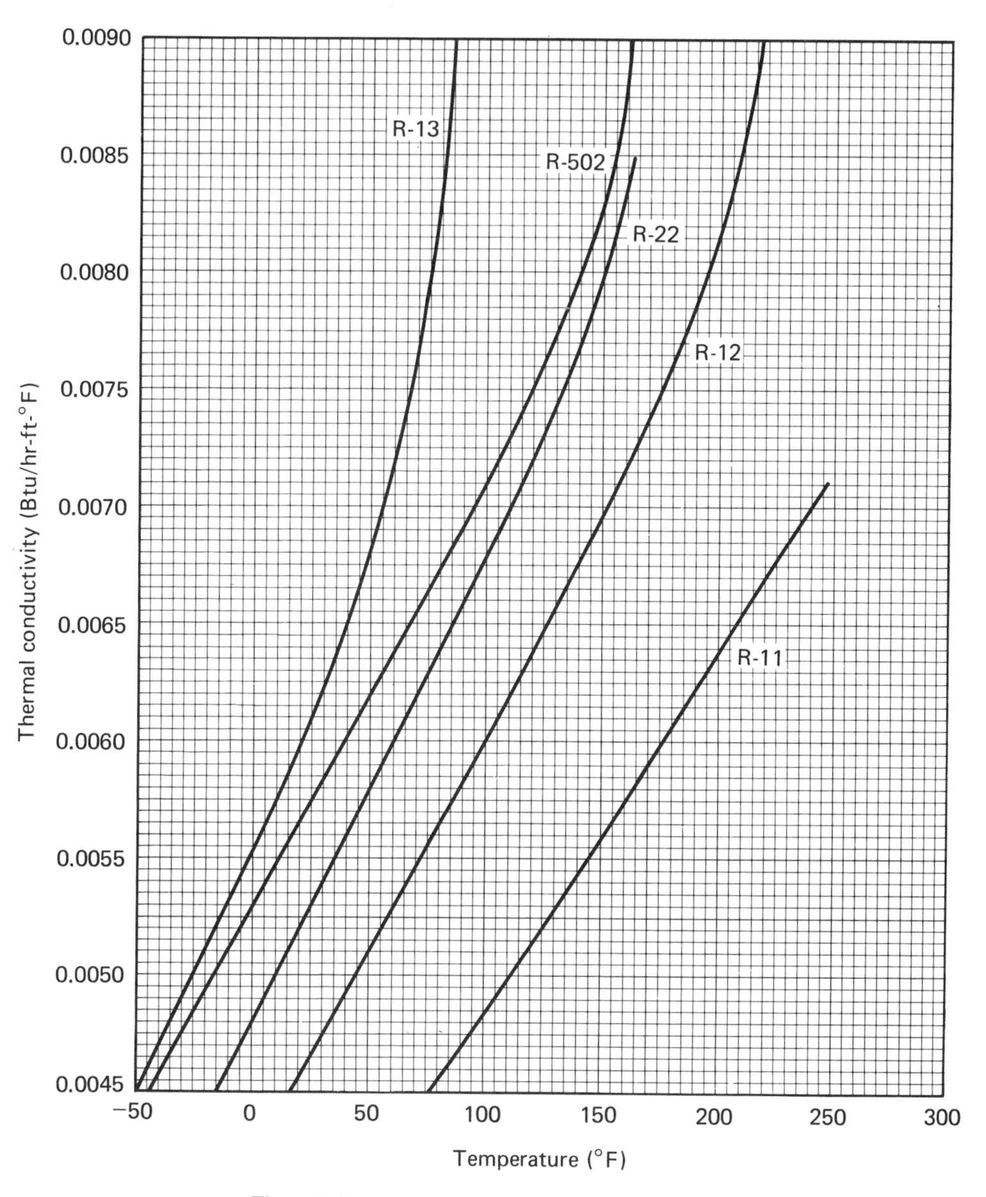

Figure 5–5 Thermal conductivity of saturated vapor.

in that region are usually small and not of vital significance for calculating refrigeration cycles. When the liquid is subcooled, the enthalpy for the saturated liquid is usually assumed. If equations were available, the heat capacity at constant pressure for the subcooled liquid could be calculated from the following relationship:

$$C_p - C_s = \left(\frac{dH}{dT}\right)_p - \left(\frac{dH}{dT}\right)_{\text{sat}} = T\left(\frac{dV}{dT}\right)_p\left(\frac{dP_{\text{sat}}}{dT}\right)$$

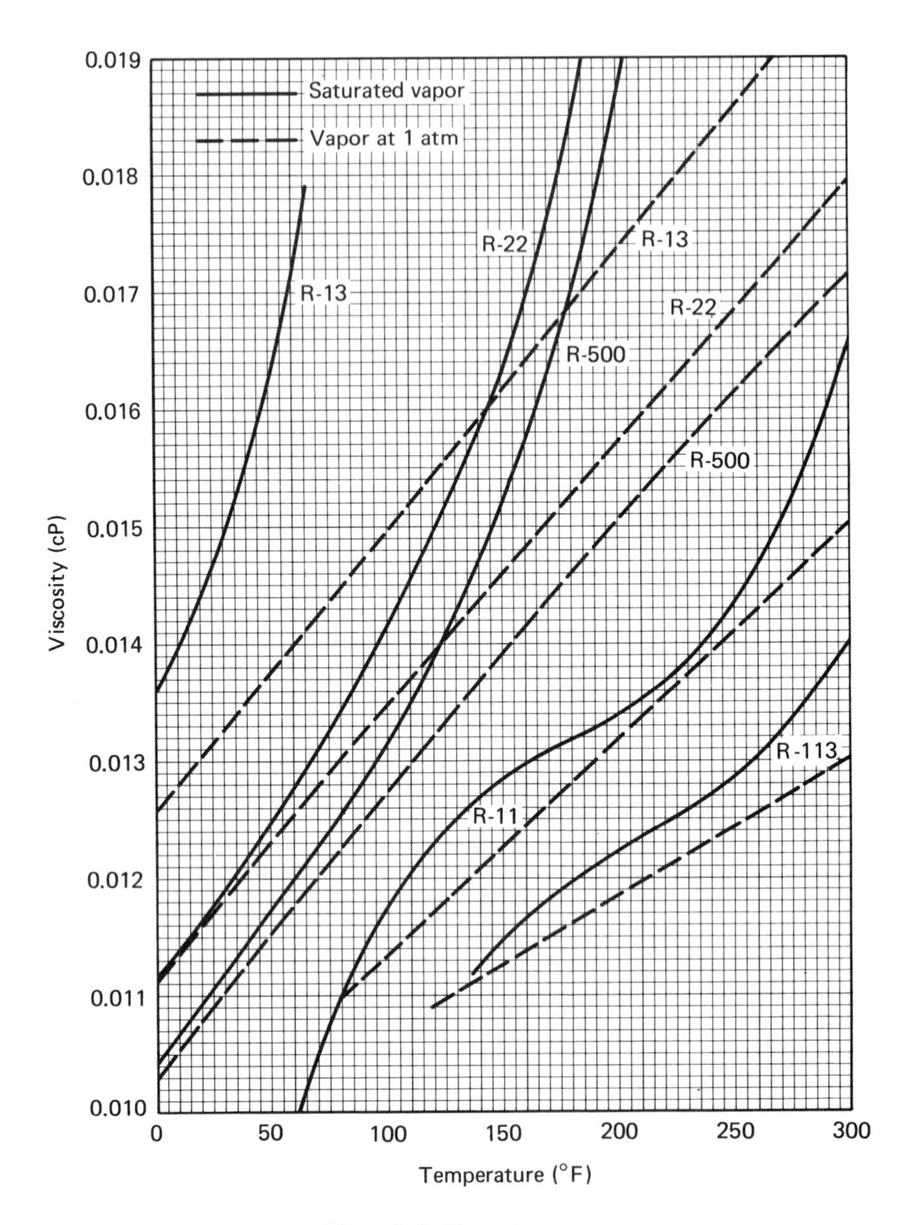

Figure 5–6 Vapor viscosity.

The following relationship between saturated liquid and saturated vapor may also be of some interest:

$$\Delta C = C_{\text{sat liq}} - C_{\text{sat vap}} = -\frac{T\,d(L/T)}{dT}$$

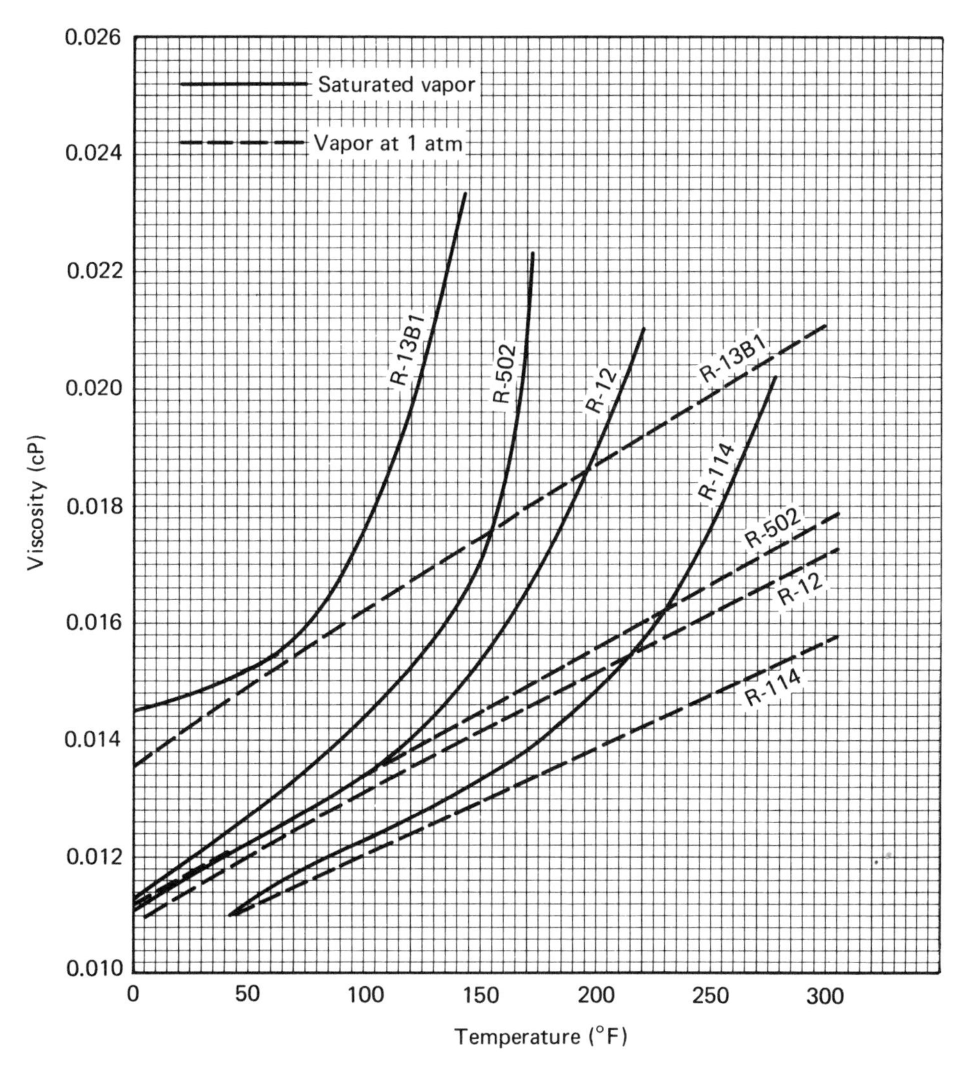

Figure 5–7 Vapor viscosity.

Vapor Heat Capacity

The specific heat of a gas can be determined either at constant pressure, C_p, or at constant volume, C_v. The ratio C_p/C_v is useful in expressions relating the compression and expansion of gases. Specific heat data for several fluorocarbon refrigerants are given in Figures 5–8 to 5–16.

For an ideal gas, the following equations relate temperature, pressure, and volume for isentropic or adiabatic compression or expansion. It is assumed that the change in the condition of the gas occurs at constant entropy. During compression

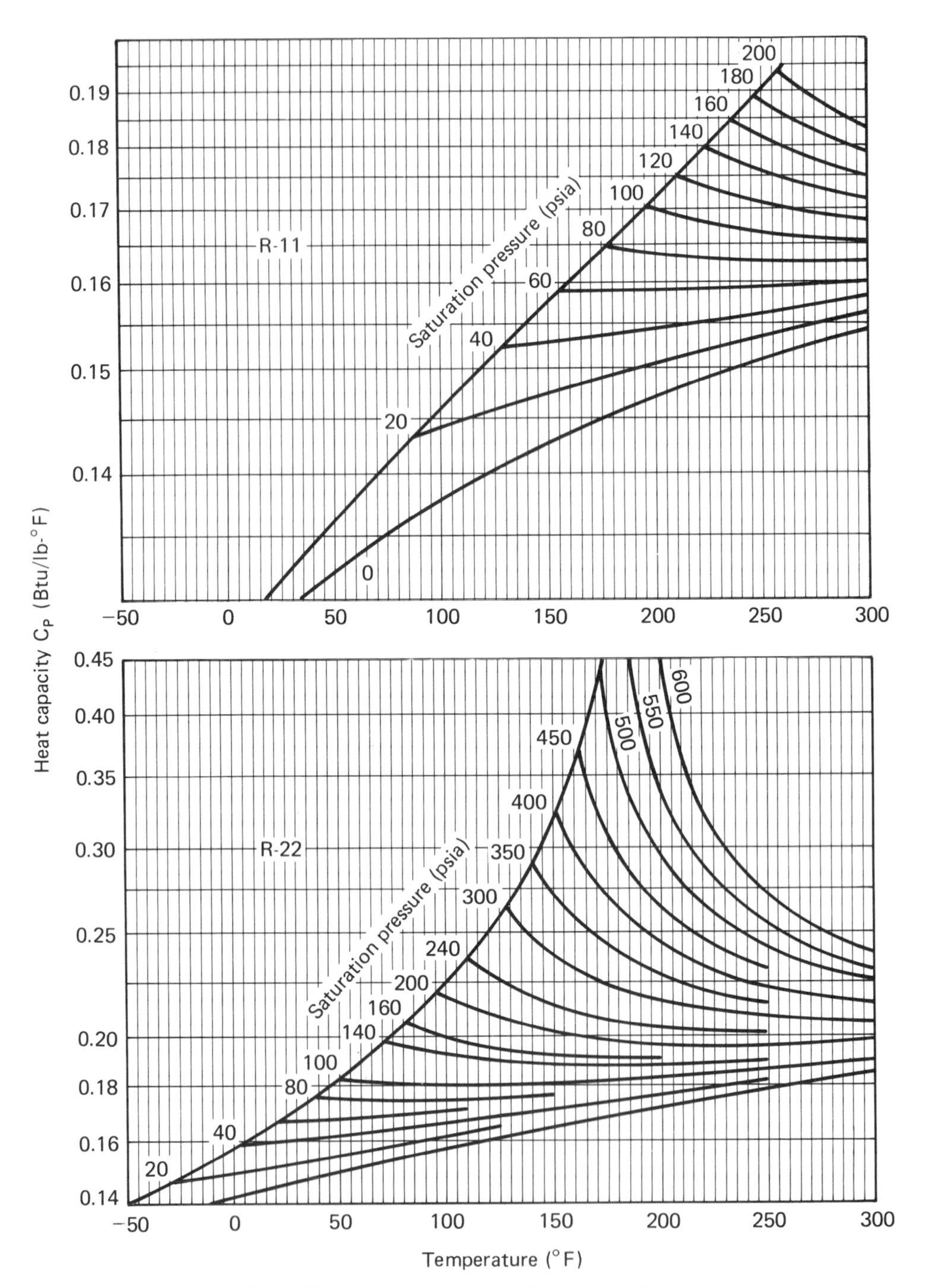

Figure 5–8 Vapor heat capacity at constant pressure.

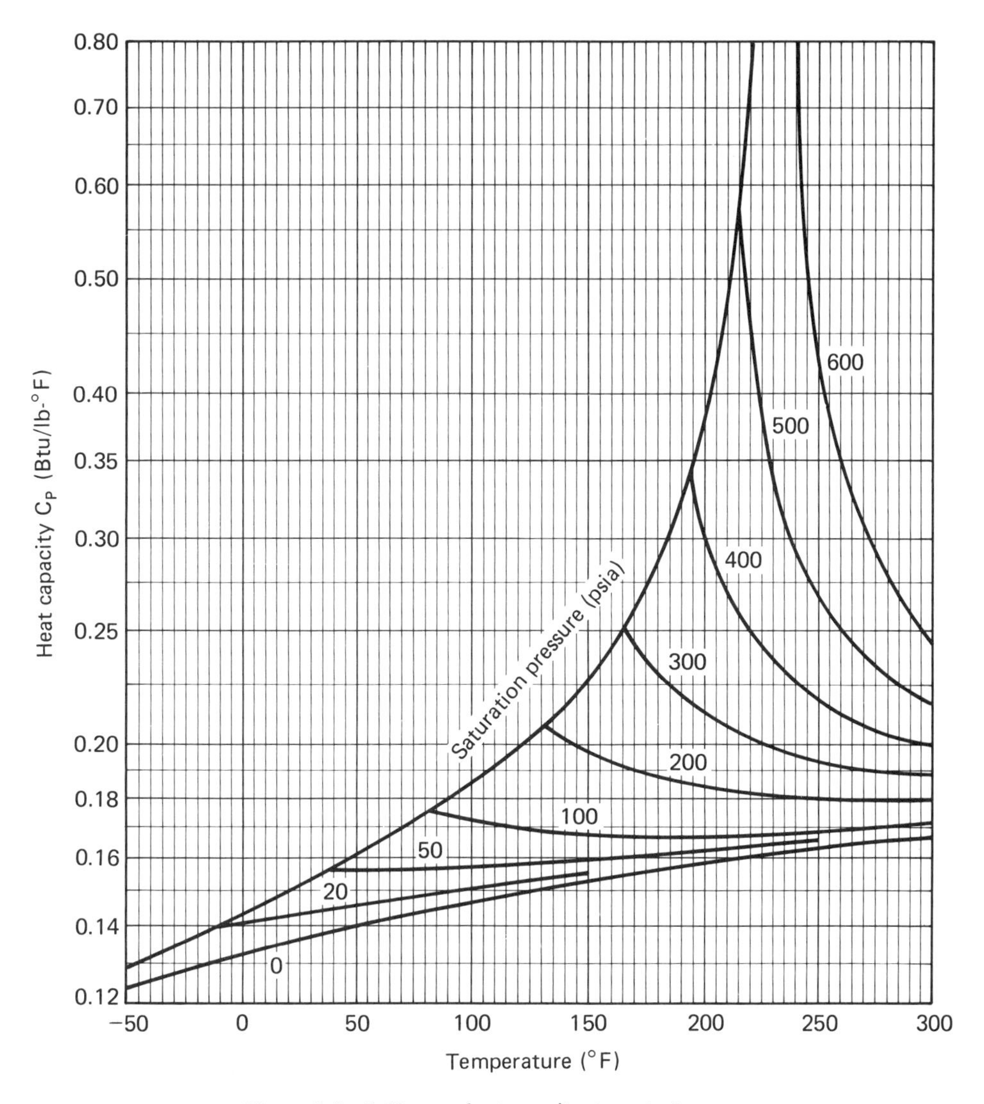

Figure 5–9 R-12 vapor heat capacity at constant pressure.

the increase in temperature is due only to the heat of compression. In practice, some heat is lost by conduction and radiation from the compressor and by increasing oil temperatures, and so on. Some heat is gained from friction in the moving parts of the compressor and by passage of a gas through a small opening. Heat gains and losses tend to offset each other. With reciprocating and rotary compressors, heat gain is often somewhat greater than heat loss and the temperature of the compressed gas somewhat higher than would be predicted.

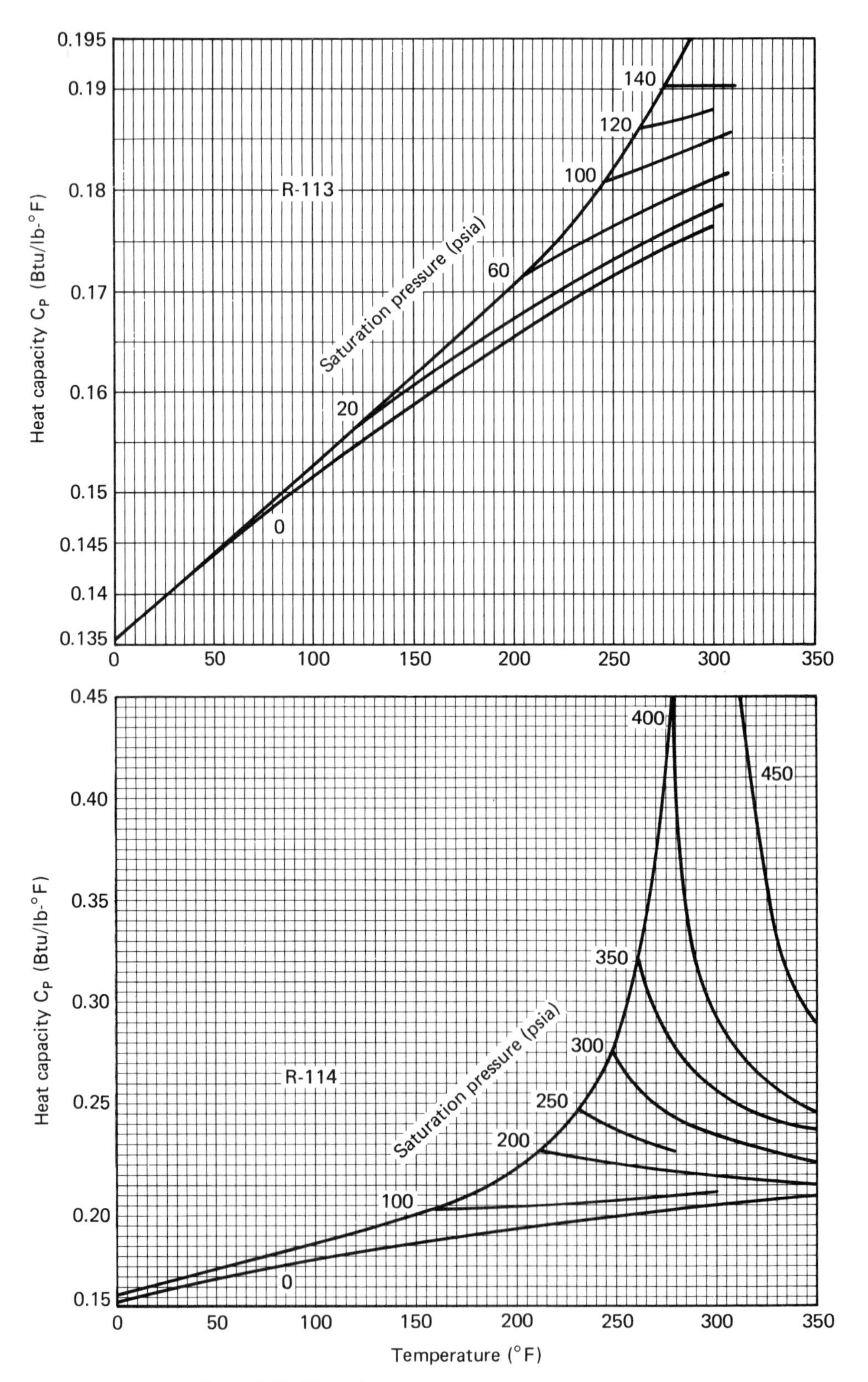

Figure 5–10 Vapor heat capacity at constant pressure.

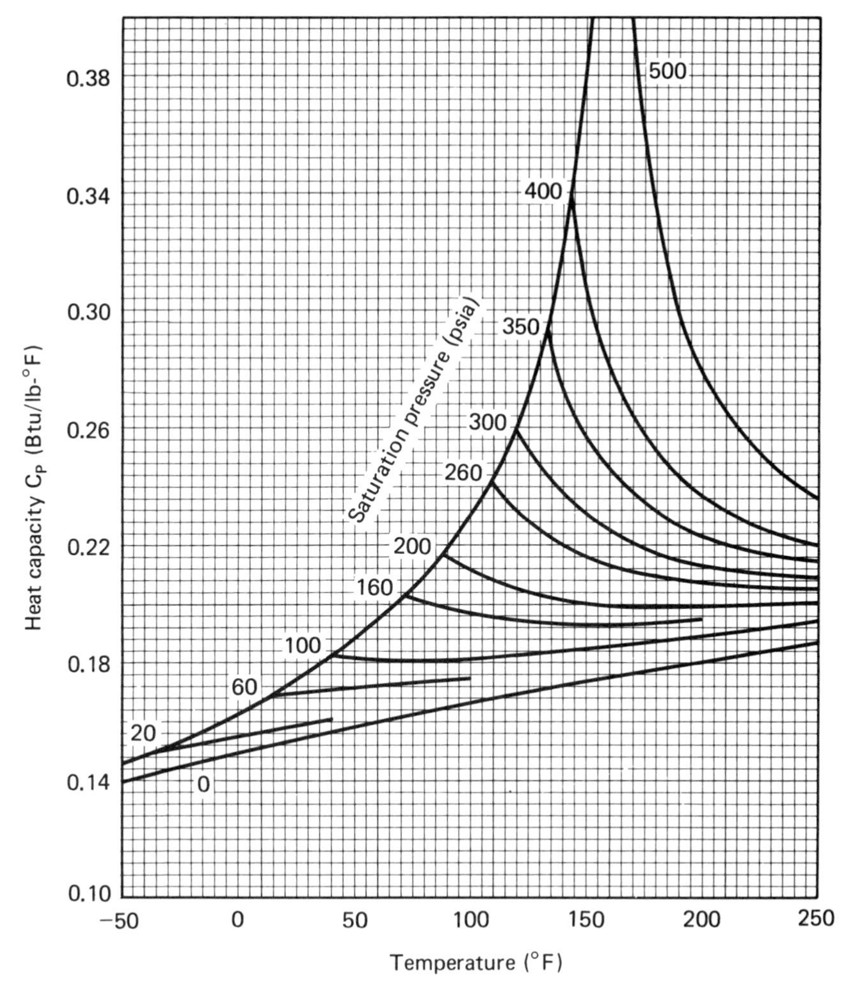

Figure 5–11 R-502 vapor heat capacity at constant pressure.

$$P_1V_1^k = P_2V_2^k \tag{5–1}$$

$$\frac{T_2}{T_1} = \left(\frac{P_2}{P_1}\right)^{(k-1)/k} \tag{5–2}$$

$$\text{work} = \frac{kP_1V_1}{k-1}\left[1 - \left(\frac{P_2}{P_1}\right)^{(k-1)/k}\right] \tag{5–3}$$

The fluorocarbon refrigerants depart somewhat from the properties of a perfect gas. Using the heat capacity ratio, k, for a given refrigerant helps increase the reliability of the calculation. In these relationships, k is assumed to be constant, although in fact it changes with both temperature and pressure.

Tables of thermodynamic properties or the equations on which they are based can be used to calculate changes in properties with accuracy and simplicity. The

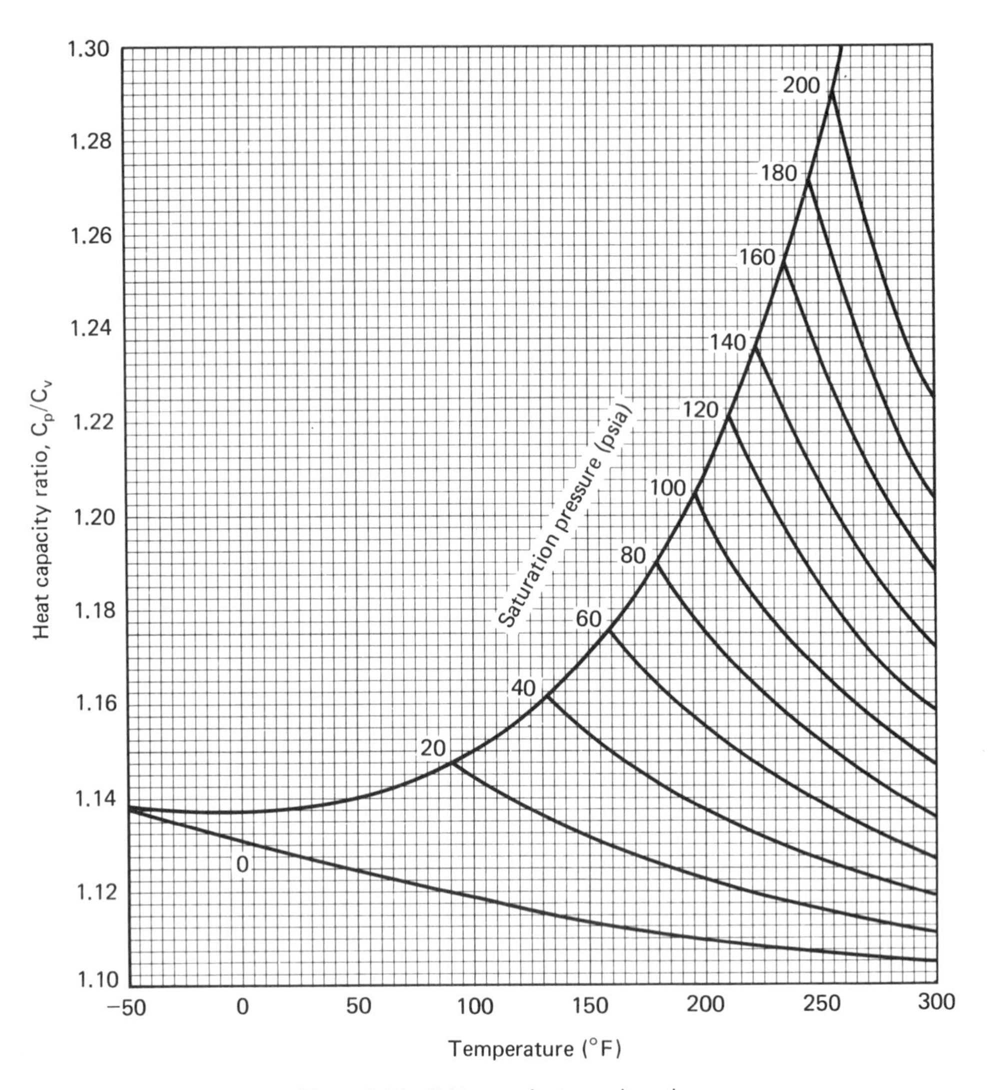

Figure 5–12 R-11 vapor heat capacity ratio.

following example illustrates the use of the equations above and compares the result with calculation from the tables.

Problem

R-12 is evaporated at 40°F and condensed at 120°F. The temperature of the gas entering the cylinder of the compressor is 80°F. The suction pressure is 51.667 psia and the discharge pressure is 172.35 psia. Calculate (a) the discharge temperature, and (b) the heat of compression.

A. *Using the ideal gas equations*

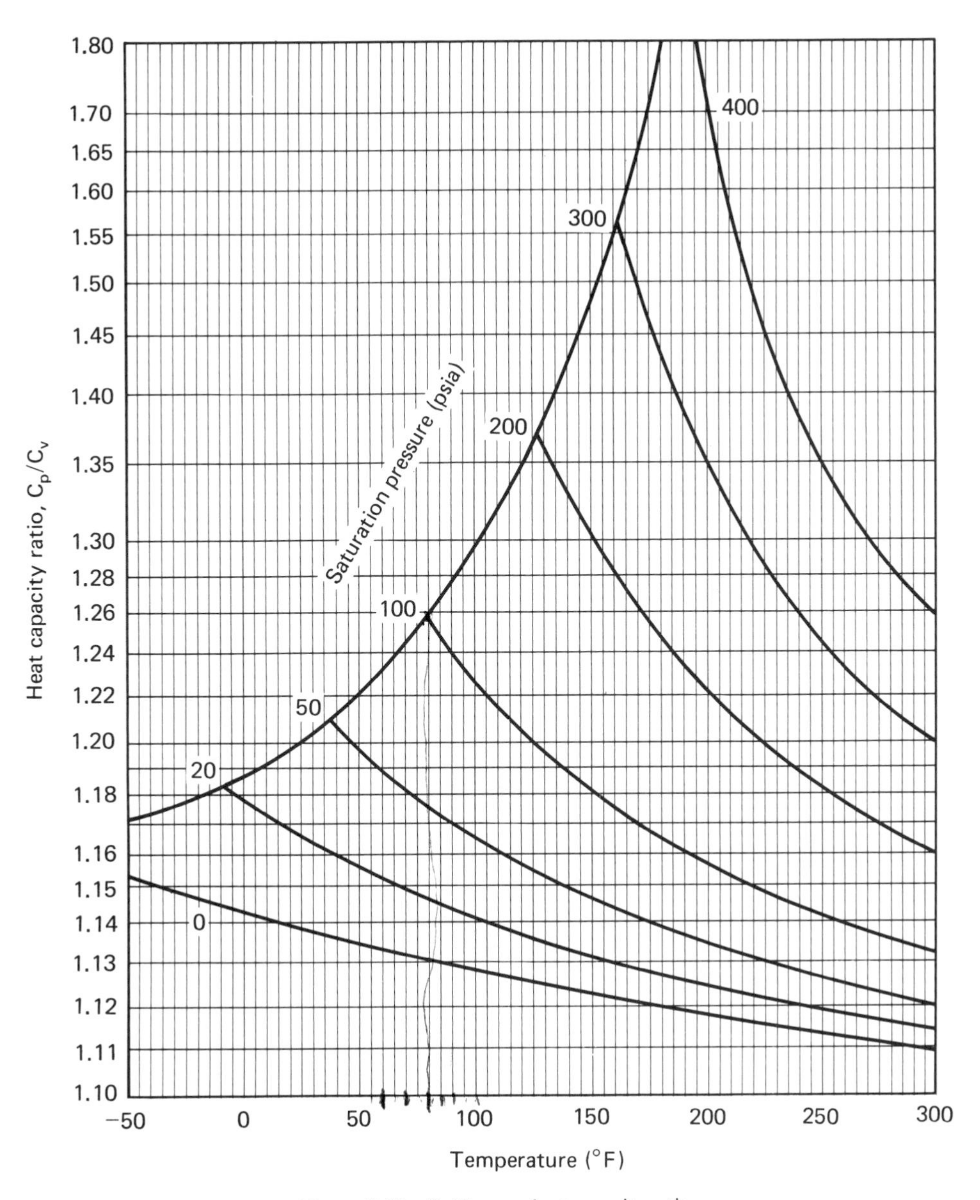

Figure 5–13 R-12 vapor heat capacity ratio.

From Figure 5–13, k at 80°F and 50 psia is 1.175 and at 175 psia and a higher temperature is about 1.25. The average is about 1.21. From equation (5–2),

$$T_2 = (80 + 459.7)\left(\frac{172.35}{51.667}\right)^{(1.21-1)/1.21}$$

$$= (539.7)(3.3358)^{0.17355} = 665.2°\text{R} = 206°\text{F}$$

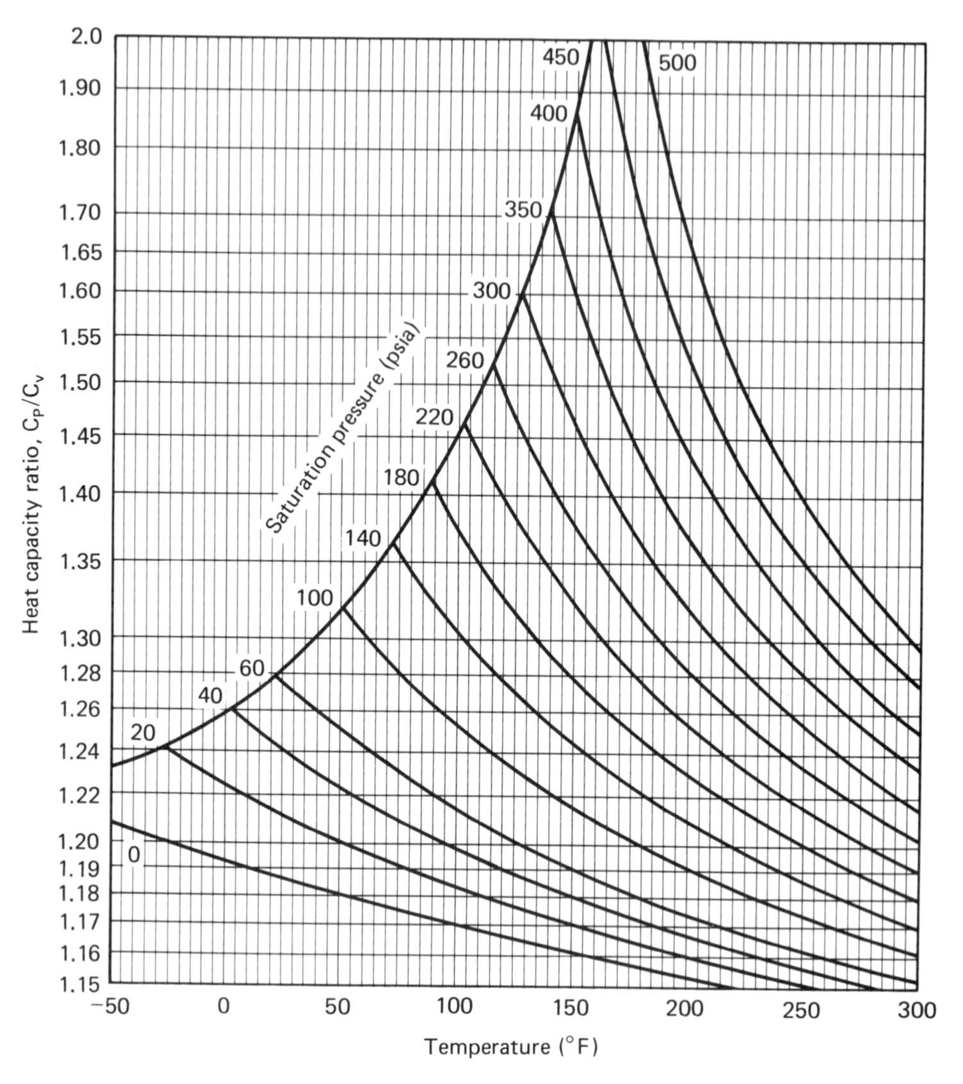

Figure 5–14 R-22 vapor heat capacity ratio.

From the gas law, calculation of volume at 80°F and 51.667 psia:

$$PV = nRT \qquad \text{or} \qquad V = \frac{(\text{wt})(R)(T)}{(P)(\text{MW})}$$

$$V_1 = \frac{(1)(10.7315)(539.7)}{(51.667)(120.93)} = 0.92697 \text{ ft}^3$$

From equation (5–1),

$$(V_2)^{1.21} = \frac{(51.667)(0.92697)^{1.21}}{172.35} = 0.27350$$

$$V_2 = 0.34251$$

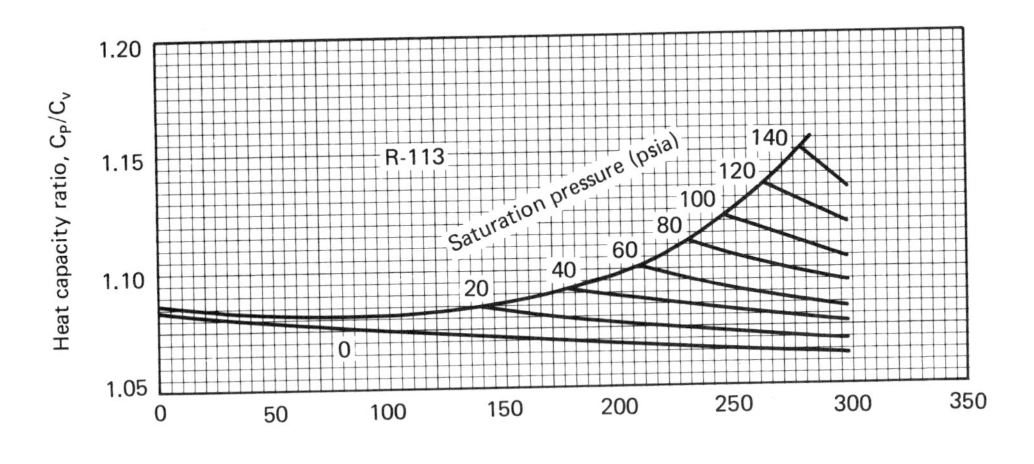

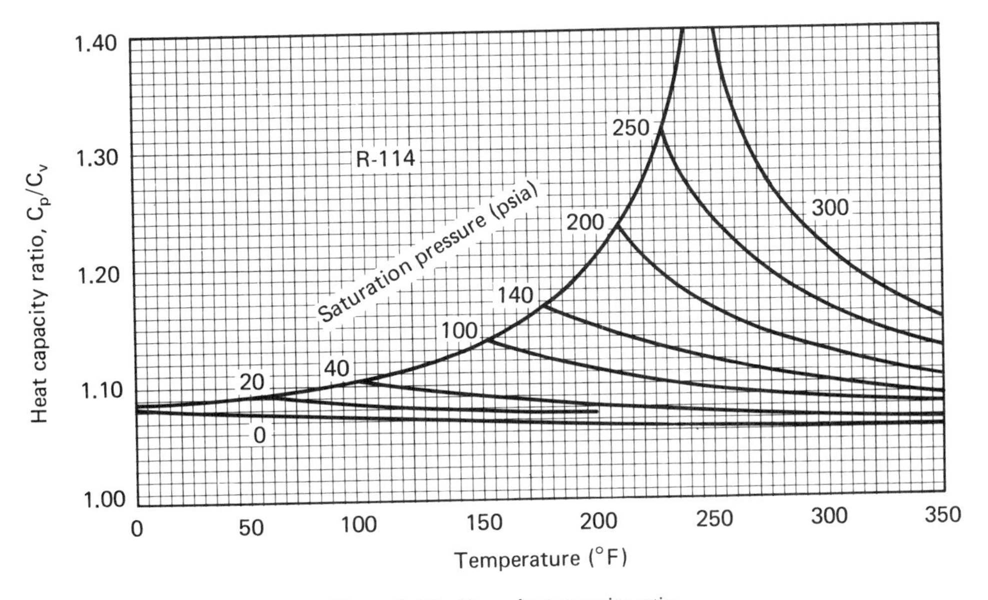

Figure 5–15 Vapor heat capacity ratio.

From equation (5–3),

$$\text{work} = \frac{(1.21)(51.667)(0.92697)}{1.21 - 1}\left[1 - \left(\frac{172.35}{51.667}\right)^{(1.21-1)/1.21}\right]$$

$$= 275.9593[1 - (3.3358)^{0.17355}] = -64.188$$

$$\text{heat absorbed in compression} = (0.185053)(64.188)$$

$$= 11.878 \text{ Btu/lb}$$

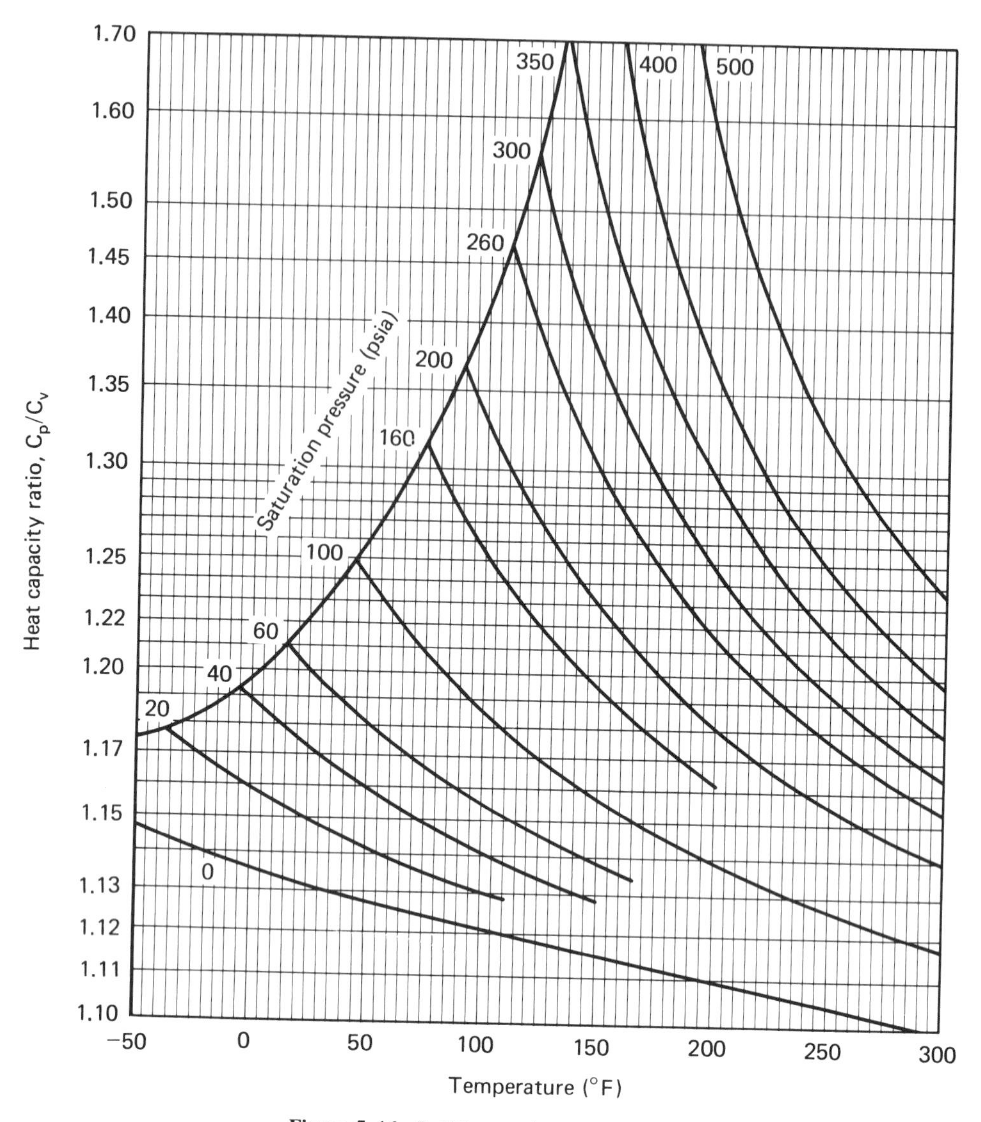

Figure 5–16 R-502 vapor heat capacity ratio.

B. *Using tables of properties*

	Gas entering compressor	Gas leaving compressor
Temperature (°F)	80	170
Volume (ft^3/lb)	0.85918	0.27428
Enthalpy (Btu/lb)	87.732	98.009
Entropy (Btu/lb-°F)	0.17798	0.17798
Change in enthalpy (Btu/lb)	10.267	

Comparison of methods of calculating properties

	Calculated from equations	From tables
Temperature (°F)		
Entering compressor	80	80
Leaving compressor	206	170
Volume (ft^3/lb)		
Entering compressor	0.927	0.859
Leaving compressor	0.343	0.274
Heat of compression (Btu/lb)	11.88	10.27

Use of the ideal gas equations will give approximate answers, but the tables not only give better results but are much easier to use.

EQUATIONS

$$C_v = a + bT + cT^2 + dT^3 + \frac{f}{T^2} + \int_{\infty}^{V} JT \left(\frac{d_2P}{dT^2}\right)_V dV$$

$$C_p = C_p^\circ + JT \int_p^0 \left(\frac{d^2V}{dT^2}\right)_P dP_T$$

$$C_p = C_v - \frac{JT(dP/dT)_V^2}{(dP/dV)_T}$$

$$C_p^\circ = C_v^\circ + R$$

NOMENCLATURE

C_p = change in heat content of liquid or vapor at constant pressure, Btu/lb-°F

C_p° = C_p for an ideal gas

C_v = change in heat content of liquid or vapor at constant volume, Btu/lb-°F

C_v° = C_v for an ideal gas

C_s = change in heat content of liquid at saturation (liquid and vapor in equilibrium), Btu/lb-°F

k = heat capacity ratio, C_p/C_v

T = absolute temperature = °F + 459.7 = °R

H = enthalpy, Btu/lb

V = volume, ft^3/lb

P = pressure, psia
P_{sat} = vapor pressure, psia
L = latent heat of vaporization, Btu/lb
R = gas constant = 10.7315 (psia-ft^3/°F-lb mol)
MW = molecular weight

$$\text{Btu/lb-°F} \times 1 = \text{cal/g-°C}$$

$$\text{Btu/lb-°F} \times 4.184 = \text{J/g-°C}$$

REFERENCES

5–1. American Society of Heating, Refrigerating and Air Conditioning Engineers, *Thermophysical Properties of Refrigerants*. Atlanta, Ga.: ASHRAE (1976).

5–2. American Society of Heating, Refrigerating and Air-Conditioning Engineers, *ASHRAE Handbook of Fundamentals*. Atlanta, Ga.: ASHRAE (1985), Chap. 17.

6

REFRIGERANT MIXTURES AND AZEOTROPES

When two liquids are mixed, a simple mixture is formed if the vapor pressure of the mixture at a given temperature is between the vapor pressures of the two components. The relationship between vapor pressure and concentration is given by Raoult's law and Dalton's law:

Raoult's law:

$$P_A = X_A(\mathrm{VP})_A \qquad \text{and } P_B = X_B(\mathrm{VP})_B$$

Dalton's law:

$$P_A + P_B = P_T$$

where P_A or P_B = partial pressure of component A or B in the solution
X_A or X_B = mol fraction of each component in the mixture
$(\mathrm{VP})_A$ or $(\mathrm{VP})_B$ = vapor pressure of pure component
P_T = vapor pressure of solution

When the vapor pressure of a mixture is equal to the sum of the partial pressures of the components as calculated by Raoult's law, the solution is said to be "ideal," as illustrated in Figure 6–1.

When the measured vapor pressure of a solution is higher than calculated as above, the deviation is said to be positive. When the deviation is so great that the measured vapor pressure is higher than that of each pure component, an azeotrope is formed.

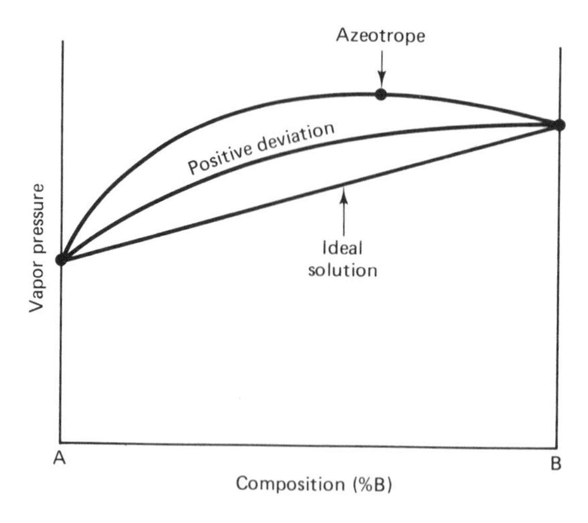

Figure 6–1 Illustration of the vapor-pressure change with composition at constant temperature for two-component mixtures.

MIXTURES

In this section, *mixture* is defined as a simple or nonazeotropic mixture containing two or more components with the properties of the mixture between those of the components.

Although it seems better to use a single refrigerant when one is available with the right properties, mixtures do have a place. For the air conditioning of overhead crane cabs in steel mills, a mixture of R-12 and R-114 is used. High ambient temperatures cause severe service problems in this application and refrigerant stability, together with moderate pressures, are important. Both R-12 and R-114 are among the more stable fluorocarbons. Using the mixture reduces the high pressures found with R-12 alone and increases the capacity compared with R-114 alone.

An interesting system was developed by Missimer [6–1] using mixtures to reach very low temperatures for use in cold traps and other applications. As many as four different refrigerants (such as R-11, R-12, R-13, and R-14) are compressed as a mixture. Through a series of condensers, continually lower-boiling liquids are removed until a final low temperature is obtained. The liquid from one condenser is evaporated to cool the next condenser, and so on.

Another example of a practical application of mixed refrigerants is with heat pumps. When a heat pump with a single refrigerant is operating in the heating cycle, the capacity is greatly reduced at low ambient temperatures. With a mixed refrigerant, the accumulator tends to store liquid. The composition of the liquid becomes rich in the higher-boiling component, so the capacity of the compressor is increased. A mixture of R-13B1 and R-152a has been studied by DuPont [6–2] with encouraging results.

Some properties of specific refrigerant mixtures are given here. Common practice is to list the lower-boiling component of a mixture first.

R-12/R-22

Various mixtures of R-12 and R-11 were widely used as aerosol propellants until environmental considerations greatly restricted their use. Some of the mixtures are still used in essential products. R-12 and R-11 are very similar in chemical structure, and mixtures follow Raoult's law very closely. Mixtures of R-12 and R-11 have been tested in refrigeration systems with good performance, as expected. However, there has been little, if any, commercial use since single refrigerants are available and there is no particular reason to use a mixture. The vapor pressure and liquid density of mixtures of R-12 and R-11 are shown in Tables 6–1 and 6–2. Liquid densities at pressures higher than the vapor pressure for R-12/R-11 (50/50) and other mixtures are recorded in Figure 6–2, page 134.

R-12/R-114

As aerosol propellants, mixtures of R-12 and R-114 were used in nearly all perfume and cosmetic products. In refrigeration, a mixture containing 60% R-12 and 40% R-114 by weight is used for air conditioning overhead crane cabs in steel mills. The volume of refrigerant used is not large, but it is a very demanding service. Condensers are air-cooled and ambient temperatures are high. A single refrigerant with sufficient stability, capacity, and reasonable pressures is not available. The mixture obeys Raoult's law. Vapor pressure and liquid density data are given in Tables 6–3 and 6–4, pages 135 and 136.

R-13B1/R-12 and R-13B1/R-22 [6–4]

Mixtures of R-13B1 with R-12 or R-22 are not used commerically but have been studied in a laboratory calorimeter [6–4]. Performance data are illustrated in Table 6–5, page 137.

R-12/R-13

Three different mixtures of R-12 and R-13 were used in reciprocating compressors, with the results recorded in Table 6–6, page 138. Normal operation was found even with refrigerants with widely different boiling points. It would appear that the condensing temperature in all three systems was controlled in some way. The actual inside temperatures probably covered a range similar to those in the evaporator.

R-152a/R-114

Mixtures of R-152a and R-114 at 60.4°F have positive deviation from Raoult's law but do not form an azeotrope (see Table 6–7, page 138).

TABLE 6–1 VAPOR PRESSURE OF R-12/R-11 SOLUTIONS (PSIA)

Temp. (°F)	R-12 (wt %)										
	100	90	80	70	60	50	40	30	20	10	0
−40	9.3	8.2	—	—	—	—	—	—	—	—	—
−30	12.0	10.8	9.6	8.2	—	—	—	—	—	—	1.0
−20	15.3	14.0	12.6	10.9	8.9	—	—	—	—	—	1.4
−10	19.2	17.7	16.0	14.1	12.3	10.4	8.7	—	—	—	1.9
0	23.9	21.9	19.9	17.9	15.9	13.8	11.9	10.0	—	—	2.6
10	29.3	27.0	24.6	22.2	19.8	17.5	15.1	12.3	9.3	—	3.4
20	35.7	32.9	30.1	27.3	24.4	21.4	18.4	15.2	11.8	—	4.3
30	43.1	39.8	36.4	33.0	29.7	26.2	22.6	18.8	14.7	9.4	5.6
40	51.7	47.7	43.7	39.7	35.7	31.6	27.3	22.7	17.6	12.0	7.0
50	61.4	56.6	51.9	47.2	42.4	37.6	32.6	27.3	21.5	15.5	8.8
60	72.4	67.0	61.5	55.9	50.2	44.6	38.6	32.3	25.6	18.5	10.9
70	84.9	78.6	72.0	63.5	56.9	52.2	45.2	37.7	30.0	21.8	13.3
80	98.9	91.3	83.7	76.2	68.7	60.9	53.0	44.5	35.8	26.4	16.2
90	115	106	97.5	88.8	79.9	70.7	61.4	51.9	42.0	31.5	19.6
100	132	122	112	102	91.9	81.8	71.5	60.6	48.9	36.6	23.5
110	151	140	129	117	106	93.9	81.6	69.3	56.5	42.8	27.9
120	172	160	147	134	121	107	93.5	79.5	65.3	49.8	32.9
130	196	182	166	151	136	122	106	91.0	74.8	57.5	38.7
140	221	205	188	172	155	137	122	104	85.0	66.1	45.1

Source: Ref. 6–3.

TABLE 6–2 LIQUID DENSITY OF R-12/R-11 SOLUTIONS (G/CM^3)

Temp. (°F)	R-12 (wt %)										
	100	90	80	70	60	50	40	30	20	10	0
−40	1.516	1.527	1.537	1.547	1.556	1.568	1.579	1.591	1.602	1.612	1.622
−30	1.501	1.512	1.522	1.533	1.543	1.555	1.567	1.578	1.589	1.600	1.610
−20	1.485	1.496	1.507	1.519	1.530	1.542	1.553	1.565	1.576	1.587	1.598
−10	1.469	1.480	1.492	1.505	1.516	1.528	1.539	1.552	1.563	1.575	1.586
0	1.452	1.465	1.477	1.490	1.502	1.514	1.526	1.539	1.551	1.563	1.574
10	1.435	1.448	1.461	1.474	1.487	1.500	1.512	1.526	1.538	1.550	1.562
20	1.418	1.432	1.446	1.459	1.472	1.485	1.498	1.513	1.525	1.537	1.549
30	1.400	1.416	1.430	1.444	1.457	1.471	1.484	1.500	1.512	1.524	1.537
40	1.382	1.397	1.413	1.429	1.442	1.457	1.471	1.486	1.498	1.511	1.524
50	1.364	1.380	1.396	1.412	1.427	1.442	1.457	1.472	1.485	1.498	1.511
60	1.345	1.362	1.379	1.395	1.412	1.427	1.442	1.458	1.471	1.484	1.498
70	1.325	1.343	1.361	1.377	1.396	1.412	1.427	1.444	1.458	1.471	1.485
80	1.305	1.324	1.343	1.360	1.379	1.396	1.412	1.430	1.443	1.458	1.472
90	1.284	1.304	1.325	1.343	1.362	1.380	1.397	1.414	1.429	1.444	1.458
100	1.262	1.284	1.305	1.325	1.346	1.364	1.382	1.399	1.415	1.430	1.445
110	1.239	1.265	1.285	1.307	1.328	1.347	1.366	1.384	1.401	1.416	1.431
120	1.216	1.243	1.264	1.289	1.311	1.330	1.350	1.369	1.387	1.402	1.417
130	1.191	1.221	1.245	1.270	1.293	1.313	1.333	1.353	1.372	1.388	1.403
140	1.165	1.194	1.223	1.250	1.274	1.295	1.316	1.337	1.357	1.374	1.389

Source: Ref. 6–3.

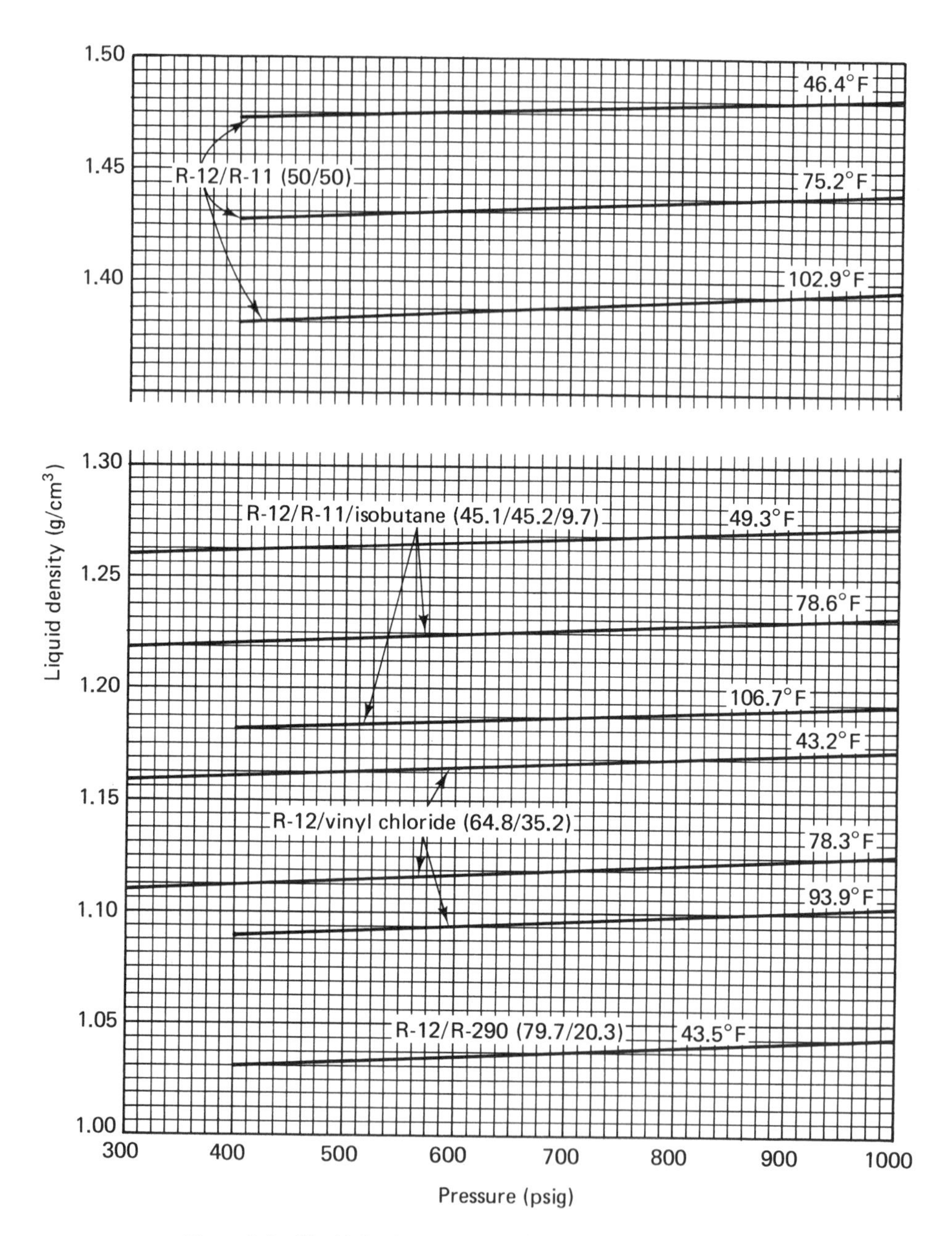

Figure 6–2 Liquid density at pressures higher than the vapor pressure.

CO_2/R-22

The addition of carbon dioxide to R-22 increases the pressure and compressor capacity at the same evaporator pressure as shown in Table 6–8, page 139. Compared with R-502, there was a slight increase in compressor discharge temperature. The CO_2/R-22 mixture did not offer sufficient advantage to receive further attention.

TABLE 6–3 VAPOR PRESSURE OF R-12/R-114 SOLUTIONS (PSIA)

Temp. (°F)	R-12 (wt %)										
	100	90	80	70	60	50	40	30	20	10	0
−40	9.3	8.8	8.2	7.6	—	—	—	—	—	—	—
−30	12.0	11.4	10.7	9.8	9.0	8.2	—	—	—	—	—
−20	15.3	14.5	13.6	12.6	11.5	10.4	9.2	7.9	—	—	—
−10	19.2	18.1	17.0	15.8	14.3	13.2	11.7	10.1	8.5	—	—
0	23.9	22.5	21.1	19.6	18.0	16.4	14.7	12.7	10.8	8.4	—
10	29.3	27.7	26.0	24.2	22.2	20.3	18.1	15.8	13.4	10.5	7.7
20	35.7	33.8	31.9	29.7	27.3	24.9	22.2	19.4	16.4	13.2	9.8
30	43.1	40.8	38.3	35.7	33.0	30.1	27.0	23.8	20.3	16.3	12.3
40	51.7	48.9	46.0	42.8	39.8	36.2	32.5	28.8	24.6	19.9	15.2
50	61.4	58.1	54.7	51.1	47.3	43.2	39.0	34.4	29.5	24.2	18.7
60	72.4	68.5	64.4	60.4	56.0	50.9	46.3	41.0	35.3	29.1	22.8
70	84.9	80.4	74.7	70.8	65.7	60.3	54.6	48.3	41.9	34.9	27.6
80	98.9	93.7	88.4	82.7	76.9	70.6	64.0	56.7	49.1	41.4	33.0
90	115	109	103	96.2	89.5	82.0	74.6	66.3	57.8	48.8	39.3
100	132	125	119	111	103	95.0	86.5	77.2	67.2	56.9	46.4
110	151	145	137	128	119	109	99.5	88.8	78.0	66.6	54.4
120	172	166	156	145	135	125	114	102	89.6	76.9	63.4
130	196	187	177	165	154	142	130	117	103	88.3	73.5
140	221	212	200	186	174	160	147	133	117	101	84.8

Source: Ref. 6–3.

TABLE 6–4 LIQUID DENSITY OF R-12/R-114 SOLUTIONS[a] (G/CM^3)

Temp. (°F)	R-12 (wt %)										
	100	90	80	70	60	50	40	30	20	10	0
−40	1.516	1.528	1.540	1.553	1.565	1.578	1.590	1.602	1.614	1.626	1.638
−30	1.501	1.512	1.524	1.537	1.549	1.562	1.574	1.586	1.599	1.611	1.624
−20	1.485	1.495	1.507	1.517	1.529	1.542	1.557	1.570	1.583	1.596	1.609
−10	1.469	1.480	1.490	1.503	1.516	1.529	1.542	1.555	1.568	1.581	1.595
0	1.452	1.462	1.473	1.487	1.500	1.513	1.527	1.540	1.553	1.566	1.580
10	1.435	1.445	1.456	1.470	1.483	1.497	1.510	1.524	1.537	1.551	1.565
20	1.418	1.427	1.439	1.453	1.467	1.480	1.494	1.507	1.521	1.535	1.549
30	1.400	1.410	1.422	1.436	1.450	1.463	1.477	1.491	1.505	1.519	1.533
40	1.382	1.393	1.405	1.419	1.433	1.447	1.461	1.475	1.489	1.503	1.517
50	1.364	1.376	1.388	1.402	1.416	1.430	1.444	1.458	1.472	1.486	1.501
60	1.345	1.358	1.371	1.385	1.399	1.413	1.427	1.441	1.455	1.469	1.485
70	1.325	1.341	1.353	1.368	1.383	1.397	1.412	1.426	1.441	1.455	1.468
80	1.305	1.321	1.335	1.350	1.364	1.379	1.386	1.406	1.422	1.437	1.451
90	1.284	1.301	1.316	1.331	1.345	1.360	1.375	1.389	1.404	1.418	1.433
100	1.262	1.279	1.295	1.310	1.325	1.340	1.355	1.371	1.386	1.401	1.416
110	1.239	1.257	1.274	1.290	1.305	1.321	1.336	1.352	1.367	1.383	1.398
120	1.216	1.235	1.252	1.268	1.284	1.300	1.315	1.331	1.347	1.363	1.379
130	1.191	1.210	1.228	1.245	1.261	1.278	1.294	1.311	1.327	1.344	1.360
140	1.165	1.186	1.203	1.220	1.237	1.254	1.271	1.289	1.306	1.323	1.340

[a] $g/cm^3 \times 62.428 = lb/ft^3$; $g/cm^3 \times 8.345 = lb/gal$.

Source: Ref. 6–3.

TABLE 6–5 MIXTURES OF R-13B1[a]

	R-13B1/R-12 mixtures (wt %)				
	0/100	10/90	25/75	63/35	100/0
Capacity					
Btu/hr	1150	1295	1480	2670	4160
Percent of R-12	100	112	129	232	362
Compression ratio	9.88	9.89	9.68	8.50	7.30
Pressure (psia)					
Suction	15.3	17.0	19.8	31.5	48.6
Discharge	151	168	192	268	355
Temperature[b] (°F)					
Suction	131	129	125	117	106
Discharge	233	235	239	247	249
	R-13B1/R-22 mixtures (wt %)				
	0/100	25/75	50/50	75/25	100/0
Capacity					
Btu/hr	2175	2850	3250	3725	4160
Percent of R-22	100	131	149	171	191
Compression ratio	9.74	9.13	8.44	7.80	7.30
Pressure (psia)					
Suction	25.0	30.2	36.6	43.6	48.6
Discharge	243	276	309	340	355
Temperature[b] (°F)					
Suction	141	134	125	115	106
Discharge	282	279	273	261	249

[a]Nominal conditions: −20/65/110; temperatures in °F, evaporating/vapor entering compressor/condensing.

[b]In compressor head $\frac{1}{16}$ in. from valve reed.

R-12/R-30 (Methylene Chloride)

Mixtures of R-12 and methylene chloride (see Table 6–9, page 139) were used to some extent in aerosol preparations, but no refrigeration application was considered.

Various Mixtures [6–6]

The atmospheric boiling points of several refrigerant mixtures have been determined (see Table 6–10, page 140).

R-22/R-12/R-11

There was some interest in this three-component mixture as an aerosol propellant. Properties of the mixture are shown in Table 6–11, page 141.

TABLE 6–6 OPERATION OF MIXTURES OF R-12 AND R-13 IN RECIPROCATING COMPRESSORS

	R-13 in mixture (wt %)		
	19	29	34
Temperature (°F)			
In evaporator			
Lowest	−61.6	−88.6	−112.0
Highest	−29.2	−43.6	−88.6
In condenser	77	77	77
Pressure (psia)			
In evaporator	14.7	14.7	14.7
In condenser	147	162	162
Refrigeration (Btu/hr)	2083	1548	909
Refrigerant circulation (lb/hr)	104.7	66.1	65.7
Coefficient of performance	1.27	1.14	0.72

Source: Ref. 6-5.

R-13B1/R-152a

Mixtures of R-13B1 and R-152a have been studied by DuPont [6–2] for use in heat pumps. A mixture containing 65 wt % of R-13B1 and 35 wt % R-152a gives about the same capacity as R-22 on a cooling cycle. However, on a heating cycle, some of the R-152a is automatically stored in the suction-line accumulator and capacity is increased. The improvement over R-22 is greater at lower ambient outdoor temperatures.

Thermodynamic properties and pressure–enthalpy charts have been prepared for R-13B1/R-152a compositions from 65/35 to 80/20 wt % in 5% increments and are available from the DuPoint Company. Some physical properties of these mixtures and performance data are given in Tables 6–12, 6–13, and 6–14, pages 141 and 142.

TABLE 6–7 SATURATED VAPOR PRESSURE FOR MIXTURES OF R-152a AND R-114 AT 60.4°F

R-152a in liquid		Measured vapor pressure (psia)	Calculated vapor pressure (psia)
Weight fraction	Mole fraction		
1.0	1.0	65.6	65.6
0.830	0.926	64.1	62.5
0.614	0.804	62.4	57.2
0.452	0.681	59.7	52.0
0.390	0.623	58.2	49.5
0	0	22.8	22.8

Source: Ref. 6–6.

TABLE 6–8 COMPRESSOR OPERATION WITH MIXTURES OF R-22 AND CARBON DIOXIDE[a]

	CO_2 (wt %)				
	0	2	4	6	8
Capacity (Btu/hr)[b]	10,660	11,530	12,460	13,072	13,894
Pressure (psig)					
Compressor suction	76	84	94.5	99	108.5
Compressor discharge	243	263	284	300	319
Expansion valve outlet	81.5	90	101	106	115
Liquid temperature near expansion valve (°F)	112.1	111.3	109.8	108.8	108.1
Compressor wattage	1,170	1,255	1,350	1,420	1,495

[a]Ambient temperature, 95°F; evaporator outlet temperature, 95°F; evaporator temperature, 47.5°F; condenser temperature, 115°F.

[b]Capacity corrected to 112°F liquid temperature.

Source: Ref. 6–6.

AZEOTROPES

Azeotropes are mixtures of two or more liquids in which the liquid and vapor have the same composition at equilibrium. In normal mixtures, the vapor is richer in the lower-boiling or more volatile component as predicted by Raoult's law and Dalton's law. Azetropes are also characterized by having either lower or higher boiling points than either component of the mixture. The vapor pressure of the azeotrope will be correspondingly either higher or lower than that of any component. Since vapor and liquid compositions are the same, an azeotrope cannot be separated by distillation. Some azeotropes involving fluocarbon products are listed in Table 6–15, page 143.

TABLE 6–9 LIQUID DENSITY OF MIXTURES OF R-12 (CCl_2F_2) AND R-30 (CH_2Cl_2) (G/CM³)

Temperature (°F)	R-12/R30 (wt %)	
	50/50	60/40
86	1.315	1.314
104	1.291	1.287
122	1.264	1.259
140	1.237	1.230
158	1.210	1.199
176	1.178	1.168
194	1.146	1.127

Source: Ref. 6–6.

TABLE 6–10 ATMOSPHERIC BOILING POINTS

R-12/R-133b		R-12/R-124		R-152a/R-11		R-142b/R-11		R-22/R-152a		R-22/R-142b	
Weight fraction R-12	Boiling point (°F)	Weight fraction R-12	Boiling point (°F)	Weight fraction R-11	Boiling point (°F)	Weight fraction R-11	Boiling point (°F)	Weight fraction R-22	Boiling point (°F)	Weight fraction R-22	Boiling point (°F)
0	42.5	0	13.5	0	−13	0	14.4	0	−13.0	0	14.4
0.1	34	0.1	5.5	0.1	−11.9	0.1	17.4	0.1	−14.3	0.1	7.7
0.2	26.5	0.2	0.5	0.2	−10.5	0.2	19.6	0.2	−16.4	0.2	0.5
0.3	18	0.3	−3.3	0.3	−9	0.3	22.1	0.3	−18.8	0.3	−7.8
0.4	10.5	0.4	−6.3	0.4	−7.2	0.4	25.2	0.4	−21.1	0.4	−15.3
0.5	3	0.5	−9.4	0.5	−4.9	0.5	29.1	0.5	−23.8	0.5	−21.5
0.6	−4	0.6	−12	0.6	−3.5	0.6	34.0	0.6	−26.6	0.6	−26.9
0.7	−9.5	0.7	−14.5	0.7	−1.3	0.7	41.5	0.7	−29.9	0.7	−31.7
0.8	−14.5	0.8	−17	0.8	2.7	0.8	51.1	0.8	−33.5	0.8	−35.7
0.9	−18	0.9	−19.3	0.9	24.8	0.9	44.4	0.9	−37.5	0.9	−38.9
1.0	−21.6	1.0	−21.6	1.0	74.9	1.0	74.9	1.0	−41.4	1.0	−41.4

R-22/R-21		R-12/R-C318[a]		R-12/R-C318[b]		R-22/R-C318[a]		R-22/R-C318[b]	
Weight fraction R-22	Boiling point (°F)	Weight fraction R-12	Boiling point at 1 atm (°F)	Weight fraction R-12	Boiling point at 39.8 psia (°F)	Weight fraction R-22	Boiling point at 1 atm (°F)	Weight fraction R-22	Boiling point at 39.8 psia (°F)
0	48.1	0	21.5	0	69.8	0	21.5	0	69.8
0.1	25.0	0.1	7.5	0.04	64.4	0.1	−5.0	0.004	62.9
0.2	8.5	0.2	−0.5	0.06	59.4	0.2	−19.5	0.10	33.2
0.3	−4.0	0.3	−6.2	0.18	45.6	0.3	−26.5	0.20	23.9
0.4	−14.0	0.4	−10.5	0.30	39.3	0.4	−30.7	0.30	13.3
0.5	−21.5	0.5	−13.5	0.49	32.0	0.5	−34.0	0.44	7.7
0.6	−27.5	0.5	−15.5	0.63	29.1	0.6	−36.8	0.64	3.7
0.7	−32.5	0.7	−17.5	1.0	24.6	0.7	−38.8	1.0	1.2
0.8	−36.2	0.8	−19.0			0.8	−40.2		
0.9	−39.3	0.9	−20.2			0.9	−41.0		
1.0	−41.4	1.0	−21.6			1.0	−41.4		

[a]From Ref. 6–6.
[b]From Ref. 6–7.

TABLE 6–11 PROPERTIES OF A MIXTURE OF R-22, R-12, AND R-11[a]

Temperature (°F)	Vapor pressure (psig)	Liquid density (g/cm^3)
0	15.0	1.432
32	42.5	1.383
70	91.0	1.312
100	145	1.250
150	203	1.127

[a]R-22, 42.8%; R-12, 28.6%; R-11, 28.6%.

Source: Ref. 6–6.

Formation

All of the factors related to the formation of azeotropes may not be fully understood, but one of the principal causes is an attraction between molecules known as hydrogen bonding. Hydrogen atoms firmly attached to one molecule become weakly attracted to other molecules of the same or different composition. In this way, small molecules may act as if they were much larger. A good example of this phenomenon is water, which has the chemical formula H_2O. Such a simple molecule with a molecular weight of 18 should have a very low boiling point. That it does not and actually boils at 212°F is due to the loose binding together or association of 30 to 40 molecules, acting as if they were one large one. Another example is ammonia, NH_3, with a molecular weight of 17. The normal boiling point is far above that expected for such a small molecule.

TABLE 6–12 PHYSICAL PROPERTIES OF R-13B1/R-152a ($CBrF_3/CH_3CHF_2$)

Weight ratio	65/35	70/30	75/25	80/20
Mole ratio	0.452/0.548	0.509/0.491	0.571/0.429	0.640/0.360
Formula weight	103.5	106.2	113.4	119.1
Estimated boiling point (°F)	−63.2	−64.4	−65.6	−66.8
Liquid density at 77°F (lb/ft^3)	76.89	79.15	84.07	81.54
Critical temperature (°F)	188[a]	183[b]	179[a]	174[a]
Specific heat, liquid at 77°F (Btu/lb-°F)	0.281	0.273	0.266	0.258
Specific heat, vapor at 77°F and 1 atm (Btu/lb-°F)	0.159	0.152	0.145	0.139
Specific heat ratio at 77°F	1.147	1.147	1.147	1.146

[a]Estimated.
[b]Experimental.

TABLE 6–13 AIR-CONDITIONING PERFORMANCE

	R-22	R-13B1/R-152a (65/35)
Indoor unit air temperature		
In/out (°F)		
Dry bulb	79.5/63.0	80.0/61.5
Wet bulb	67/58	69/60
Compressor pressure (psia)		
Suction	91.7	99.2
Discharge	294.7	312.7
Compressor temperature (°F)		
Suction	67.5	54.0
Discharge	226.0	186.0
Expansion inlet temperature (°F)	108.0	102.0
Average evaporator temperature (°F)	55.0	52.0
Average condenser temperature (°F)	123.0	121.0
Btu removed per lb dry air	6.5	6.8
Watts	2661	2735
Charge size (lb)	6.5	8.75

Some relationships of molecular weight and boiling point are illustrated in Figure 6–3, page 144. In general, the higher the molecular weight, the higher the boiling point in the same series of related compounds. For example, the hydrocarbons (methane, CH_4; ethane, C_2H_6; propane, C_3H_8; etc.) increase in a regular way. The completely halogenated compounds (not containing hydrogen) also increase regularly in boiling point with molecular weight.

TABLE 6–14 HEATING PERFORMANCE

	R-22			R-13B1/R-152a		
Secondary refrigerant temperature (°F)	47	17	0	47	17	0
Average condenser temperature (°F)	101	87	78.5	105	94	86
Compressor pressure (psia)						
Suction	65	48	36.5	78	52	44
Discharge	217	177	156	304	237	228
Compressor temperature (°F)						
Suction	62	37.5	41	53	25	31
Discharge	218	210	233.5	201.5	154	208
Expansion inlet temperature (°F)	86	78.5	73	76	34.5	38
Indoor unit air (ΔT)	25.5	15	8	27.5	21	14
COP	3.0	2.1	1.3	2.8	2.7	2.1
Initial charge (wt %)				70/30	60/40	70/30
Circulating composition (wt %)				76/24	71/29	80/29
Charge size (lb)	6.25	6.25	6.25	14.0	14.0	14.0

TABLE 6–15 AZEOTROPES

Composition	Wt %	Boiling point (°F)	Reference
R-11/R-123, $CCl_3F/CHCl_2CF_3$	22/78	74.5	6–6
R-11/methyl formate, $CCl_3F/HCOOCH_3$	82/18	68	6–15
R-11/acetaldehyde, CCl_3F/CH_3CHO	55/45	60.1	6–15
R-12/R-152a, CCl_2F_2/CH_3CHF_2	73.8/26.2	−28.3	6–20
R-12/R-31, CCl_2F_2/CH_2ClF	78/22	−21.3	6–29
R-12/R-227, CCl_2F_2/C_3HF_7	86.5/13.5	32 (45.1 psia)	6–9
R-12/R-40, CCl_2F_2/CH_3Cl	72.5/27.5	−31	6–10
R-12/hexafluoropropene, CCl_2F_2/C_3F_6	42/58	−27	6–11
R-12/dimethyl ether, CCl_2F_2/CH_3OCH_3	90/10 (at 32°F)		6–12
R-12/R-134, CCl_2F_2/CHF_2CHF_2	68.5/31.5	−28.5	6–40
R-12/propyne, CCl_2F_2/C_3H_4	14/86	90	6–27
R-13B1/R-32, $CBrF_3/CH_2F_2$	80/20	−83	6–40
R-22/R-12, $CHClF_2/CCl_2F_2$	75/25	−42.5	6–22
R-22/R-115, $CHClF_2/CClF_2CF_3$	48.8/51.2	−49.8	6–21
R-22/R-290, $CHClF_2/C_3H_8$	68/32	−49	6–13
R-22/hexafluoropropene, $CHClF_2/C_3F_6$	75/25		6–11
R-22/R-218, $CHClF_2/C_3F_8$	46/54	32 (89 psia)	6–9
R-23/R-13, $CHF_3/CClF_3$	40.1/59.9	−127.6	6–7
R-23/R-116, CHF_3/CF_3CF_3			6–23
R-31/R-114, $CH_2ClF/CClF_2CClF_2$	55.1/44.9	9.6	6–30
R-32/R-115, $CH_2F_2/CClF_2CF_3$	48.2/51.8	−71	6–7, 6–31
R-113/propylene oxide, CC_2FCClF_2/C_3H_6O	25/75	94.3	6–6
R-113/methyl alcohol, CCl_2FCClF_2/CH_3OH	90/10	103.1	6–24
R-113/acetone, CCl_2FCClF_2/CH_3COCH_3	88.9/11.1	113	6–25
R-113/R-30, CCl_2FCClF_2/CH_2Cl_2	50.5/49.5	98.6	6–26
R-113/ethyl alcohol, CCl_2FCClF_2/CH_3CH_2OH	96.2/3.8	113	6–6
R-114/R-133, $CClF_2CClF_2/CHClFCHF_2$	63/37	32.9	6–27
R-114/R-21, $CClF_2CClF_2/CHCl_2F$	70/30	104 (at 54.8 psia)	6–9, 6–13
R-114/R-600, $CClF_2CClF_2/C_4H_{10}$	59/41	28	6–15
R-114/vinyl chloride, $CClF_2CClF_2/CH_2{=}CHCl$	28/72	6.4	6–27
R-114/vinyl methyl ether, $CClF_2CClF_2/CH_2{=}CHOCH_3$	60/40	24.4	6–27
R-115/R-152a, $CClF_2CF_3/CH_3CHF_2$	69/31	−42.3	6–16

TABLE 6–15 (Continued)

Composition	Wt %	Boiling point (°F)	Reference
R-115/R-290, $CClF_2CF_3/C_3H_8$	68.4/31.6 (at 32°F)	−51.9	6–28
R-143a/R-115, $CH_3CF_3/CClF_2CF_3$	60/40	−56	6–8
R-218/R-32, C_3F_8/CH_2F_2	32/68	−72	6–19
R-C318/R-124a, c-$C_4F_8/CClF_2CHF_2$	20/80 (by vol)	11.3	6–17
R-C318/R-21, c-$C_4F_8/CHCl_2F$	68/32	16.5	6–6
R-C318/R-600a, c-C_4F_8/i-C_4H_8	70/30 (at 70°F)		6–18

Another factor relating to boiling point is the great reactivity of fluorine compared with other halogens, such as chlorine and bromine. Fluorine atoms become very strongly attached to carbon atoms. This strong attachment and short carbon-to-fluorine bond distance is responsible for the stability and low toxicity of the fluorocarbon molecules and for the unusually low boiling points. In some ways, fluorine behaves somewhat like hydrogen in spite of a much greater atomic weight (19 to 1). For example, methane, CH_4, has a boiling point of −259°F and carbon tetrafluoride, CF_4, boils at −198°F—only 61°F higher, even though the molecular weight is 88 compared with 16 for methane.

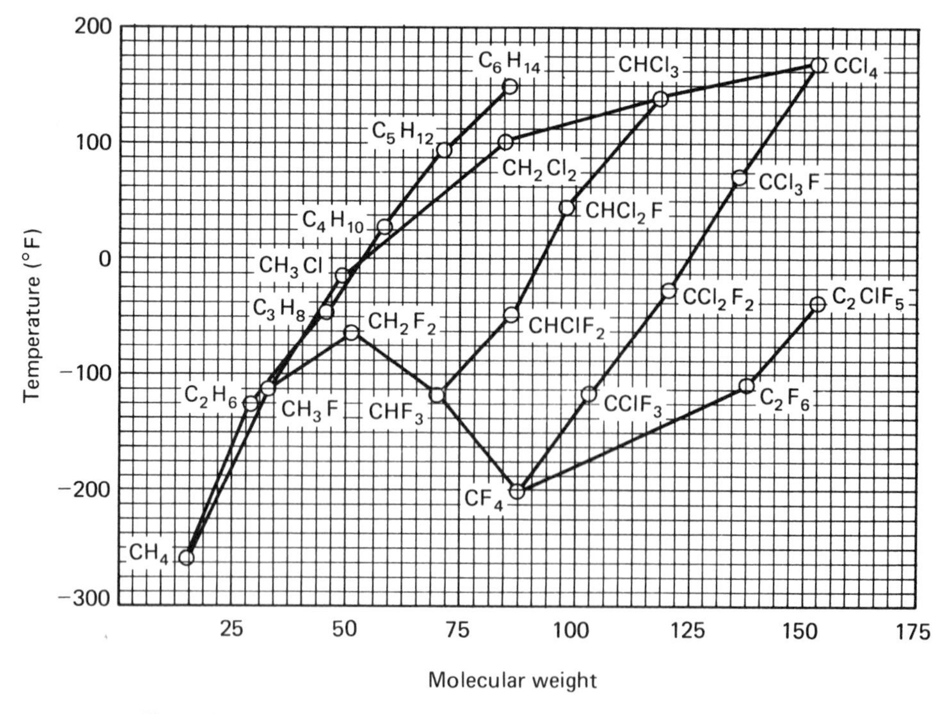

Figure 6–3 Relationship of molecular weight and atmospheric boiling point.

One way to look at azeotropes containing halogenated compounds is from the standpoint of association.

1. Compounds containing only C and H atoms are not associated.
2. Compounds without H are not associated.
3. Compounds containing both hydrogen and halogen atoms may be associated and the boiling point is higher than expected. A good example is CH_2F_2, with a relatively high boiling point and a rather low molecular weight.

Azeotropes found in this group of related compounds nearly always contain one component capable of association and one that is not. The simple or unassociated molecule tends to interfere with the association of the other component, thus reducing the apparent size of the molecule and lowering its boiling point. Azeotropes formed in this way are lower-boiling than the individual components. So, if looking for an azeotrope, as a start choose one associated molecule and one nonassociated molecule and mix. If nothing happens, try another combination. Association is undoubtedly not the whole story. For example, R-11 (CCl_3F) and R-21 ($CHCl_2F$) should form an azeotropic mixture, but none has been found.

Most known azeotropes have higher boiling points than the components. The association theory also applies here if the components of the mixture tend to increase the degree of association. For example, water (H_2O) and ethanol (C_2H_5OH) are each associated in the liquid phase. When mixed, an azeotrope is formed with a boiling point higher than that of either alone. Association of water and alcohol molecules has apparently increased the effective molecular weight and boiling point of the mixture.

Effect of Temperature

The refrigerant nomenclature standard developed by ASHRAE reserved the 500 series for azeotropes. A number is assigned in historical sequence as the azeotropes become commercially important. So far, seven products have been recognized, although only three, R-500, R-502, and R-503, have survived and are in commercial use. Each azeotrope is identified as a definite composition with a definite boiling point. However, the azeotropic composition changes with temperature and in most cases, the assigned composition is not that found at the atmospheric boiling point. Generally, the difference is not great and is not of practical significance. It does seem, however, that in describing an azeotrope, both composition and temperature should be recorded. The temperature corresponding to the designated composition for some azeotropes is shown in Table 6–16.

When the search for new and perhaps better refrigerants first pointed toward azeotropes, the change of composition with temperature was not fully appreciated. Different investigators used different conditions and methods and there was some controversy about exact numbers. The differences were of little more than academic

TABLE 6–16 AZEOTROPIC COMPOSITION AND TEMPERATURE

R	Composition	Wt %	Azeotrope temperature (°F)	Boiling point (°F)	Reference
500	R-12/R-152a	73.8/26.2	32	−28.3	6–34
501	R-22/R-12	75/25	−40[a]	−42.5	6–22
502	R-22/R-115	48.8/51.2	66	−49.8	5–32
503	R-23/R-13	40.1/59.9	−127.6	−127.6	—
505	R-12/R-31	78/22	240	−21.3	6–35
	R-22/R-218	68/32	20	−54	6–32

[a]There is some controversy about the temperature corresponding to the official composition.

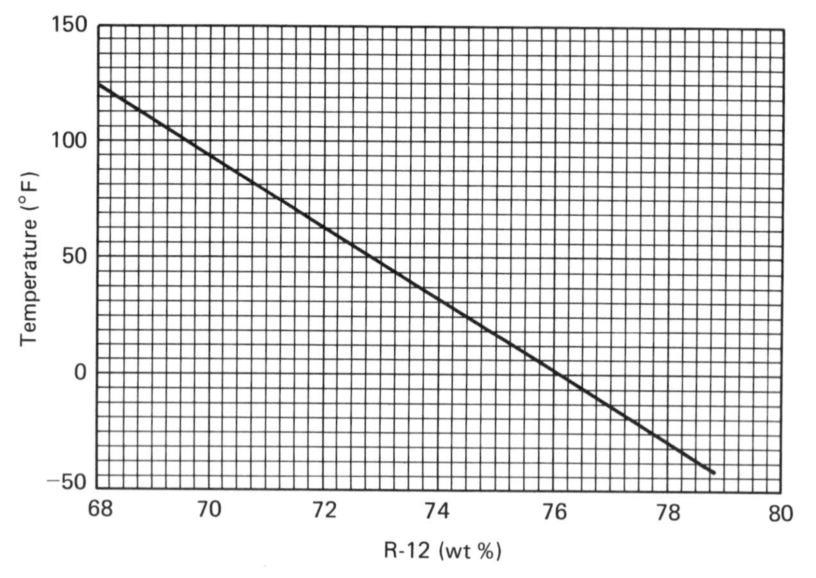

Figure 6–4 R-500 azeotrope (CCl_2F_2/CH_3CHF_2): change in composition with temperature. (From Ref. 6–34.)

interest but did tend to increase interest in the subject and the discovery of many new azeotropes.

The composition change with temperature for R-500 is shown in Figure 6–4, for R-501 in Figure 6–5, for R-502 in Figure 6–6, and for R-22/R-290 in Figure 6–7, page 149. In some cases, the associated pressure is also given. For the azeotrope, R-114/R-21, a composition of 70/30 and a pressure of 54.8 psia is reported at 104°F [6–14] and 75/25 (14 psia) at 32°F [6–9]. As can be seen, the rate of change varies with the azeotrope, but in no case is it large enough to affect the operation of a refrigeration system.

	Change in composition per °F
R-501	0.111
R-500	0.091
R-502	0.055
R-22/R-290	0.050

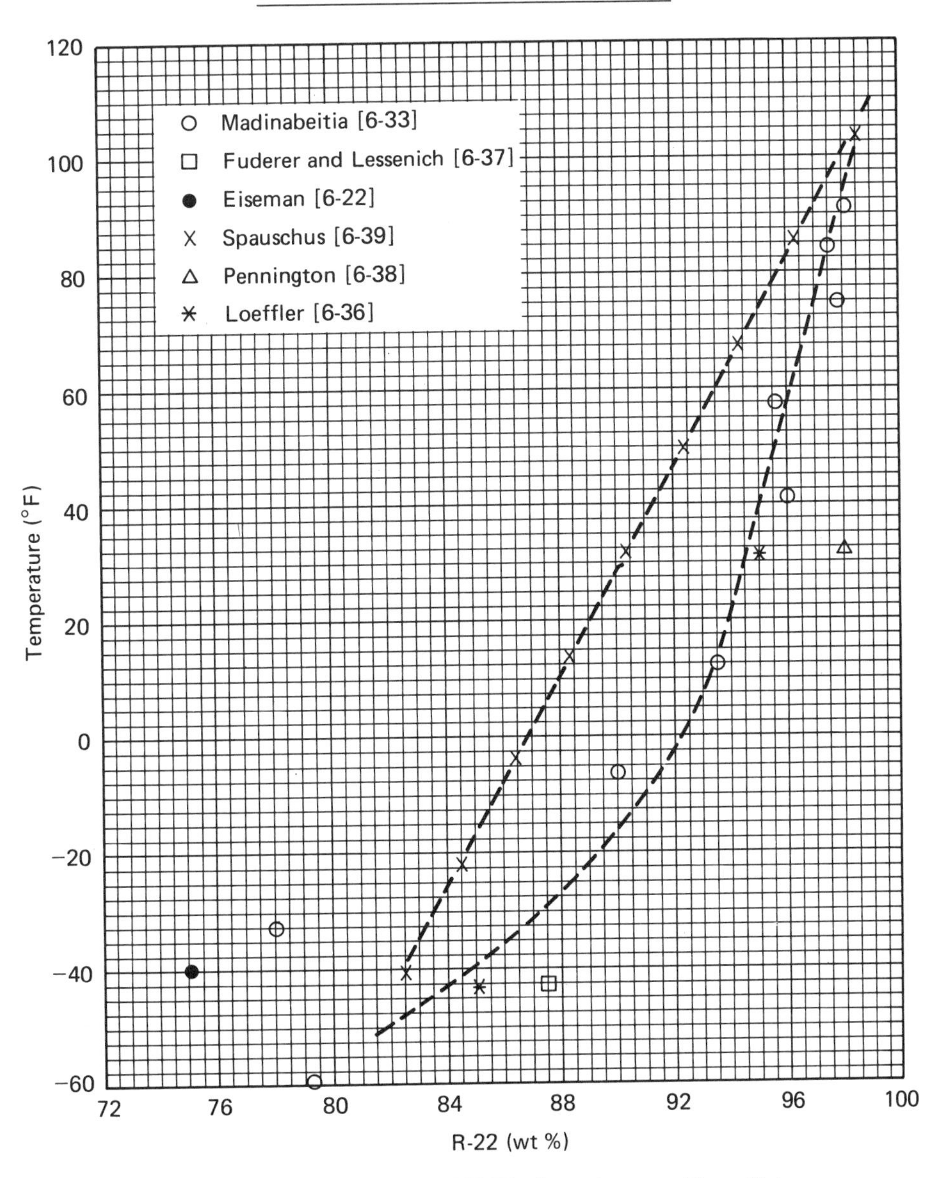

Figure 6–5 R-501 azeotrope ($CHClF_2/CCl_2F_2$): change in composition with temperature.

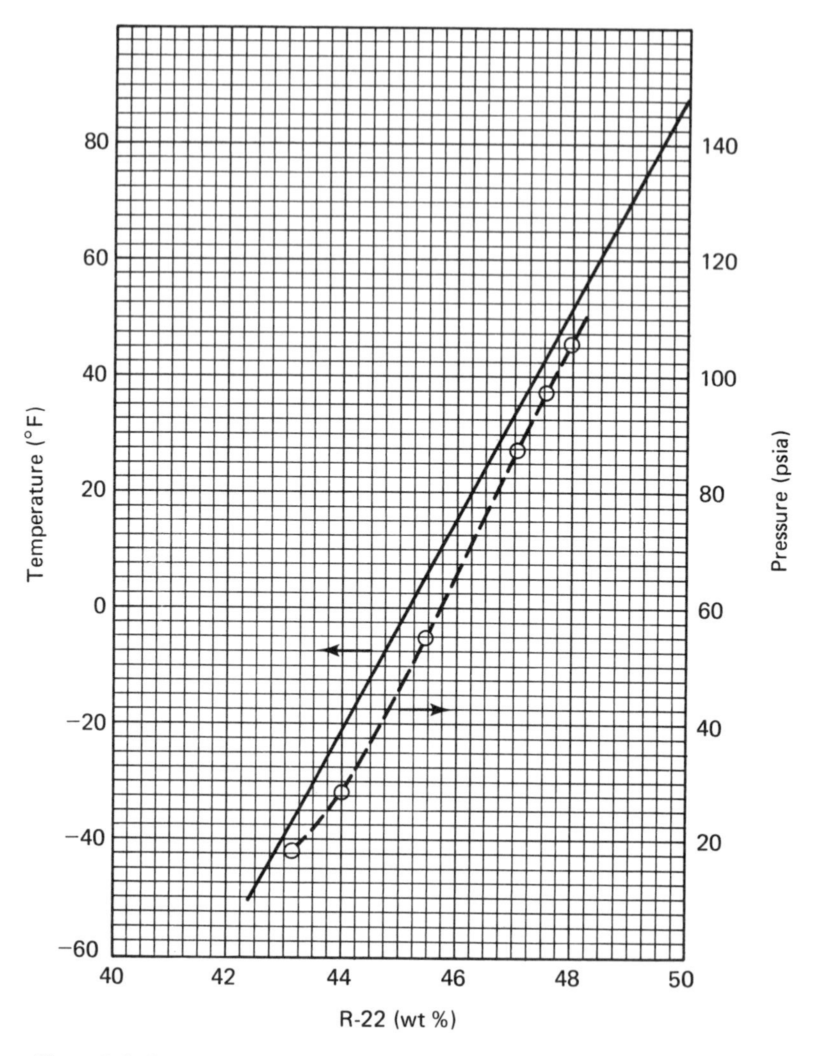

Figure 6–6 R-502 azeotrope ($CHClF_2/CClF_2CF_3$): change in composition with temperature and pressure.

One method of determining the existence and composition of an azeotrope is to measure the equilibrium boiling point of several different mixtures at a constant pressure, usually atmospheric. A curve is drawn through the measured points and the boiling points of the individual components. If an azeotrope exists, the curve will have a minimum and the composition at that point can be graphically determined. Such curves are shown for R-500 in Figure 6–8, R-501 in Figure 6–9, and R-502 in Figure 6–10.

An interesting aspect of the boiling-point curve at constant pressure is the effect of small changes in composition. This point is illustrated in Table 6–17, page 151. For example, with R-501, the concentration of R-22 can vary by 50% with only a 1-degree change in boiling point.

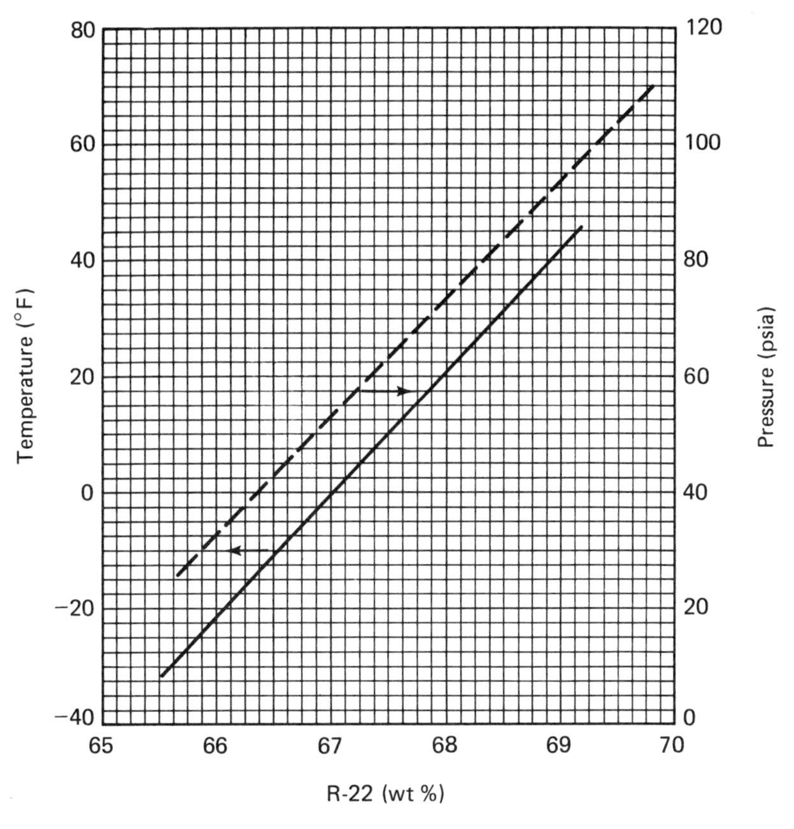

Figure 6–7 Azeotrope of R-22 and R-290 ($CHClF_2/C_3H_8$. (From Ref. 6–32.)

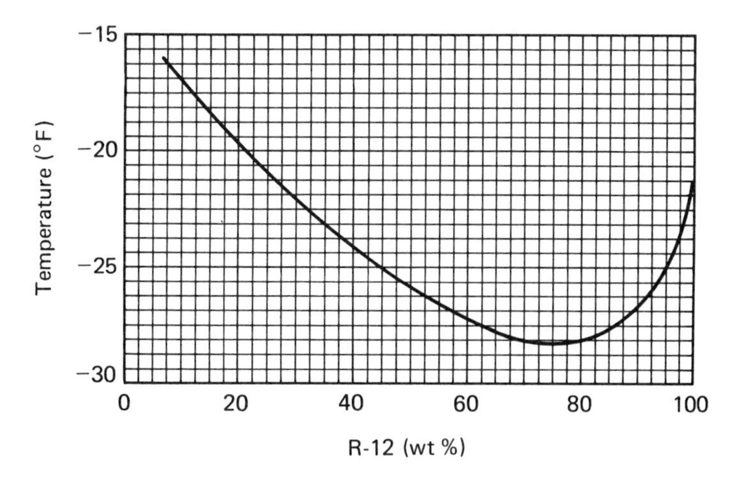

Figure 6–8 R-500: change of equilibrium boiling point with composition at a constant pressure of 1 atm. (From Ref. 6–5.)

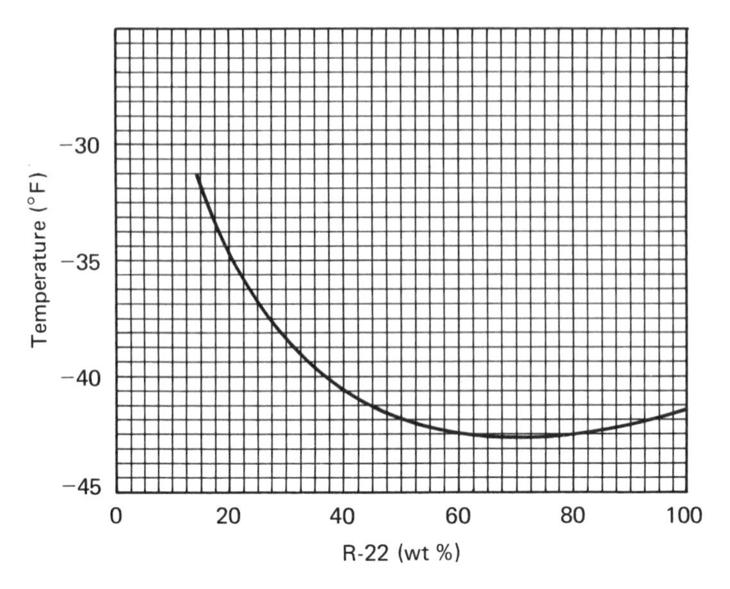

Figure 6–9 Azeotrope R-501: change of equilibrium boiling point with composition at a constant pressure of 1 atm. (From Ref. 6–22.)

Predicting Critical Temperatures

Li [6–43] has proposed the following empirical relationship for calculating critical temperatures.

$$\frac{T_b}{T_c} = X_1 \left(\frac{T_{b1}}{T_{c1}}\right) + X_2 \left(\frac{T_{b2}}{T_{c2}}\right)$$

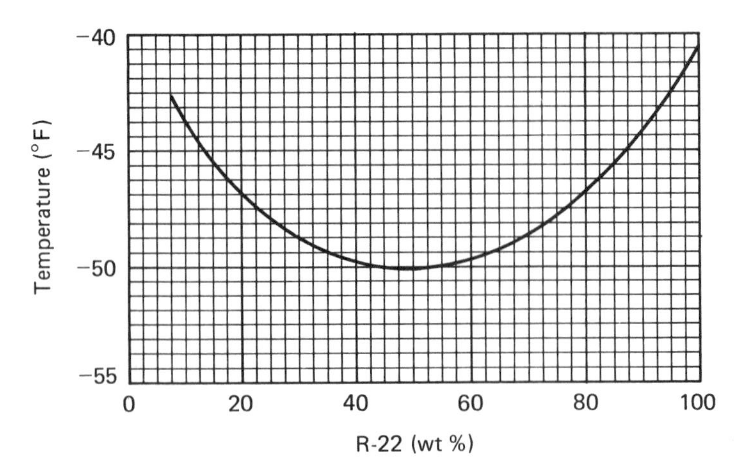

Figure 6–10 Azeotrope R-502: change of equilibrium boiling point with composition at a constant pressure of 1 atm. (From Ref. 6–5.)

TABLE 6–17 COMPOSITION CHANGE RELATED TO BOILING POINT (WT %)

	Boiling-point change	
	1°F	0.5°F
R-500	26	17
R-501	50	36
R-502	35	25

where T_b = boiling point of the azeotrope
T_c = critical temperature of the azeotrope
X_1 = mole fraction of component 1
X_2 = mole fraction of component 2
T_{c1} = critical temperature of component 1
T_{c2} = critical temperature of component 2
The relationship can be extended to three or more components.

Estimated critical temperatures are compared with measured values for several azeotropes in Table 6–18.

TABLE 6–18 CRITICAL TEMPERATURES OF AZEOTROPES

	Composition at 1 atm (wt %)	Critical temperature (°F)		
		Estimated	Observed	Difference
R-12/R-152a	73.6/26.4	218.2	221.3	3.1
R-22/R-115	60.2/39.8	182.4	179.8	−2.6
R-23/R-13	40.1/59.9	67.5	67.1	−0.4
R-31/R-114	47.8/52.2	285.0	285.3	0.3
R-170/R-23	34.4/63.6	52.8	53.3	0.5
R-31/R-12	11.1/88.9	233.4	235.3	1.9
R-115/R-152a	82.8/17.2	174.8	176.9	2.1
R-32/R-218	41.6/58.4	138.0	137.4	−0.6
R-13B1/R-32	80.5/19.5	136.6	142.0	5.4
R-22/R-115/R-218	30/44/26	172.0	174.0	2.0

The relationship can also be used to estimate the boiling point at a given pressure from data for the individual components. Some comparisons with measured data are given in Table 6–19.

The list of azeotropes in Table 6–15 gives a minimum of information. Additional data and comments for some of the azeotropes attaining commercial status and others of general interest follow.

R-500. The azeotrope of R-12 and R-152a was patented by the Carrier Corporation and included in their family of refrigerants as Carrene 7. The objective was to find a refrigerant to use in standard refrigerators that would produce the same

TABLE 6–19 ESTIMATED BOILING POINTS (°F)

	Composition at 1 atm (wt %)	Critical temperature (°F)	Pressure (psia)	Boiling point (°F)	
				Estimated	Observed
R-32/R-115	48.3/51.7	152.2	14.7	−71.1	−69.2
			63.4	−8.1	−6.9
			138.8	32.0	34.1
			286.4	79.6	79.9
R-32/R-12	6.0/94.0	170.6	14.7	−64.5	−62.7
			35.6	−29.9	−27.4
			69.4	1.3	3.5
			122.3	32.0	33.7
			219.5	68.9	69.7
R-23/R-13	40.1/59.9	67.1	14.7	−126.2	−124.5
			177.0	−19.4	−17.5
			222.0	−5.4	−5.6

capacity when used with 50-Hz current (common in Europe) as R-12 with 60-Hz current. R-500 gives about 17% more capacity than R-12 and thus was an excellent solution to the problem. Although the 50-Hz market did not develop, R-500 has found a useful place as a refrigerant with both reciprocating and centrifugal compressors, offering capacity between that of R-12 and R-22. The patent was offered to the public by Carrier some years ago and has since expired.

R-501. The azeotrope of R-22 and R-12 has been known for many years but has not really developed commercial stature. It has the least change in boiling point for a wide change in composition of any known refrigerant. At atmospheric pressure, the boiling point of the azeotrope is about 1°F lower than that of R-22. Up to about 25 wt % of R-12 can be added to the R-22 without any significant change in boiling point or in the operation of equipment designed for R-22. In practical low-temperature operations, some R-12 is often added to the R-22 to improve oil return. So R-501 (or rather a mixture of R-501 and R-22) is being used, although not intentionally and not as an original charge. The use of R-501 as a low-cost R-22 has been considered and some testing performed. The results were not outstanding and apparently not sufficient to compensate for the commercial disadvantage of dealing with another refrigerant or refrigerant mixture.

The performance of several different mixtures of R-22 and R-12 has been measured in a laboratory calorimeter [6–44]. The results are illustrated in Table 6–20 and confirm the slight effect on capacity for mixtures containing up to about 25% R-12.

R-502

The azeotrope of R-22 and R-115 had been discovered and patented for several years before a need for its special properties developed. This need was in the severe operating conditions for frozen-food and ice cream display cabinets in supermarkets.

TABLE 6–20 PERFORMANCE OF MIXTURES OF R-22 AND R-12

	R-22/R-12 mixtures (wt %)					
	0/100	25/75	50/50	75/25	85/15	100/0
Nominal conditions: −20/65/110[a]						
Capacity						
Btu/hr	1,250	1,625	2,135	2,315	2,355	2,360
% of R-22	53	69	90	97	100	100
Compression Ratio	9.88	9.54	9.40	9.46	9.51	9.74
Discharge temperature (°F)	236	249	262	273	275	282
Nominal conditions: 40/65/110[a]						
Capacity						
Btu/hr	8,660	11,060	12,555	13,310	13,545	13,660
% of R-22	63	81	92	97	99	100
Compression ratio	2.93	2.83	2.83	2.86	2.86	2.90
Discharge temperature (°F)	170	173	178	190	187	191

[a]Temperature in °F, evaporating/vapor entering compressor/condensing.

Source: Ref. 6–44.

Here the compressors were usually located in a machinery room often 80 or 100 ft or more distant from the evaporator in the front part of the store. Long return-gas or suction lines were necessary, allowing the cold refrigerant vapor to warm somewhat before it entered the compressor. Higher temperatures for refrigerant vapor entering a compressor mean much higher temperatures on leaving. R-22 was the refrigerant in many systems and compressor discharge temperatures became very high. Service problems became intolerable [6–45,6–46]. High temperatures in the compressor cause many problems, including deterioration of motor-winding insulation, loss of lubricating ability of the oil, bearing and piston seizure, broken valve reeds, refrigerant and oil decomposition, and the formation of carbon and oil sludge.

After extensive laboratory evaluation, R-502 was used to replace R-22 in thousands of existing low-temperature systems in supermarkets. In all cases, compressor discharge temperatures were lowered—sometimes by as much as 100°F—and service problems were greatly reduced. Today, a majority of new frozen-food display cases are designed to use R-502.

The properties of R-502 responsible for relatively low compressor discharge temperatures include:

High vapor heat capacity. The gas is able to absorb more Btu/lb for each degree rise in temperature because of the more complex nature of the R-115 part of the azeotrope.

High vapor flow rate. Refrigeration calculations and thinking are based on the weight of the refrigerant—which has little to do with the operation of a system. The higher molecular weight of R-502 compared with R-22 means a lower latent heat in terms of Btu/lb and so more pounds of refrigerant are circulated—providing more refrigerant for soaking up heat with a lower rise in temperature.

Low compression ratio. Compressed gas leaves the compressor at a lower temperature.

The refrigeration capacity of R-502 is 20 to 30% greater than R-22 at low evaporating temperatures (−40°F) but is about the same or perhaps a trifle less than R-22 in the air-conditioning range. Since many of the desirable properties of R-502 are contributed by the R-115, why not use it alone? It can be used, but at a considerable sacrifice in capacity. The performance in a laboratory calorimeter is illustrated in Table 6–21. In field operations, compressor discharge temperatures are often much higher than those found in the laboratory. R-502 is sometimes described as a refrigerant with the thermal properties of R-12 and the capacity of R-22.

R-503

The azeotrope of R-23 and R-13 has an atmospheric boiling point almost exactly the same as ethane, R-170, and so is a direct competitor for the very-low-temperature market. R-13 is also used for this application, but R-503 produces about 10 to 15% more capacity.

TABLE 6–21 PERFORMANCE OF VARIOUS REFRIGERANTS

	Refrigerant			
	12	115	22	502
Nominal conditions: −20/65/110[a]				
Capacity				
Btu/hr	1,250	1,675	2,360	2,650
% of R-22	53	77	100	112
Compression ratio	9.88	9.15	9.74	8.88
Discharge temperature (°F)	236	194	282	238
Nominal conditions: 40/65/110[a]				
Capacity				
Btu/hr	8,660	10,040	13,660	14,090
% of R-22	63	73	100	103
Compression ratio	2.93	2.79	2.90	2.77
Discharge temperature (°F)	170	139	191	162

[a]Temperature in °F, evaporating/vapor entering compressor/condensing.
Source: Ref. 6–44.

R-505, R-506

These azeotropes containing R-31 with R-12 and with R-114 were withdrawn from the market when it was discovered that R-31 is more toxic than was thought originally. R-505 was being used to replace R-12 in some applications, especially in centrifugal compressors. R-506 was used as the refrigerant for air conditioning in overhead crane cabs in steel mills. It was replaced with a mixture of R-12 and R-114 (60/40 by weight).

R-143a/R-115 (60/40)

The azeotrope of R-143a and R-115 was patented by DuPont [6–8] and claimed to operate with slightly greater capacity than R-502 and lower discharge and motor temperatures. Although R-143a is flammable, the azeotropic mixture is nonflammable in air at room temperature and 212°F. Vapor pressure data are shown in Table 6–22 and performance data in Table 6–23 for various compositions.

TABLE 6–22 VAPOR PRESSURE OF MIXTURES OF R-143a AND R-115 (PSIA)

	Composition, R-143a/R-115 (wt %)					
Temperature (°F)	0/100	25/75	40/60	60/40	75/25	100/0
53	90	122.6	128.6	132.0	131.1	127.0
4				56.7	56.2	54.0
−40	13.9	20.5	21.4	21.9	21.5	20.5
−60	8.2	12.2	12.7	12.9	12.9	12.4

TABLE 6–23 PERFORMANCE DATA FOR R-143a/R-115 (60/40)[a]

	Evaporator temperature (°F)	R-143a/R-115	R-22	R-502
Capacity (Btu/hr)	20	44,700	40,510	42,250
	−20	15,900	12,990	15,660
	−40	7,230	5,700	7,000
Compression ratio	20	3.90	4.20	—
	−20	8.75	9.70	—
	−40	13.70	15.80	—
Discharge temperature (°F)	20	190	226	195
	−20	253	320	266
	−40	284	356	302
Motor temperature (°F)	20	115	111	107
	−20	120	148	130
	−40	149	—	158

[a]Condensing temperature, 110°F; temperature of gas entering compressor, 65°F.

R-115/R-152a (69/31) and R-12/Dimethyl Ether (90/10)

These azeotropes have not been used commercially but have been studied in the laboratory with the results given in Table 6–24.

TABLE 6–24 PERFORMANCE OF R-115/R-152a (69/31) AND R-12/DIMETHYL ETHER (90/10)[a]

	Evaporator temperature (°F)	R-115/R-152a	R-12/DME	R-22
Capacity (Btu/hr)	20	35,222	26,658	40,514
	−20	12,474	7,204	12,986
	−40	4,705	2,726	5,700
Discharge temperature (°F)	20	187	225	226
	−20	245	294	320
	−40	267	279	356
Motor temperature (°F)	20	—	120	111
	−20	179	188	148
	−40	216	—	—

[a]Condensing temperature, 110°F; temperature of gas entering compressor, 65°F.
Source: Ref. 6–5.

R-115/R-290 (68.4/31.6)

The azeotrope of R-115 and R-290 (propane) was studied by Valbjorn [6–28] and compared with R-502. He found that the refrigerating capacity is about the same with either refrigerant but that the temperature of the motor winding is about 30°F lower with R-115/R-290. The compressor discharge temperature is also estimated to be lower. The azeotropic mixture is flammable, but precise explosion limits have not been determined. The following equation was developed to describe the vapor pressure of R-115/R-290:

$$\ln P = A + \frac{100B}{T} + C \ln\left(\frac{T}{100}\right) - D\left(\frac{T}{100}\right)^6$$

where $A = 14.0259$
$B = -27.8232$
$C = -2.1036$
$D = -0.00004669$
ln = log to the base e
$P = \text{kPa/cm}^2$, abs
$T = \text{K} = °\text{C} + 273.15$

Other properties are given in Table 6–25.

TABLE 6–25 PROPERTIES OF R-115/R-290 (68.4/31.6)

Normal boiling point	−46.6°C (−51.9°F)
Molecular weight	86.2
Critical temperature	80.7°C (177.3°F)
Critical pressure	35.30 atm (518.8 psia)
Latent heat of vaporization	48.0 kcal/kg (86.4 Btu/lb)

R-124a/R-C318 (60/40 by Weight)

The azeotrope of R-124a and R-C318 was patented by DuPont [6–16] but found no commercial use. A shortage in the supply of R-12 in the USSR, however, led to an evaluation of the azeotrope as a substitute [6–47]. It was reported to operate very well. The following equations were developed:

Vapor Pressure:

$$\log P = A - \frac{B}{T} + C \log T - DT$$

where $A = -5.5437732$
$B = 1080.52926$
$C = 4.5835157$
$D = 0.00529580678$
$P = \text{kg/cm}^2$, abs
$T = \text{K} = °\text{C} + 273.15$

Equation of State

$$P = \frac{RT(V + B)}{V^2} - \frac{A}{V^2}$$

where

$$A = A_0\left(1 - \frac{a}{V}\right) \qquad \text{and} \qquad B = B_0\left(1 - \frac{b}{V}\right)$$

where $A_0 = 99.0025958$
$B_0 = 0.07096519$
$R = 5.2485$
$a = -0.02116791$
$b = -0.013497367$
$P = \text{kg/m}^2$
$V = \text{m}^3\text{/kg}$
$T = \text{K}$

REFERENCES

6–1. D. J. Missimer, "Mechanical System Can Reach −140°C," *Res. Dev.* (July 1972), 40.

6–2. DuPont Company, Freon Products Division, "Freon 13B1/152a Refrigerant Mixture for Heat Pumps," Bulletin RT-74 (1981).

6–3. DuPont Company, Freon Products Division, "Vapor Pressure and Liquid Density of Freon Propellents," Bulletin FA-22 (1970).

6–4. R. C. McHarness and D. D. Chapman, "Refrigerating Capacity and Performance Data for Various Refrigerants, Azeotropes, and Mixtures," *ASHRAE J.*, 4 (1962), 49.

6–5. V. F. Chaikovsky and A. P. Kuznetsov, "The Use of Mixtures of Refrigerants in Compression Refrigerating Machines," *Kholod. Tekh.*, 1 (1963), 9; abstract in *J. Refrig.*, May–June (1963), 66.

6–6. DuPont Company, Freon Products Division, unpublished data.

6–7. G. H. Whipple, "Vapor-Liquid Equilibria of Some Fluorinated Hydrocarbon Systems," *Ind. Eng. Chem.*, 44 (July 1952), 1664.

6–8. A. Piacentini and F. P. Stein, "An Experimental and Correlative Study of the Vapor-Liquid Equilibria of the Tetrafluoromethane-Trifluoromethane System," Chem. Eng. Prog. Symp. Ser., *Phase Equilibria and Related Properties*, No. 81, vol. 63, p. 28.

6–9. D. E. Kvalnes, "Refrigerant Compositions," U.S. Patent 3,337,287 to DuPont Company (Apr. 9, 1968).

6–10. W. A. Pennington, "Progress in Refrigerants: 1," *World Refrig.* (Feb. 1957), 85; 2 *World Refrig.* (Mar. 1957), 151.

6–11. H. G. Brant, "Methyl Chloride-Expedient Substitute," *Refrig. Eng.*, 46 (1943), 303.

6–12. W. A. Pennington and W. H. Reed, "Azeotrope of 1,1-Difluoroethane and Dichlorodifluoromethane as a Refrigerant," *Chem. Eng. Prog.*, 46 (1950), 464; "The Evolution of a New Refrigerant," *Mod. Refrig.*, 53 (1950), 123.

6–13. W. H. Reed, "Azeotropic Mixture for Use as a Refrigerant," U.S. Patent 2,511,993 to Carrier Corp. (1950).

6–14. W. H. Reed and W. A. Pennington, "Refrigeration System Containing a Novel Refrigerant," U.S. Patent 2,630,686 to Carrier Corp. (1953).

6–15. J. Fleischer, "Refrigerants and Methods of Transferring Heat," U.S. Patent 2,191,196 to General Motors Corp. (1940).

6–16. H. Lewis, ''Azeotropic Composition of 1,1-Difluoroethane and Monofluoropentachloroethane,'' U.S. Patent 2,641,580 to DuPont Company (1953).
6–17. A. F. Benning and J. D. Park, ''Separating Fluorine Compounds,'' U.S. Patent 2,384,449 to DuPont Company (1945).
6–18. W. H. Reed, ''Propellant Composition,'' U.S. Patent 2,968,628 to Shulton Inc. (1961).
6–19. NACA TN 3226.
6–20. W. A. Pennington and W. H. Reed, ''Process for Producing Increased Refrigeration,'' U.S. Patent 2,479,259 to Carrier Corp. (1949).
6–21. A. F. Benning, ''Azeotropic Refrigerant Composition of Monochlorodifluoromethane and Chloropentafluoroethane,'' U.S. Patent 2,641,579 to DuPont Company (1953).
6–22. B. J. Eiseman, Jr., ''The Azeotrope of Monochlorodifluoromethane and Dichlorodifluoromethane,'' *J. Am. Chem. Soc.*, 79 (1957), 6087.
6–23. E. H. Hadley and L. A. Bigelow, ''Action of Elementary Fluorine upon Organic Compounds,'' *J. Am. Chem. Soc.*, 62 (1940), 3302.
6–24. E. Bennett and H. M. Parmelee, ''Azeotropic Composition,'' U.S. Patent 2,999,816 to DuPont Company (1961).
6–25. B. J. Eiseman, Jr., ''Azeotropic Composition,'' U.S. Patent 2,999,815 to DuPont Company (1961).
6–26. F. A. Bower and H. M. Parmelee, ''Azeotropic Composition,'' U.S. Patent 2,999,817 to DuPont Company (1961).
6–27. J. R. Haase and R. J. Scott, ''Azeotropic Aerosol Propellants,'' *Soap Chem. Spec.*, (Aug. 1964), 105.
6–28. K. V. Valbjorn, ''The R-115/R-290 Azeotrope as a Refrigerant,'' *ASHRAE J.*, 10 (Apr. 1968), 47.
6–29. K. P. Murphy and S. P. Orfeo, ''Novel Fluorocarbon Composition,'' U.S. Patent 3,634,255 to Allied Chemical Company (1972).
6–30. K. P. Murphy and S. P. Orfeo, ''Azeotropic Mixture,'' U.S. Patent 3,505,232 to Allied Chemical Company (1970).
6–31. ''Mixtures of Fluorinated Hydrocarbons,'' French Patent 1,350,327 to Allied Chemical Company (1963).
6–32. L. J. Long, ''The Effect of Pressure and Temperature on Azeotropic Composition in the Monochlorodifluoromethane/Monochloropentafluoroethane and Monochlorodifluoromethane/Propane Systems,'' thesis, University of Delaware (June 1962).
6–33. D. Madinabeitia, ''The Effect of Pressure and Temperature on Azeotropic Composition in the Monochlorodifluoromethane and Dichlorodifluoromethane Binary System,'' thesis, University of Delaware (June 1960).
6–34. W. A. Pennington, ''Effect of Temperature on Azeotropy in 1,1-Difluoroethane and Dichlorodifluoromethane,'' *Ind. Eng. Chem.*, 44 (1952), 2397.
6–35. Allied Chemical Corp., private communication.
6–36. H. J. Loeffler, ''Some Properties of Binary Systems Frigen 12-Frigen 22 and the Ternary System F-12–F-22–Basic Naphthene Mineral Oil,'' *Kaltetechnik*, 12 (1960), 256.
6–37. A. Fuderer and W. Lessenich, ''Azeotropic and Other Frigen Mixtures as Refrigerants,'' *Kaltetechnik*, 15 (1963), 235.
6–38. W. A. Pennington, ''Refrigerants,'' *Air Cond. Heat. Vent.*, 55 (Nov. 1958), 71.

6–39. H. O. Spauschus, "Vapor Pressures of Mixtures of Refrigerants 12 and 22," *ASHRAE J.*, 4 (Sept. 1962), 49.

6–40. B. J. Eiseman, Jr., "Multicomponent Refrigerants," International Institute of Refrigeration, meeting, Czechoslovakia (Sept. 1965).

6–41. H. Skolnik, "Effect of Pressure in Azeotropy," *Ind. Eng. Chem.*, 43 (1951), 172.

6–42. S. B. Lippincott and M. M. Lyman, "Vapor Pressure–Temperature Nomographs," *Ind. Eng. Chem.*, 38 (1946), 320.

6–43. C. C. Li, "Predicting Critical Temperatures of Azeotropes," *ASHRAE J.*, 14 (May 1972), 51.

6–44. R. C. McHarness and D. D. Chapman, "Refrigerating Capacity and Performance Data for Various Refrigerants, Azeotropes and Mixtures," *ASHRAE J.*, 4 (1962), 49.

6–45. R. C. Downing, "Comparison of Freon 502 Refrigerant with Freon 22 and Freon 12," *Refrig. Serv. Contract.* (May 1962), 18.

6–46. R. C. Downing, "Freon 502," *Mod. Refrig.*, 69 (1966), 777.

6–47. I. I. Perelstein, "The Thermodynamic Properties of the Azeotropic Mixture of R 124 and RC 318," *Kholodilnaya Tekhnika*, No. 2 (1962), 76. Abstract in Journal of Refrigeration, Nov./Dec. (1964), 134.

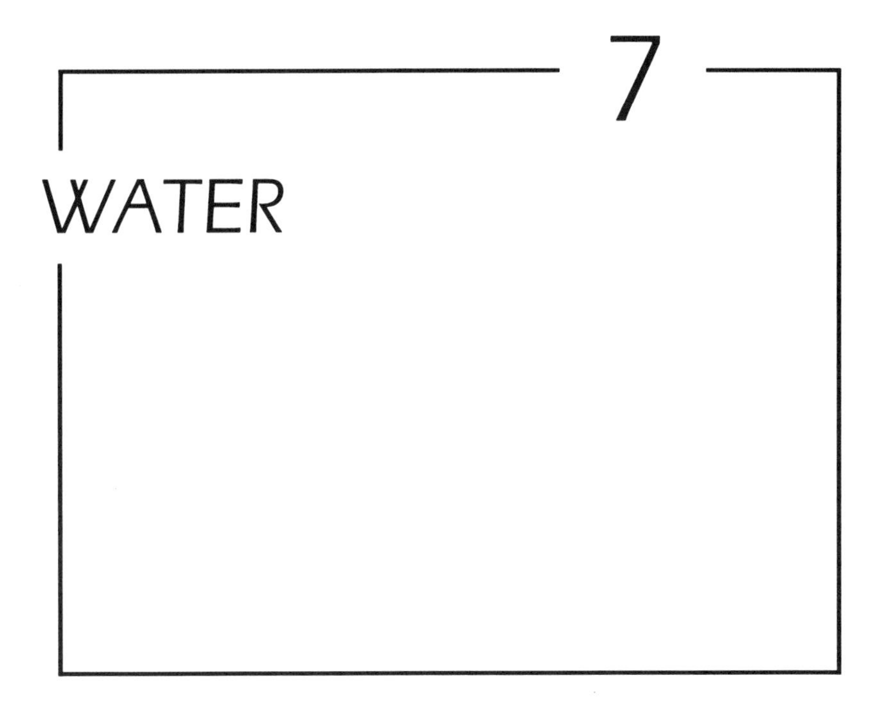

7 WATER

Manufacturers of refrigeration and air-conditioning equipment make special efforts to dry the systems before the refrigerant and oil are added—and for good reason. Water is a most undesirable contaminant. By itself or in combination with air it may cause rusting, corrosion, copper plating, refrigerant decomposition, valve damage, oil sludging, and general deterioration of the system. If water solubility in the refrigerant is exceeded at low temperatures, ice may form in the expansion valve or capillary tube and restrict the flow of refrigerant, or in severe cases, stop it altogether. Pinhole corrosion of aluminum parts has been traced to the presence of water.

No one knows for sure what limitation on the amount of water present in a system should be set, but there is general agreement that the less water present, the better. From sealed-tube studies, Parmelee concluded that dissolved moisture has no significant effect on stability [7–1]. Tests were made with R-12 and oil up to 80 ppm by weight based on the total liquid content (equal parts of refrigerant and oil by volume). Similar tests with R-22 and oil contained up to 570 ppm water. However, there is some question as to how these static tests relate to the dynamic conditions in an operating system. Oil concentration at the critical hot spots in the system may be very small and undissolved water vapor available for undesirable reactions. A survey of manufacturers [7–2] regarding acceptable moisture limits in systems lists 10 to 60 ppm for R-12 and 50 to 200 ppm for R-22, depending on the application. These numbers are apparently parts per million by weight based on the refrigerant. Presumably, experience with water in these ranges has been good, but it seems reasonable to keep the water content as low as possible.

Over the years, the specification limit for water in new refrigerant has been progressively lowered from about 50 ppm to the present level of 10 ppm. Some water is also present in the lubricating oil. A specific limitation is not usually listed, but a dielectic-strength requirement corresponds to a water content of about 25 ppm.

SOLUBILITY

Solubility of Water in Liquid Refrigerant [7–3]

The solubility of water in the liquid phase of a number of refrigerants as a function of temperature is given in Figure 7–1. It is apparent that water is far less soluble in some refrigerants than in others and that the solubility is directly affected by the temperature. For example, the solubility of water in R-12 at low temperatures is so low compared with R-22 that icing problems with R-12 would be expected to be much more severe than with R-22. It can be seen that R-114, R-13B1, and R-11 would be similar to R-12 in this respect, while R-500 and R-502 would have more tolerance for water as far as ice formation is concerned.

Solubility of Refrigerant in Water [7–3, 7–4]

In all cases, the solubility of fluorinated refrigerants in water is quite low and ordinarily is of little concern in refrigeration work. There should be no liquid water in operating systems. However, in some applications there is a possibility of refrigerant coming into contact with water through equipment failure in condensers, water chillers, drinking fountains, heat reclaim systems, and so on. If this happens, the amount of refrigerant that might dissolve in the water may be of some importance. The solubility of refrigerants in water is shown in Figures 7–2 to 7–10, pages 165–173, as a function of temperature and pressure based on measurements by Parmelee [7–4].

Stepakoff and Modica measured the solubility of R-114 in water in connection with its use in purifying seawater [7–5]. They found a somewhat lower solubility than reported by Parmelee, as shown in Table 7–1, page 174, using the following equation to describe the results:

$$\ln S\ (\text{ppm/atm}) = \frac{3321}{T\ (\text{K})} - 6.48$$

Stepakoff and Modica also calculated heats of solution as shown here in Table 7–2, page 174. Henry's constant is followed by the temperature in degrees Celsius.

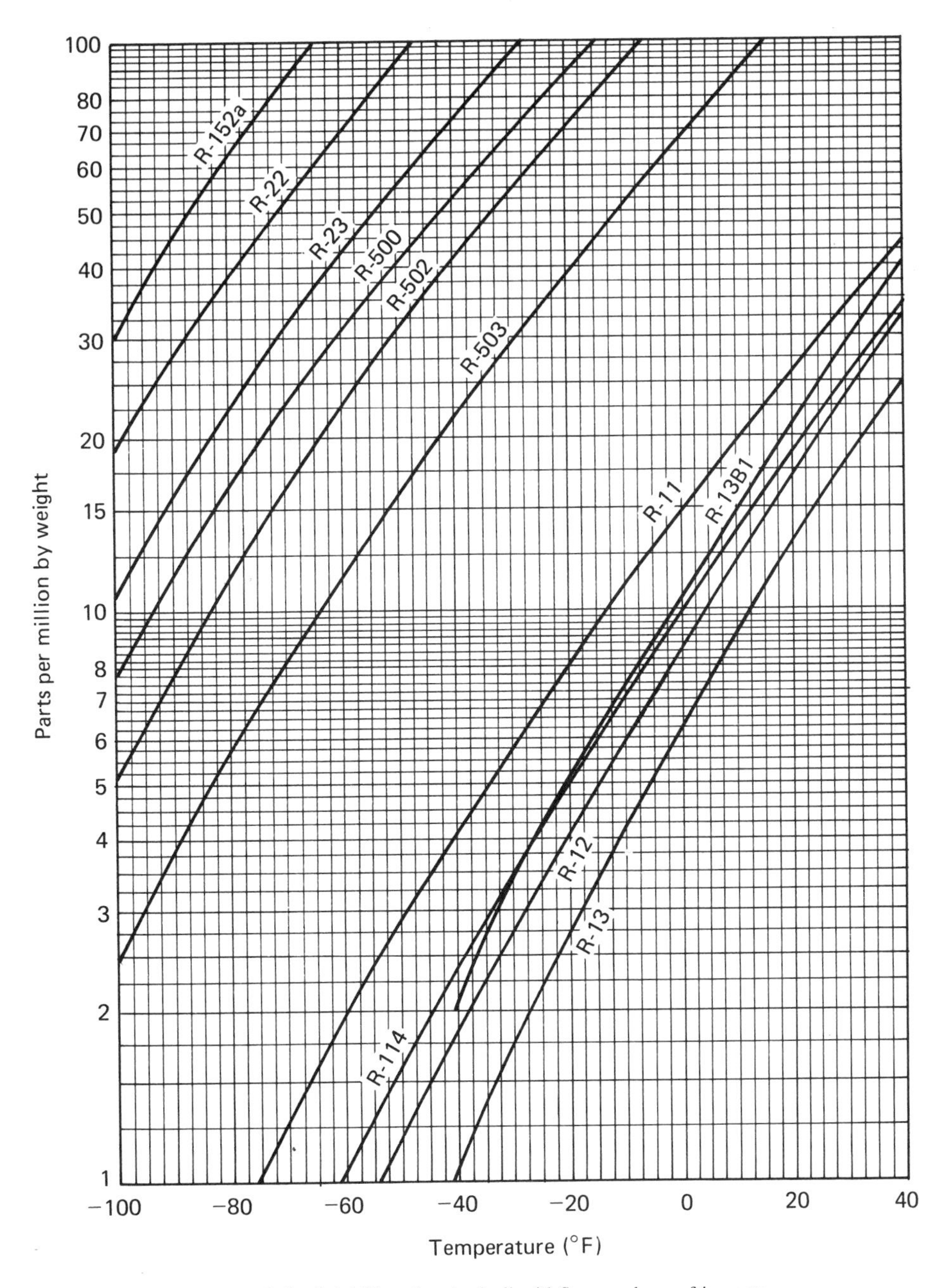

Figure 7–1 Solubility of water in liquid fluorocarbon refrigerants.

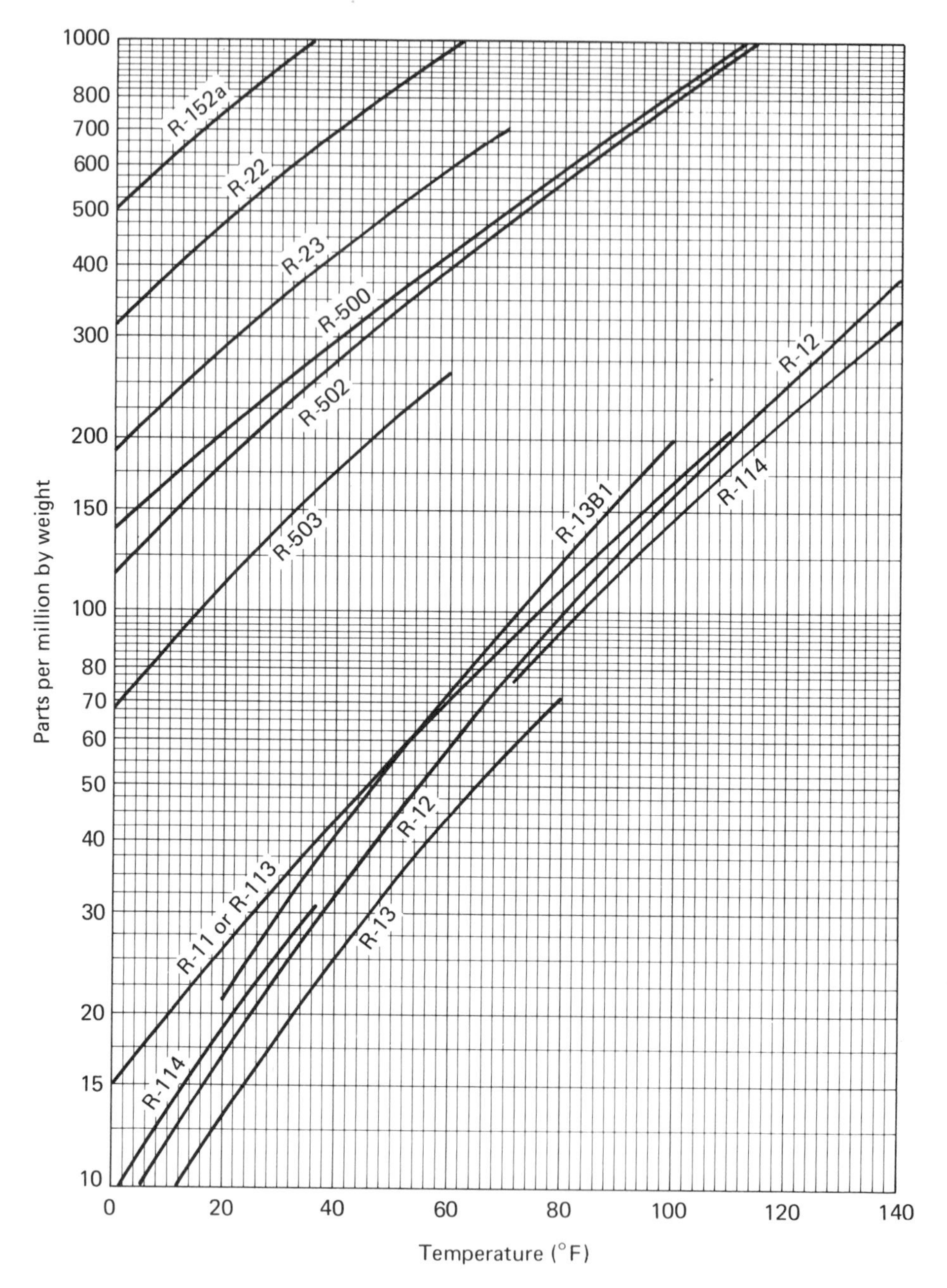

Figure 7–1 (Continued)

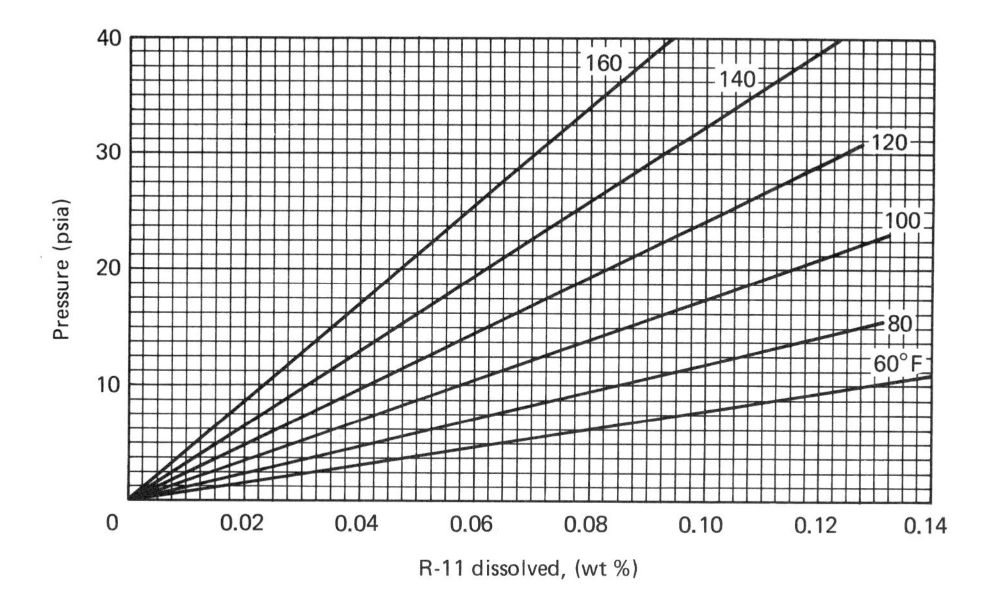

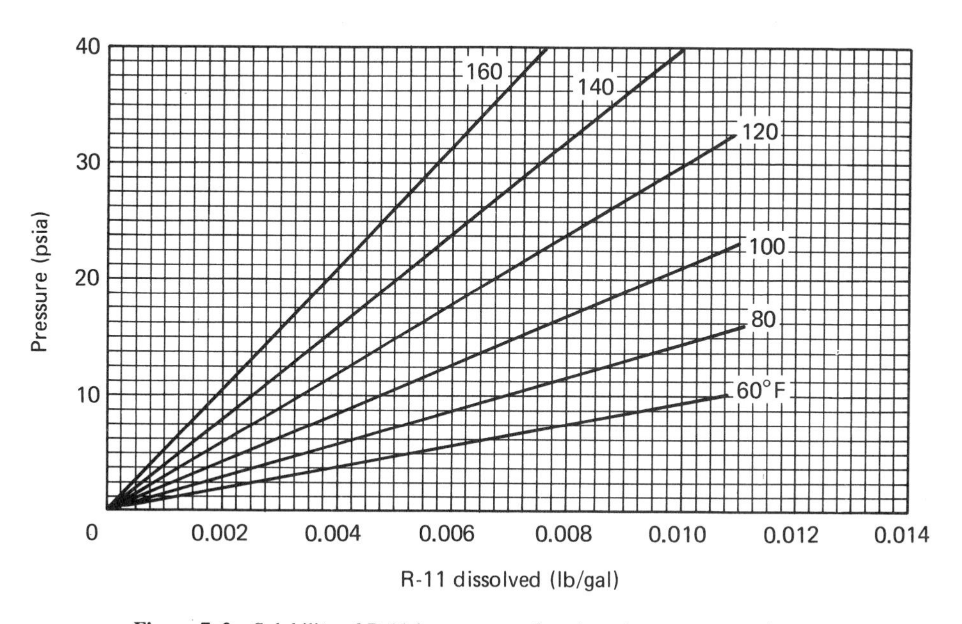

Figure 7–2 Solubility of R-11 in water as a function of temperature and pressure.

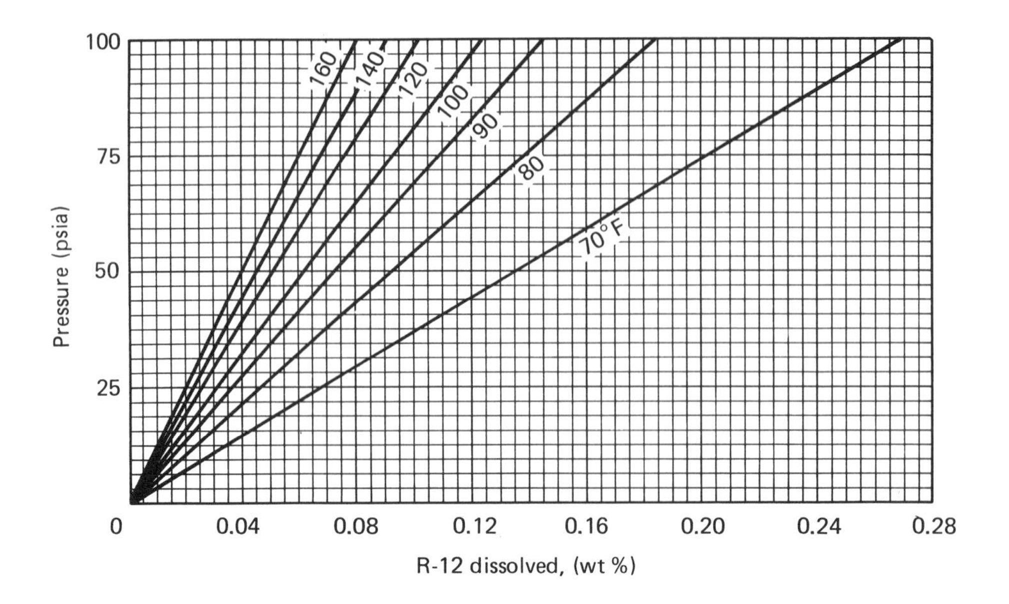

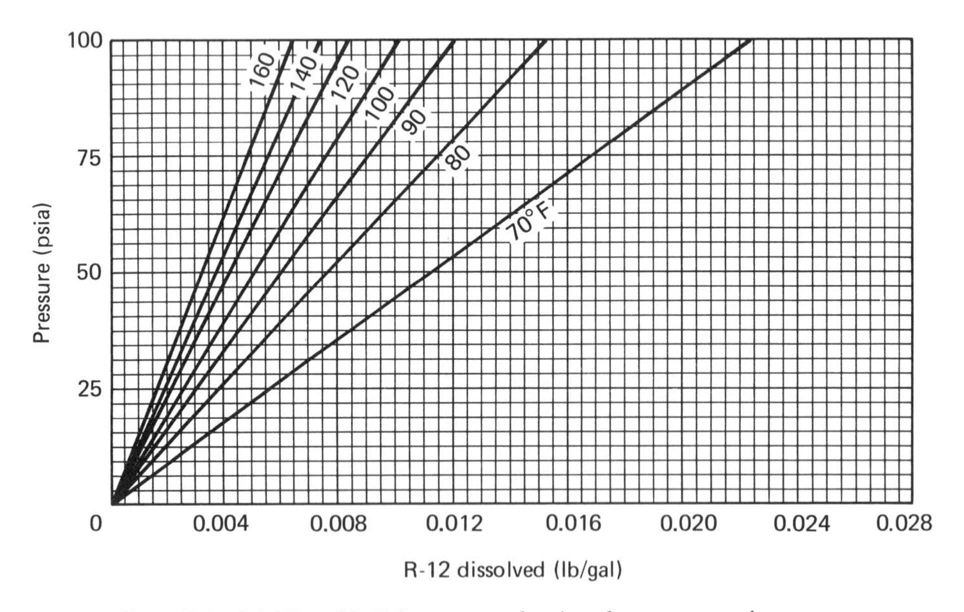

Figure 7–3 Solubility of R-12 in water as a function of temperature and pressure.

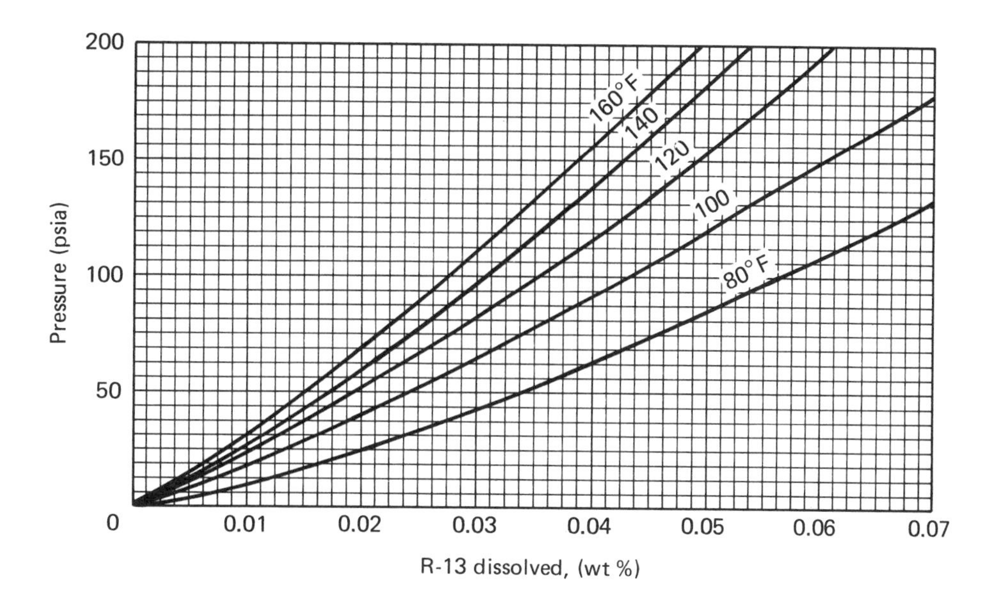

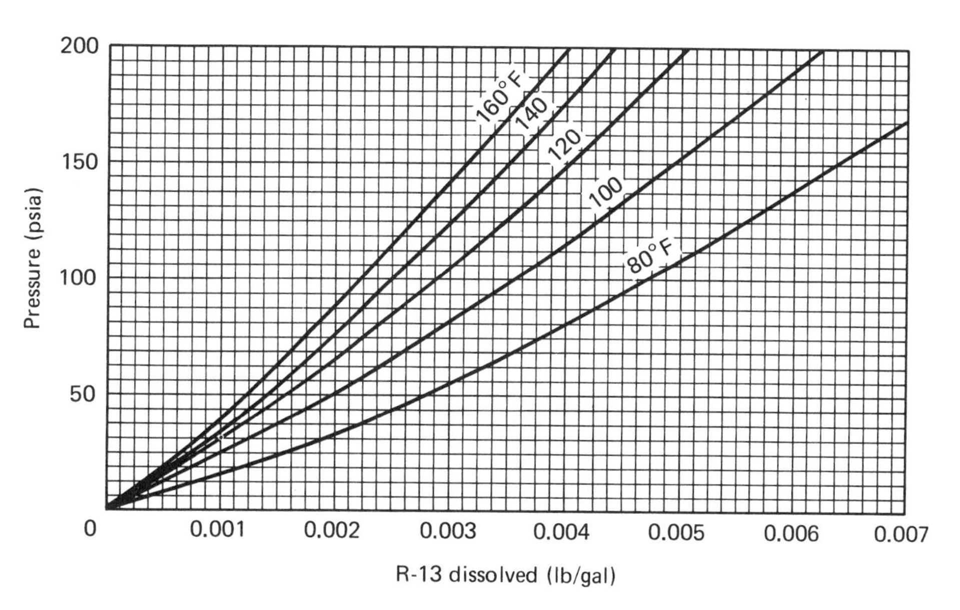

Figure 7–4 Solubility of R-13 in water as a function of temperature and pressure.

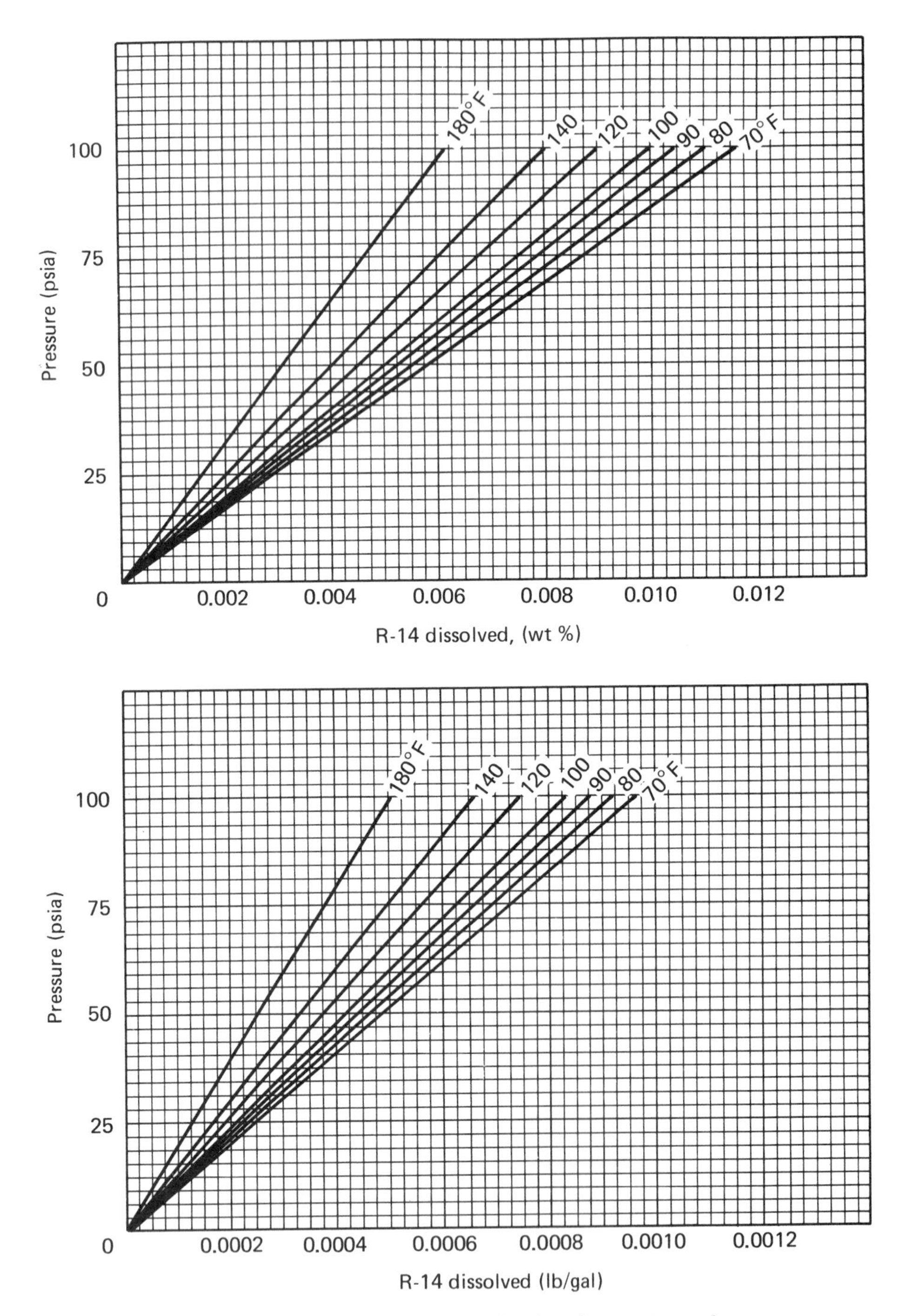

Figure 7–5 Solubility of R-14 in water as a function of temperature and pressure.

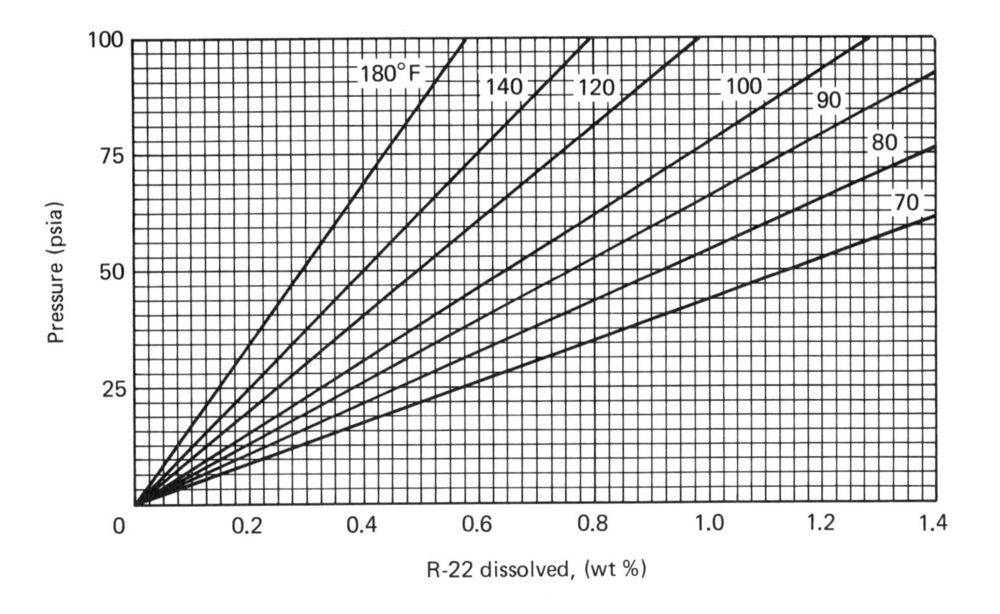

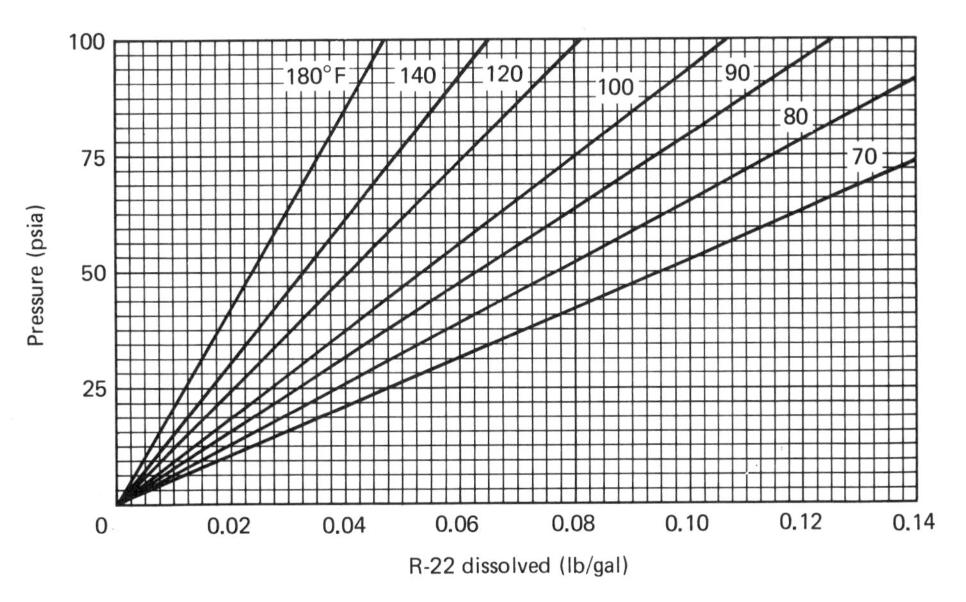

Figure 7–6 Solubility of R-22 in water as a function of temperature and pressure.

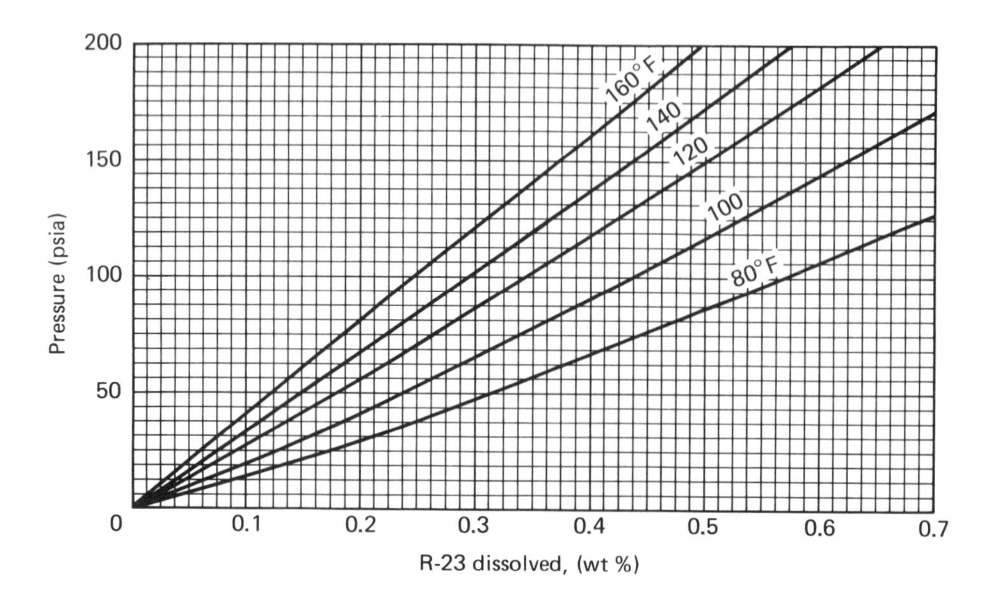

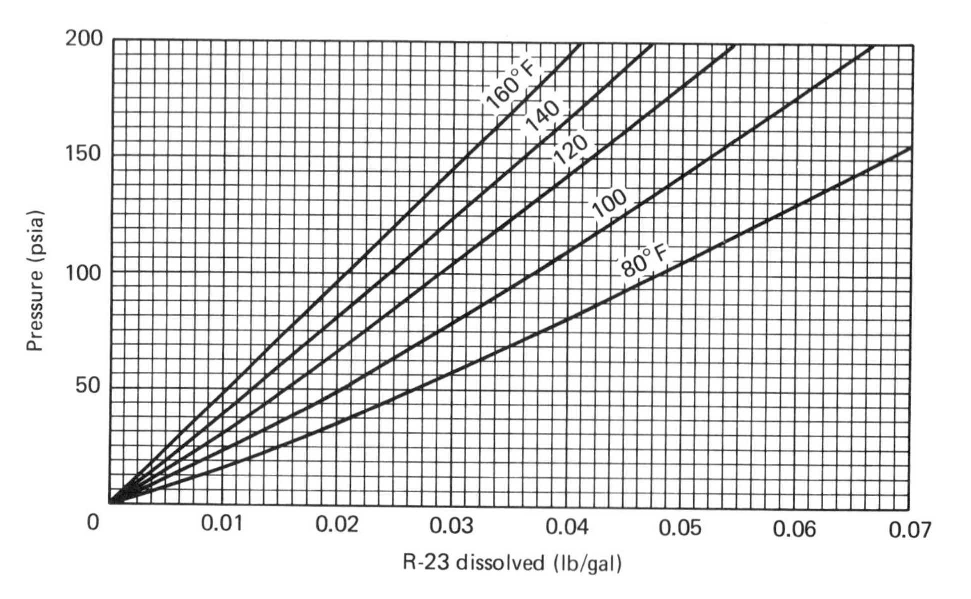

Figure 7–7 Solubility of R-23 in water as a function of temperature and pressure.

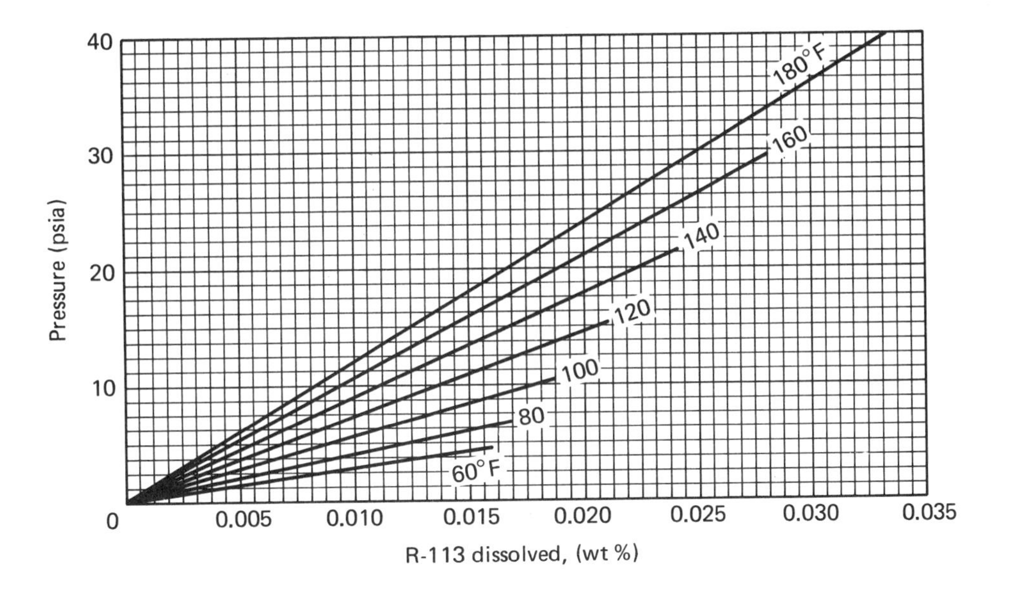

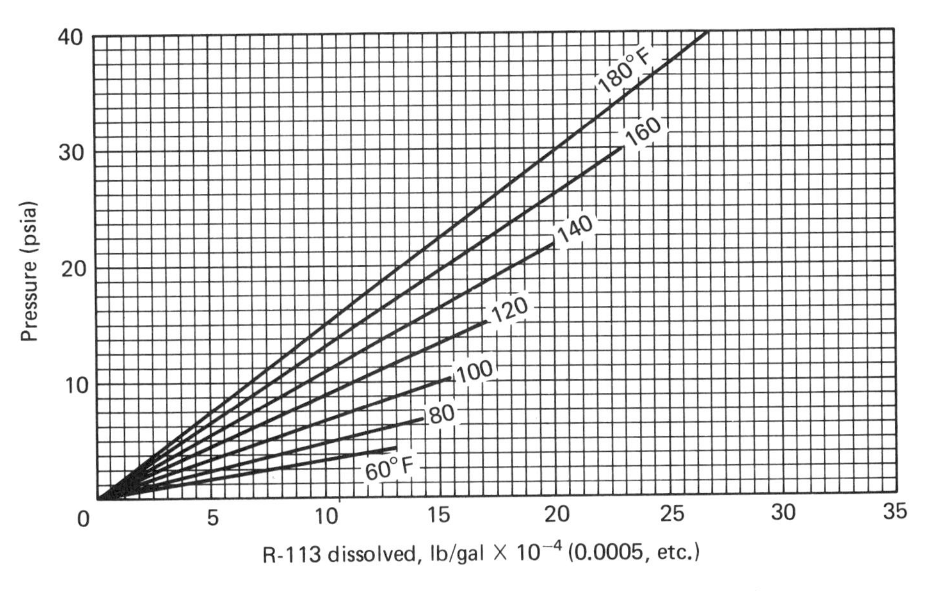

Figure 7–8 Solubility of R-113 in water as a function of temperature and pressure.

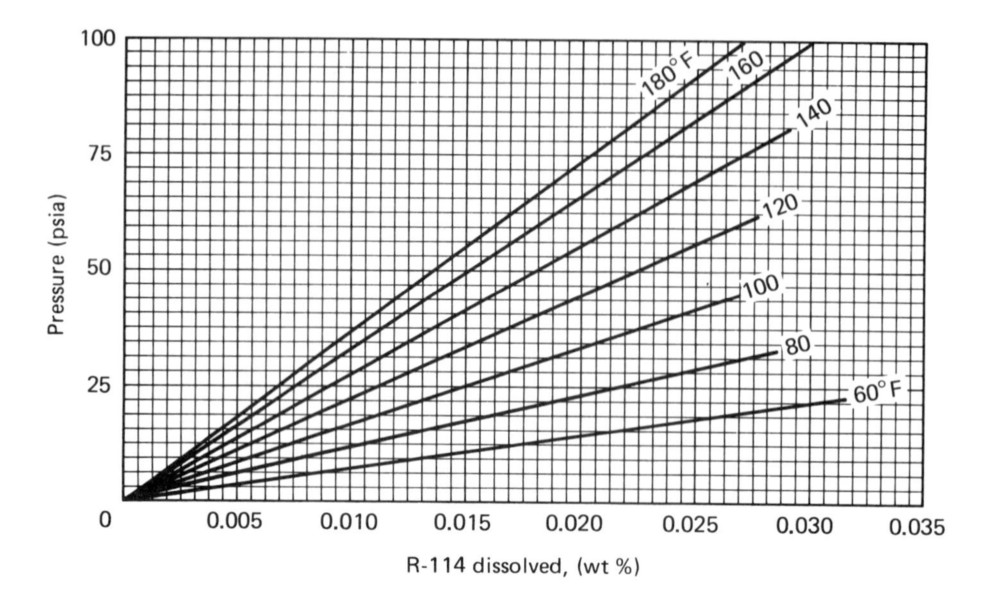

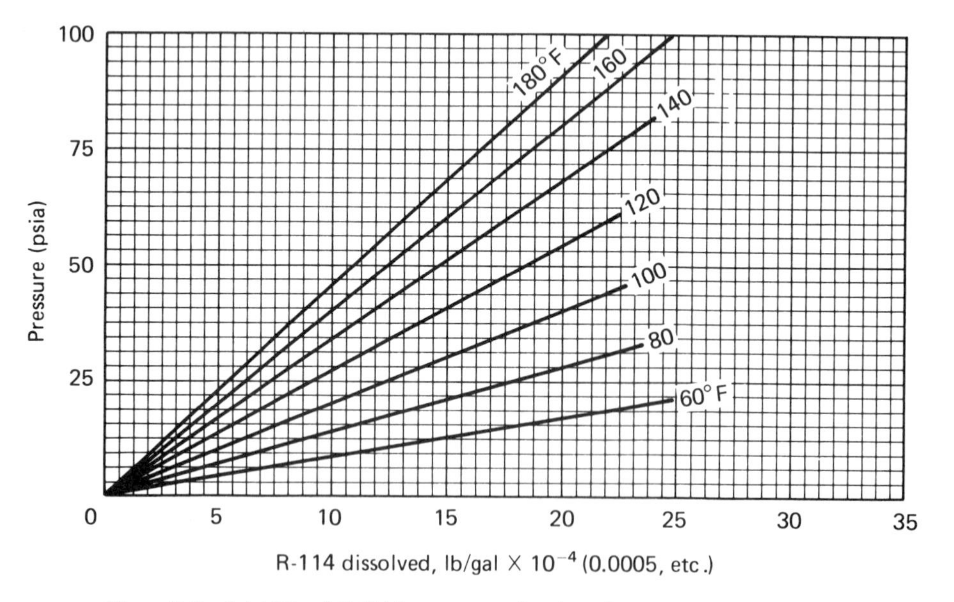

Figure 7–9 Solubility of R-114 in water as a function of temperature and pressure.

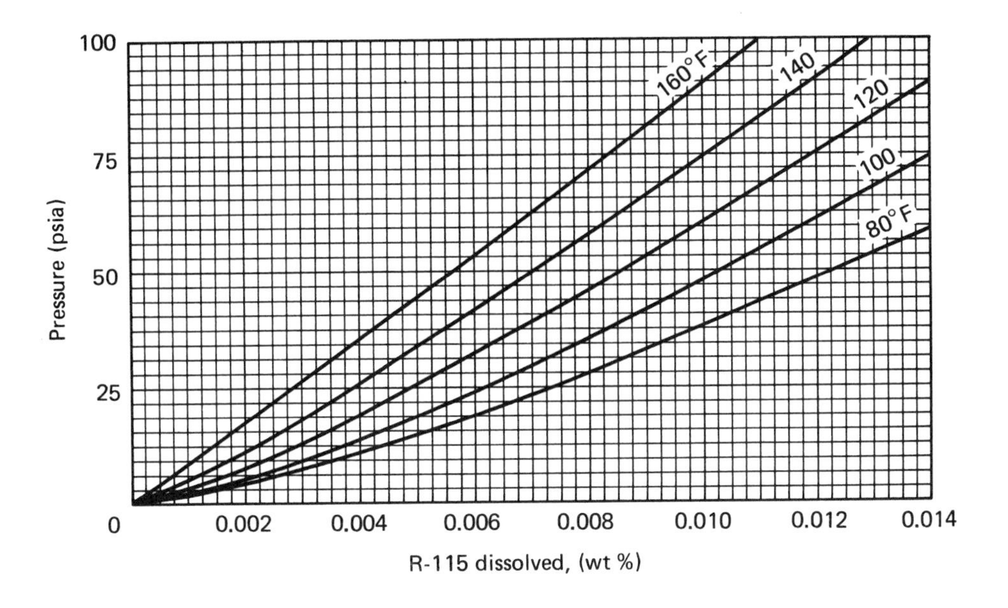

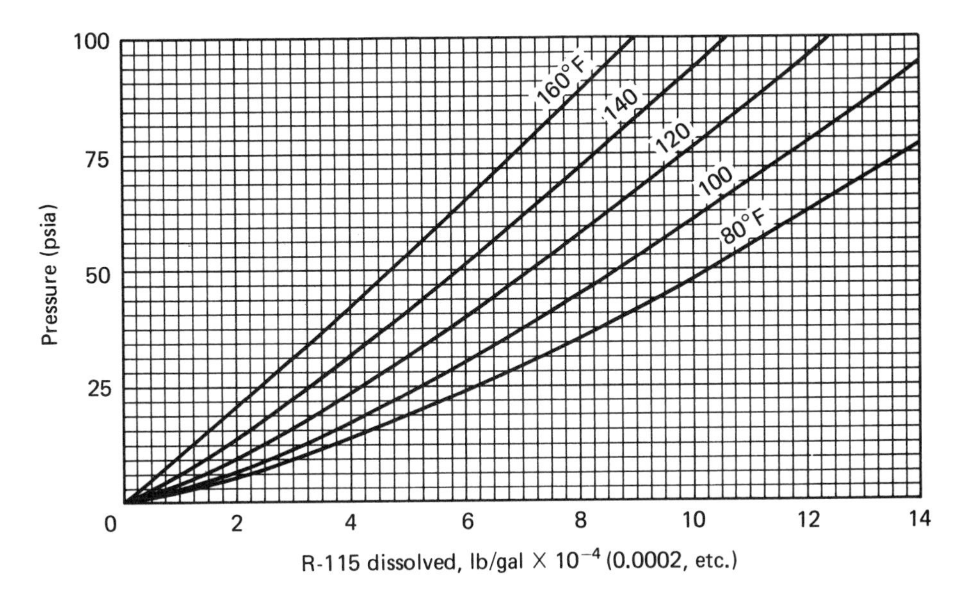

Figure 7–10 Solubility of R-115 in water as a function of temperature and pressure.

TABLE 7–1 COMPARISON OF SOLUBILITY DATA FOR R-114

Temperature		From Figure 7–9, ppm by weight at 15 psia	From the equation above (ppm/atm)[a]
°F	K		
60	288.7	208	152
80	299.8	128	99
100	310.9	90	67
120	322.0	67	46
140	333.2	46	33

[a] ppm by weight per atmosphere of pressure

TABLE 7–2 AVERAGE HEATS OF SOLUTION

R	Formula	Henry's constant (atm) (Temperature, °C)		Heat of solution (cal/mol)
12	CCl_2F_2	25,500 (25)	79,900 (75)	4720
13	$CClF_3$	60,400 (25)	161,000 (75)	4042
14	CF_4	257,000 (25)	1,630,000 (75)	7608
22	$CHClF_2$	1,590 (25)	5,010 (75)	4868
23	CHF_3	4,060 (25)	32,400 (75)	8565
115	ClF_2CF_3	147,000 (25)	660,000 (75)	6199
C318	cC_4F_8	222,000 (0)	336,000 (25)	4050
114	$CClF_2CClF_2$	91,400 (0)	207,000 (50)	6598
31	CH_2ClF	189 (10)	341 (25)	6475
21	$CHCl_2F$	341 (10)	702 (25)	8608
142b	CH_3CClF_2	2,240 (10)	4,370 (25)	6970

DISTRIBUTION OF WATER BETWEEN LIQUID AND VAPOR

In a closed container of refrigerant, such as a shipping cylinder, any water present will be distributed between the liquid and vapor phases in a definite ratio. Distribution ratios for several refrigerants are given in Table 7–3. The numbers are the ratios of water concentration in each phase. The amount of water present in each phase will depend on the amount of that phase present. In a nearly full cylinder, most of the water will be dissolved in the liquid even though the concentration of water in the liquid is very small. On the other hand, for a cylinder containing a small liquid phase, most of the water may be in the vapor phase. Elsey and Flowers [7–6] studied this relationship and showed that the ratios applied when the water content of the liquid was small as well as at saturation.

In most cases, the concentration of water is higher in the vapor phase than in the liquid phase and the ratio is greater than 1. The reverse is true for R-22, R-502, and for R-500 at lower temperatures. If the ratio is less than 1, the vapor is drier than the liquid. As vapor is removed from the cylinder or other container, the remaining liquid will become wetter. If the ratio is more than 1, the vapor is

TABLE 7-3 DISTRIBUTION OF WATER BETWEEN VAPOR AND LIQUID PHASES[a]

	Temperature (°F)				
	−40	0	40	80	100
R-11	81.7	62.3	50.6	34.8	30.1
R-12	17.1	13.3	10.1	6.9	5.5
R-13B1	3.14	2.74	2.18	1.50	1.21
R-22	0.21	0.30	0.39	0.41	0.41
R-113	—	141	98.9	61.0	50.2
R-114	52.2	32.1	24.5	15.9	13.3
R-500	0.63	0.88	1.12	1.13	1.11
R-502	0.40	0.53	0.66	0.65	0.63

[a] $\dfrac{\text{Water in vapor (wt \%)}}{\text{Water in liquid (wt \%)}}$.

wetter than the liquid and when removed will leave the liquid drier. In both cases equilibrium must be reestablished each time vapor is removed. It is always good practice to discharge refrigerant from a cylinder through a drier in order to remove any water that may be present in the valve fittings or tubing.

MOISTURE INDICATORS

Moisture indicators are valuable aids in servicing and maintaining operating systems and are widely used with all types of refrigeration and air-conditioning equipment.

Moisture indicators can be used with all fluorinated refrigerants, including R-500 and R-502. However, the water content indicated by a dry or wet reading will be different for each depending on the solubility of water in each refrigerant. The active ingredient in the indicator is usually a cobalt salt with a different color when it contains water of hydration than when it is dry. The water in the salt is in equilibrium with the water dissolved in the refrigerant.

The temperature of the liquid refrigerant affects the color of the indicator. For example, with a "dry" color at 125°F there may be two or three times as much water present as at 75°F. A temperature calibration is provided by the manufacturer. A precise temperature measurement is not needed. If the liquid line feels cool, the calibration at about 75°F can be used; if warm, about 125°F.

Oil may saturate the paper in the indicator and temporarily produce a tan color that will obscure the color of the indicator. Liquid refrigerant will usually flush the oil out of the paper without affecting its use for indicating water.

Excess water—that is, undissolved liquid water—may wash the colored salt out of the paper and destroy its usefulness. Alcohol may do the same thing.

Eight hours should be allowed for the indicator to reach equilibrium and give the right color. The time depends on the amount of charge, the circulation rate and moisture level of the refrigerant, and the condition of the drier.

The indicator should be located in the liquid line upstream from the drier so

TABLE 7–4 COMPARISON OF COLOR AND WATER CONCENTRATION

R	Parts per million at 75°F	
	Dry color	Wet color
12	Below 5	Above 15
22	Below 30	Above 100
502	Below 15	Above 50

that bubbles in the liquid will be meaningful. Liquid refrigerant must flow through the indicator. It cannot be used to estimate the water content in a cylinder by attaching to the valve since equilibrium may never be established under these conditions. The best place to install a moisture indicator is in a bypass in the liquid line. Only a small part of the liquid need flow through the indicator in order to maintain equilibrium conditions. With the indicator in a bypass line, it will be less affected by impurities than it might be in the main liquid stream.

A dry reading in the moisture indicator will show that the water content of the system is at an acceptably low level for all fluorinated refrigerants even though the actual water concentration will be different for each refrigerant. The concentration indicated by the color may be slightly different for indicators made by different manufacturers, but the comparison of color and concentration in Table 7–4 will illustrate the general range.

DRIERS

Liquid-line driers are highly recommended for all field-serviced equipment. Not only are they good insurance in case the system drying procedures were not completely successful, but they also act as filters and will remove solid contaminants that might otherwise be harmful. Driers should be located as close to the expansion valve or capillary tube as is convenient, for two reasons:

1. Drying agents are more effective at low temperatures than at high temperatures, so the drier should be located in the coolest part of the liquid line.
2. An appreciable amount of time is required for a drier to remove the maximum amount of water. At all times the driest part of the system will be immediately following the drier, so the expansion valve and capillary tubing will have the greatest possible protection against icing and plugging.

Combined filters and driers have been developed for use in the suction line to protect the motors in hermetic compressors from solid contaminants—especially following a motor burnout. In this application, the filtering action is probably more important than the drying action, although drying does occur. Studies of the effectiveness of vapor and liquid drying have been made by Jones [7–7] and others. With vapor-phase drying, equilibrium is established between the water absorbed on the drying agent and the water vapor in the gas passing over or through the agent. It does not make much difference what the gas is. Air, refrigerant vapor, or any other gas would be dried to about the same point (on a volume basis).

In liquid-phase drying, equilibrium is established between the water adsorbed on the drying agent and the water dissolved in the liquid. The equilibrium point will depend on the solubility of water in the liquid refrigerant. A liquid in which water is more soluble will not be dried as well by a given drier as a liquid in which water is less soluble. For example, R-22 will not be dried as well as R-12.

There is some indication that vapor-phase drying is almost as effective as liquid-phase drying for refrigerants with a high solubility for water, such as R-22. On the other hand, drying in the liquid phase is several times more effective than in the vapor phase for refrigerants, such as R-12, with rather low solubility for water. The drying agent in the suction line is more likely to be coated with oil, and its effectiveness thereby reduced, than in the liquid line. High refrigerant velocities in the suction line lead to poor contact between the water vapor and the drying agent. From all of these considerations, it seems that for best results, a suction line drier should be used in addition to, not in place of, a liquid-line drier.

Drying Agents

The drying agents or desiccants in current use are all quite effective in removing water. The most common types in use are silica gel, alumina gel, and synthetic silicates. Frequently, two or more of these agents are used together in commercial driers to take advantage of the good characteristics of each. The physical form of the drying agent is important since it must be porous enough so that the flow of refrigerant is not hindered, but at the same time it must not produce powder or otherwise deteriorate when water is adsorbed.

Do not use older drying materials such as calcium chloride or calcium oxide. They tend to become powdery as time goes on and the powder is carried out of the drier into other parts of the system, where it could be harmful. They also tend to become caked and less effective. It is especially important not to use alkaline or basic materials such as these or any other alkaline agent with R-22. This refrigerant is sensitive to alkaline conditions and tends to be decomposed by basic materials. This is of no concern under ordinary conditions since refrigeration systems tend to be slightly acidic from the presence of organic acids in oils, the effect of slight amounts of oxygen that may be present, and slight decomposition of refrigerant, if this occurs. Under acidic conditions R-22 is very stable. Only when alkaline materials are purposely introduced into a system will R-22 be affected.

The use of antifreeze materials may be tempting under some conditions, but they are not recommended. It is better to remove the cause of trouble rather than to provide relief by curing the symptoms temporarily. Adding anything to the basic refrigeration system often leads to trouble. The least contaminated system will be the best performing system with less repair work needed—and it will last longer.

Antifreeze agents may react directly with component metals and contaminate the system, especially at high temperatures. The products of this reaction may in turn catalyze further deterioration of the system. As a general rule, the most stable system contains the least number of dissimilar materials. Tests with R-12 illustrate the effect of other materials. For example, R-12 by itself in a quartz tube is quite stable in the range 800 to 1000°F. The addition of steel reduces this temperature range to the 500°F level. The addition of oil further reduces the temperature at which the combination is thermally stable to the 300°F range. The presence of other materials, such as antifreeze agents, metal oxides, soldering fluxes, and air, may lead to a further reduction in the temperature range at which refrigerants will operate properly. It is much better to remove water rather than merely to lower its freezing point by the addition of another chemical substance.

At one time methanol was used quite widely as an antifreeze until its harmful effect was gradually recognized. It was found that the presence of large amounts of methanol will cause severe corrosion of aluminum. Small amounts cause less corrosion but even a little is objectionable. Addition of methanol does not remove water. Both the water and methanol remain in the system. Over a period of time they can react chemically not only with the refrigerant, metals, or oils but also with insulation materials. Water and methanol (together with some air) contribute to the formation of metal oxides, which in turn help to decompose the refrigerant.

Methanol can react with aluminum regardless of its effect on a refrigerant or other metals present. This reaction produces a gray gelatinous mass and hydrogen. In one test [7–8], R-12 containing 4% methanol with oil, aluminum, copper, and steel was stored for 60 days at 250°F. At the end of the test, a vapor sample contained over 50 vol % hydrogen. In similar tests, containing 2% and 4% methanol with R-12 and R-22, there was a greater degree of copper plating and attack on the aluminum than in control tests without methanol. Even small quantities of methanol or other alcohols will degrade Mylar polyester film and make it brittle.

FIELD DRYING

All refrigeration and air-conditioning systems should be thoroughly dried in the field before charging with refrigerant. Whenever any part of the system is opened so that air and water can enter, it is necessary to remove these contaminants to ensure that subsequent operation of the equipment will be as free of trouble as possible. Even with units charged and sealed at the factory, it is desirable to evacuate and dry connecting lines and other parts that may have been open to the air, if only for a short time.

The best method of drying equipment in the field is by using a good vacuum pump to reduce the pressure in the system below the vapor pressure of water, thus causing it to boil or evaporate and the vapors to be withdrawn by the pump [7–9]. If the deep vacuum is broken by introducing clean, dried refrigerant or dry nitrogen to a pressure of 0 psig and the system is again evacuated, the time required for good dehydration can be reduced and the same degree of dryness produced. When water evaporates it is cooled and the vapor pressure is reduced; for this reason, a deep vacuum is needed. The vapor pressure of water (or ice) is given in Table 7–5.

TABLE 7–5 VAPOR PRESSURE OF WATER

Temperature (°F)	In. Hg at 32°F	psig	μm
212	29.92	14.696	759,968
200	23.46	11.52	595,884
180	15.29	7.51	388,370
160	9.65	4.74	245,110
140	5.88	2.89	149,352
120	3.45	1.69	87,630
100	1.93	0.95	49,022
80	1.03	0.51	26,162
60	0.52	0.26	13,208
40	0.25	0.12	6,350
32	0.18	0.089	4,572
20	0.10	0.049	2,540
0	0.037	0.018	940
−20	0.013	0.0064	330
−40	0.0037	0.0018	94
−60	0.0010	0.00049	25.4
−80	0.00024	0.00012	6.1

CONDENSATION

Although not directly connected with refrigerants, the condensation of water from the air on cold surfaces can be an aggravating problem. It may cause rusting of metal, rotting of wood, damage to paper cartons, formation of unpleasant stains, and other disagreeable and expensive difficulties. It is the result of a condition that is simple to correct in theory, but is sometimes more difficult in practice.

Condensation can occur only when the relative humidity of the air is 100% with respect to a cold surface. A degree of humidity that might be acceptable with an object at 40°F might permit the condensation of free water on an object with a temperature of 0°F. In Table 7–6, temperatures at which condensation will begin in ambient air at various levels of humidity are illustrated.

To avoid the consequences of condensation, three possibilities are:

1. Raise the temperature of the object.
2. Lower the relative humidity.
3. Apply an impervious coating.

Lowering the relative humidity with air conditioning will solve most condensation problems, but there are limitations. For example, reducing the relative humidity to 30% with a dry bulb temperature of 70°F will prevent condensation on objects at temperatures higher than about 40°F. For lower temperatures, insulation would help since the surface temperature in contact with the air is the important factor.

TABLE 7–6 CONDENSATION TEMPERATURES[a]

Relative humidity (%)	Ambient dry bulb temperature (°F)	
	70	80
100	70	80
90	67	77
80	64	73
70	60	69
60	56	65
50	51	60
40	45	54
30	37	46
20	28	35

[a]Temperature of an object on which water will begin to condense, °F.

LIQUID-PHASE SEPARATION

In large open-type centrifugal compressor systems operating with low-side pressures below atmospheric, water and air may tend to leak in. With the large amounts of refrigerant involved, it is difficult to use drying agents in the ordinary way. Instead, purging units are used to eliminate these troublesome impurities. A small portion of liquid refrigerant from the condenser is diverted to a small tank cooled by a separate refrigeration system or by using refrigerant from the condenser of the large machine. For removal of the water, it is necessary to cool the liquid refrigerant to a low temperature so that the solubility of water in the refrigerant will also be low. If the amount of water present is greater than the solubility at the low temperature, the water will separate as a liquid phase on top of the liquid refrigerant. A sight glass is provided so that the boundary between the liquid layers can be seen.

The liquid refrigerant layer can be drained down into the evaporator and the liquid water layer can be discarded. Since the solubility of water in refrigerant goes down quite rapidly with temperature, it is obvious that the lower the temperature in the separation container, the more water will be removed.

WATER ENTERING REFRIGERATION SYSTEMS [7–10, 7–11]

Water may be present in refrigeration systems from several sources. It may be present in a system because of inadequate drying before charging. It may be introduced in the refrigerant and oil. It may be formed from materials in the system that degenerate as a result of severe operating conditions. However, it cannot enter by forcing its way in against a stream of refrigerant leaking out at a pressure higher than atmospheric.

At one time there was a widespread belief that water could by some means creep into a system when refrigerant was leaking out. One explanation offered was Dalton's law of partial pressures, which says that at equilibrium the total pressure of a mixture of gases will be equal to the sum of the partial pressures of each gas present. This "law" is approximately true, but applies only at equilibrium. If two containers filled with different gases are separated by a small (or large) orifice, the composition of gases will be the same in each container at equilibrium. Refrigerant leaking from a system is not an equilibrium condition. One demonstration has been reported in which a moisture indicator showed "wet" after it had been operated for some time with a refrigerant leak. However, parts of the system were made of plastic and it is quite possible that water did enter the system by diffusion through the plastic.

In order to test the inward migration of water theory, several analytical studies were conducted [7–10].

1. Eight 25-lb cylinders of R-12, equipped with needle valves leaking at different rates, were placed in an atmosphere saturated with water vapor. Four cylinders were upright, so the leak was from the vapor phase, and four were inverted, so the leak was from the liquid phase. In all cases the water content of the refrigerant at the end was essentially the same as at the beginning (see Table 7–7).
2. Leaks were simulated by attaching small needle valves to two 145-lb cylinders of R-12 and three 125-lb cylinders of R-22. After 167 days there was essentially no change in the water content of the refrigerant, as shown in Table 7–8.
3. Cylinders of R-12 and R-22 with known leaks were analyzed at intervals during storage covering a 9-month period from March through November 1955 (see Table 7–9). The water content of the refrigerant ranged from 3 to 6 ppm. The rate of refrigerant loss was not checked.

The evidence against moisture entering through a leak from which refrigerant is leaving is convincing. Whether the leak is from a cylinder or a refrigeration system would not seem to make any difference.

TABLE 7–7 MOISTURE CONTENT OF R-12 CYLINDERS LEAKING THROUGH THE VALVE

Water content (ppm)		R-12	Rate of leakage	Duration of	Type of
Start	Finish	Phase	(lb/month)	test (days)	valve
16	5	Vapor	10.7	43	Packed
4	7	Vapor	1.1	146	Packed
5	3.5	Vapor	17.0	25	Bellows
3	3	Vapor	4.1	92	Bellows
5.5	4.5	Liquid	2.9	117	Packed
4	4.5	Liquid	0.9	145	Packed
5	3.5	Liquid	8.0	39	Bellows
3	2.5	Liquid	0.9	93	Bellows

TABLE 7–8 VAPOR-PHASE LEAKAGE WITH R-12 AND R-22 FOR 167 DAYS

	Water content (ppm)		Rate of leakage (lb/month)
	Start	Finish	
R-12	5	3	4.8
	5	5	5.0
R-22	4	4	3.6
	5	5	1.4
	6	5	3.8

TABLE 7–9 MOISTURE ANALYSIS IN LEAKING CYLINDERS AFTER SEVERAL MONTHS

	Number of cylinders	Location of leak	Moisture (ppm)
R-12	1	Valve stem	4
	1	Valve leak	4
	14	Fuse metal	3–6
	5	Weld lock	3–6
	1	Fuse plug	3
	2	Valve threads	3–4
R-22	1	Fuse metal	5

HYDROLYSIS [*7–12, 7–13*]

The hydrolysis rate for the flourocarbon refrigerants as a group is very low compared with other halogenated compounds. Within the group, however, there is considerable variation. Conditions of temperature and pressure and the presence of other materials greatly affect the rate. Typical hydrolysis rates are illustrated in Table 7–10. Under neutral or acidic conditions, the presence of hydrogen in the molecule has little effect but under alkaline conditions hydrolysis may be rapid.

Hydrolysis is affected not only by conditions but also by the solubility of the fluorocarbon in water or aqueous solution and the nature of the compound. Completely halogenated compounds seem not to hydrolyze by direct attack of hydroxyl ions on the molecule but rather by free-radical formation [7–12]. This mechanism is illustrated for R-11.

$$CCl_3F \longrightarrow CCl_2F\cdot + Cl\cdot$$

What happens then depends on conditions and what else is present. The following reaction may occur with the formation of equal moles of R-21 and acid:

$$CCl_2F\cdot + Cl\cdot + 2H_2O \longrightarrow CHCl_2F + HOCl$$

TABLE 7–10 HYDROLYSIS IN AQUEOUS SYSTEMS[a,b]

System	Refrigerant				
	R-11	R-12	R-22	R-113	R-114
Water alone	<0.005	<0.005	<0.01	<0.005	<0.005
Water and copper	0.02	<0.005	0.02	<0.005	<0.005
Water and steel	3	0.8	0.12	50	1.4
Water and aluminum	0.008	1.6	0.087	40	0.002
Water and zinc	2	1.3	0.097	300	1.3
Water and tin	0.05	<0.002	0.009	<0.002	0.7
Sodium carbonate (Na_2CO_3), 1%	0.12	0.04	220	< 0.01	<0.01
Sodium carbonate (Na_2CO_3), 1% and steel	0.11	0.03	220	—	—
Sodium hydroxide (NaOH), 10%, 140°F	100	40	Very rapid	80	50

[a]Temperature, 86°; pressure, 1 atm.
[b]Grams of refrigerant hydrolyzed per liter of solution, saturated with the gas, per year.

The R-21 may then hydrolyze in the usual way to form three more moles of acid.

$$CHCl_2F + 2H_2O \longrightarrow HCOOH + 2HCl + HF$$

On the other hand, compounds containing hydrogen such as R-22 may hydrolyze directly.

$$CHClF_2 + 2H_2O \longrightarrow HCOOH + HCl + 2HF$$

Compounds containing hydrogen are especially sensitive to pH—being relatively stable in neutral or acidic solutions but rapidly hydrolyzed under alkaline conditions. This difference is illustrated in Table 7–11, page 185. The alkaline hydrolysis of R-22 has been used as a means of distinguishing it from R-12 in field usage [7–12].

R-11

Church and Mayer [7–14] studied the effect of water on the decomposition of R-11. Metal bombs 12 in. long with internal diameters of 3 in. and a 1-in. wall thickness were half filled with metal shavings. Measured amounts of water, air, and R-11 were added and the bombs rotated at a temperature of 150°F for 108 hr or longer.

In one series of tests, R-11 containing various amounts of water from 0.001 to 0.123 wt % and 4 vol % air was heated at 150°F for 108 hr. The results were about the same in cast steel and cast iron. The following decomposition was observed.

Water (%)	R-11 decomposition (%)
0.001	0.3
0.08	1.0
0.12	1.3
1.0	3.8

At the same conditions with 0.12% added water, the effect of various metals on the decomposition was as follows:

Metal	R-11 decomposition (%)
Bronze (copper, lead, tin)	2.13
Lead sheet	1.99
Cast iron	1.56
Aluminum	1.46
Cast steel	1.32
Bronze (copper, tin)	0.07
Copper	Trace

Most of the metals in combination behaved independently except for copper and aluminum. The decomposition with this pair was more than twice that of aluminum alone. The effect of air was negligible, at least up to 4 vol %. Various mechanisms for the decomposition of R-11 were considered, including hydrolysis, but direct reaction with the metal was thought to predominate.

R-114

Two interesting studies of the hydrolysis of R-114 include (1) the work of Faloon and Farrar [7–15] in connection with the use of R-114 as a heat transfer agent in the gaseous diffusion process for separating the isotopes of uranium at Oak Ridge, Tennessee; and (2) the work of Stepakoff and Modica [7–5] relating to the use of R-114 in the desalination of seawater by direct contact.

Faloon and Farrer study. The tests were carried out in 10-in. tubes of 1-in. heavy-walled Pyrex tubing. Metal test pieces, 20 ml of the chlorofluorocarbon, and 10 ml of distilled water were added. The tubes were heated at 100°F. The extent of hydrolysis was determined by measuring the amount of chloride ion developed. The results are given in Table 7–12.

R-114 contains 92% of the symmetrical isomer, $CClF_2CClF_2$, and 8% of the unsymmetrical isomer, CCl_2FCF_3. R-114a has 87.5% unsymmetrical isomer. When

TABLE 7–11 HALOCARBON VAPOR SOLUBILITY IN WATER AND HYDROLYSIS RATES AT 15°C

	Concentration in water with 0.1 part per billion by volume of halocarbon in vapor in 1 atm of air (g/g water)	Hydrolysis rate constant, K in water[a]	
		pH 7	pH 8[b]
CCl_4	0.86×10^{-9}	3.0×10^{-4} (2)	1.0×10^{-4} (2)
CH_3CCl_3	4.2×10^{-9}	8.0×10^{-6} (1)	—
CH_3Cl	0.7×10^{-9}	2.5×10^{-5} (1)	0.22 (2)
R-11	0.15×10^{-9}	$<3.0 \times 10^{-6}$ (2)	2.25×10^{-4} (2)
R-12	0.033×10^{-9}	$<4.0 \times 10^{-6}$ (2)	2.75×10^{-4} (2)
R-22	0.39×10^{-9}	$<3.0 \times 10^{-8}$ (1)	1.75 (2)
R-113	0.045×10^{-9}	$<1.2 \times 10^{-4}$ (2)	4.8×10^{-4} (2)
R-114	0.018×10^{-9}	$<2.2 \times 10^{-4}$ (2)	—
R-115	0.008×10^{-9}	1.0×10^{-8} (2)	—

[a]The order of the reaction is in parentheses after the K value. The units for the hydrolysis rates are:

$$\text{First order:} \quad \frac{dx}{dt} = kc = \left(\frac{\text{liter}}{\text{hr}}\right)\left(\frac{\text{mol}}{\text{liter}}\right)$$

$$\text{Second order:} \quad \frac{dx}{dt} = kc = \left(\frac{\text{liter}}{\text{mol-hr}}\right)\left(\frac{\text{mol}}{\text{liter}}\right)$$

[b]Estimated from alkaline hydrolysis studies. For pH 8, the OH^- concentration is 10^{-6} mol/liter.

Source: Ref. 7–13.

TABLE 7–12 Hydrolysis Rates of Chlorofluorocarbons[a]

	Aluminum	Copper	Steel
R-114	<0.01	0.21	3.2
R-114a	3.8	0.42	29
R-113	10.1	0.38	24

[a]Rate of chloride evolution [micromoles per square inch of metal surface per $(\text{day})^{1/2}$].

the great difference in the hydrolysis rates of the two isomers was found, various mixtures were prepared and tested with the results shown in Table 7–13.

The differences in the stability of the two compounds indicate the following order of halogen reactivity:

$$\text{least stable} \quad =CCl_2 < -CFCl_2 << -CF_2Cl \quad \text{most stable}$$

TABLE 7–13
HYDROLYSIS OF R-114 VERSUS R-114a

R-114a (%)	Decomposition (%)
0	0.002
20	0.011
40	0.06
60	0.35
80	2.0
100	10.2

The mechanism of the reaction in these tests was considered. Direct hydrolysis to produce hydrochloric acid, which then attacked the metal, seemed reasonable except that both isomers of dichlorotetrafluoroethane were hydrolyzed at about the same rate when metal was not present.

Reactive metals such as aluminum are effective in dechlorination, and this possibility was examined and discarded. The evidence against this mechanism is that water must be present for the production of chloride ion and corrosion of aluminum and that dechlorination is envisaged to remove chlorine from adjacent carbon atoms. However, the reactive center seems to be two chlorine atoms on one carbon atom.

The authors finally settle on a mechanism of reaction on the surface of the aluminum with hydrolysis occurring catalytically on a ''barrier film surface.''

Stepakoff and Modica study. The use of R-114 in desalination involved evaporation of the R-114 while in intimate contact with saline water. The resulting ice was separated by filtration and recovered as essentially salt-free water. In addition to data on the solubility of R-114 in water it was desirable to know the amount that might be lost by hydrolysis. Some data are shown in Table 7–14.

For comparison, at 300 hr in water at 25°C (77°F) and 1.67 atm. of R-114 pressure, the fluoride concentration was 4×10^{-7} mol/liter. Assuming that the ''mol'' refers to the moles of R-114 undergoing hydrolysis, we have

$$4 \times 10^{-7} \times 170.9 \text{ (mol. wt. of R-114)} = 683.7 \times 10^{-7} \text{ g/(liter) (300 h)}$$

$$683.7 \times 10^{-7} \text{ divided by } 300 = 2.28 \times 10^{-7} \text{ g/liter-h}$$

$$= 54.7 \times 10^{-7} \text{ g/liter-day}$$

$$= 2 \times 10^{-7} \text{ g/liter-year}$$

$$= 0.002 \text{ g/liter-year}$$

The value reported in Table 7–10 is <0.01 g/liter-year.

TABLE 7–14 HYDROLYSIS OF R-114[a]

	Water: 25°C, 1.67 atm	Water: 52°C, 1 atm	7% NaCl solution: 25°C, 1.67 atm	3.4% NaCl solution and 316 stainless steel: 25°C, 1 atm	3.4% NaCl solution and aluminum: 25°C
Exposure					
100 h		55		4.7	
200 h	1.5	92	0.3	5.6	
300 h	4.0	132	0.7	6.7	4.0
400 h	6.4		1.0		10
450 h					17
500 h			1.4		
600 h			1.8		
Slope (mol/liter-hr)	0.0244	0.381		0.0107	0.0635
Solubility (mol/liter)	0.00914	0.00242		196 ppm	0.00547 at 24.7 psia
k[b] (h)	4.46 at 1 atm	157 at 1 atm		2.35 at 1 atm	11.6 at 1 atm

[a]Fluoride concentration (mol/liter $\times 10^7$).

[b]k, specific heat constant $= \dfrac{\text{slope of plot of fluoride concentration vs. time}}{\text{solubility at the same temperature and pressure}}$

R-22

The hydrolysis of R-22 in alkaline solutions is very rapid. Some measurements of solubility and hydrolysis of hydrogen-containing compounds are given in Table 7–15. The very low rate of hydrolysis of R-32 is noteworthy.

The hydrolysis of R-22 can be described by the following equation:

$$\text{Rate of hydrolysis} = [K]\,[\text{R-22 conc.}]\,[\text{OH}^-]$$

where the rate of hydrolysis (actually the rate of disappearance of R-22) is in mol/liter-min.

The hydrolysis constant, K, is in liters/mol-min and is in fact not constant but a function of temperature. A plot of log K versus $(1 \times 10^3)/T$ (K) is a straight line. Some values are

$$20°\text{C} = 0.045$$

$$40°\text{C} = 0.24$$

$$60°\text{C} = 1.05$$

TABLE 7–15 SOLUBILITY AND HYDROLYSIS DATA

Temperature (°C)	Conditions	Solubility (mol/liter × 10)	Rate of hydrolysis (mol/liter-min × 10)
		R-21	
10	10% NaOH	4.52	—
		5.18	—
		4.18	0.59
35	Water	6.22	—
35	2.77 *M* NaCl	2.31	4.55
35	1% NaOH	7.63	0.83
35	2% NaOH	7.37	—
		6.22	—
		6.42	1.25
		7.73	
35	5% NaOH	5.85	2.37
35	10% NaOH	4.54	4.19
		4.71	3.88
50	10% NaOH	3.66	14.0
		R-22	
10	10% NaOH	1.13	56.6
25	10% NaOH	—	171
35	1 *N* Na_2CO_3	1.85	1.77
35	10% NaOH	—	280
		—	260
		R-23	
35	10% NaOH	0.96	0.336
		R-32	
35	10% NaOH	1.23	0.1

Source: Ref. 7–13.

The concentration of R-22 in water is given by

$$\log k = \frac{B}{C + t} - A$$

where k = solubility in water, lb/gal-psia
t = °F mol. wt. of R-22 = 86.47
A = 4.0136 1 gal (U.S.) = 3.7853 liters
B = 160.97 1 lb = 453.6 g
C = 54.58

The hydroxyl ion concentration $[OH^-]$ is 1×10^{-7} g mol/liter for neutral water.

Using the equation above, some values for the hydrolysis of R-22 are given in Table 7–16.

TABLE 7–16 HYDROLYSIS OF R-22 IN WATER

Temperature (°F)	Solubility (g mol/liter-atm)	Hydrolysis (g/liter-year)
60	0.0501	0.0068
80	0.0310	0.0113
100	0.0217	0.0197
120	0.0165	0.0351
140	0.0133	0.0623
160	0.0111	0.108

Source: Ref. 7–12.

REFERENCES

7–1. H. M. Parmelee, "Sealed-Tube Stability Tests on Refrigeration Materials," *ASHRAE Trans.*, 71, pt I (1962), 154.

7–2. American Society of Heating, Refrigerating and Air-Conditioning Engineers, *ASHRAE Systems Handbook*. Atlanta, Ga.: ASHRAE (1984), Chap. 28.

7–3. DuPont Company, Freon Products Division, "Solubility Relationships between Fluorocarbons and Water," Bulletin B-43.

7–4. H. M. Parmelee, "Water Solubility of Freon Refrigerants," *Refrig. Eng.*, 61 (Dec. 1953), 1341.

7–5. G. L. Stepakoff and A. P. Modica, "The Hydrolysis of Halocarbon Refrigerants in Freeze Desalination Processes: Pt I. Solubility and Hydrolysis Rates of Freon 114 ($CClF_2CClF_2$)," *Desalination*, 12 (1973), 85; "Pt II. Theoretical Prediction of Hydrolysis Rates and Comparison with Experimental Data," *Desalination*, 12 (1973), 239.

7–6. H. M. Elsey and L. C. Flowers, "Equilibria in Freon 12–Water Systems," *Refrig. Eng.*, 57 (Feb. 1949), 153.

7–7. Evan Jones, "Liquid or Suction Line Drying," *Air Cond. Refrig. Bus.* (Sept. 1969).

7–8. DuPont Company, Freon Products Division, *Refrigerants and Service Pointers Manual*, (1970).

7–9. Robinair Manufacturing Corporation, "Fundamentals of Dehydrating a Refrigerant System," (1969).

7–10. DuPont Company, Freon Products Division, "Test Results Showing That Moisture Does Not Diffuse into Leaking Refrigerant Cylinders," Bulletin B-24 (1964).

7–11. R. C. Downing, "More on Moisture," *Refrig. Serv. Contract.* 52 (July 1984), 29.

7–12. DuPont Company, unpublished information.

7–13. DuPont Company, Freon Products Division, "Freon Fluorocarbons, Properties and Applications," Bulletin B-2 (1971).

7–14. J. M. Church and J. H. Mayer, "Stability of Trichlorofluoromethane in the Presence of Moisture and Certain Metals," *J. Chem. Eng. Data*, 6 (July 1961), 449.

7–15. A. V. Faloon and R. L. Farrar, Jr., "Hydrolysis of Water-Saturated Chlorofluorocarbons in the Presence of Metals and the Associated Corrosion," Report No. K-1461, Union Carbide Nuclear Co., Oak Ridge, Tenn. (Feb. 1961).

8

SOLUBILITY OF AIR AND OTHER GASES

AIR

In 1951, Parmelee published comprehensive data on the solubility of air in R-12 and R-22 [8–1]. Later he studied the relationship of air and R-11 [8–2].

Air is an unwelcome impurity in refrigerants and in refrigeration machines. When air is present, pressure readings will be higher than for the refrigerant alone. Diagnosing service problems is more difficult and efficiency suffers. Corrosion, oil sludging, and acid formation tend to increase. The sensitivity and reliability of control devices are affected when air is present in refrigerant used in the device. Defining the air/refrigerant solubility relationship was a big step forward for the refrigeration and air-conditioning industry.

The total amount of air present in a cylinder or machine is usually quite small, but most of it is in the vapor phase. Solubility in the liquid is slight but important in understanding air concentrations. Good analytical technique and sampling are essential. For these studies, refrigerant samples were carefully collected. The refrigerant was frozen in liquid nitrogen and the residual gas was measured with due care for temperature, volume, and pressure corrections to assure valid answers. The refrigerant was liquefied and refrozen several times to be sure that no gas was occluded in the solid. Modern instruments have greatly simplified analytical procedures, but the methods developed by Parmelee and performed with skill and patience produced valuable information.

Henry's Law

The solubility of air in the liquid phase of R-11, R-12, and R-22 was found to obey Henry's law, which can be expressed as

$$X_2 = K_2 P_2 \tag{8–1}$$

where X_2 = mol fraction of air in the liquid phase
K_2 = solubility constant with the dimensions of reciprocal pressure
P_2 = partial pressure of air

Experimental measurements were made at −40°F, 0°F, and 75°F and equations developed to represent the data. The solubility of air in the liquid phase as a function of partial pressure is given in Figure 8–1 for R-11, Figure 8–2 for R-12, and Figure 8–3 for R-22. The fact that straight lines through the experimental data points extend to the origin in this type of chart shows that Henry's law is followed. The equations are used to extrapolate to somewhat higher temperatures than actually measured. The measurements were made at atmospheric pressure and reported as percent by volume—cubic centimeters of air per 100 cm^3 of vapor or vaporized liquid. At this low pressure, volume fraction or percent is essentially the same as mole fraction or percent. At higher pressures the refrigerant gas departs from ideality and suitable corrections must be made.

The charts in Figures 8–1, 8–2, and 8–3 were constructed by assigning values for the partial pressure of air and using values for K_2 calculated with equations (8–3), (8–4), and (8–5), respectively. If the total pressure (air plus refrigerant) is measured, the partial pressure of the air can be calculated by

$$P_2 = \frac{Y_2 P}{a + Y_2(1 - a)} \tag{8–2}$$

where P_2 = partial pressure of air, psia
Y_2 = mole fraction air in the vapor
P = total pressure, psia
$a = PV/nRT$, the compressibility factor for the refrigerant
P = pressure, psia
V = volume, ft^3
n = lb mol = weight in pounds divided by molecular weight
T = °R = °F + 459.67
R = the gas constant = 10.7318 for the units above

For other units (based on 10.7318):

$$\frac{(\text{atm})(\text{ft}^3)}{(\text{lb mol})(°\text{R})} \qquad R = 0.730257$$

$$\frac{(\text{atm})(\text{cm}^3)}{(\text{g mol})(°\text{K})} \qquad R = 82.0593$$

$$\frac{(\text{ft})(\text{lb})}{(\text{lb mol})(°\text{R})} \qquad R = 1545.42$$

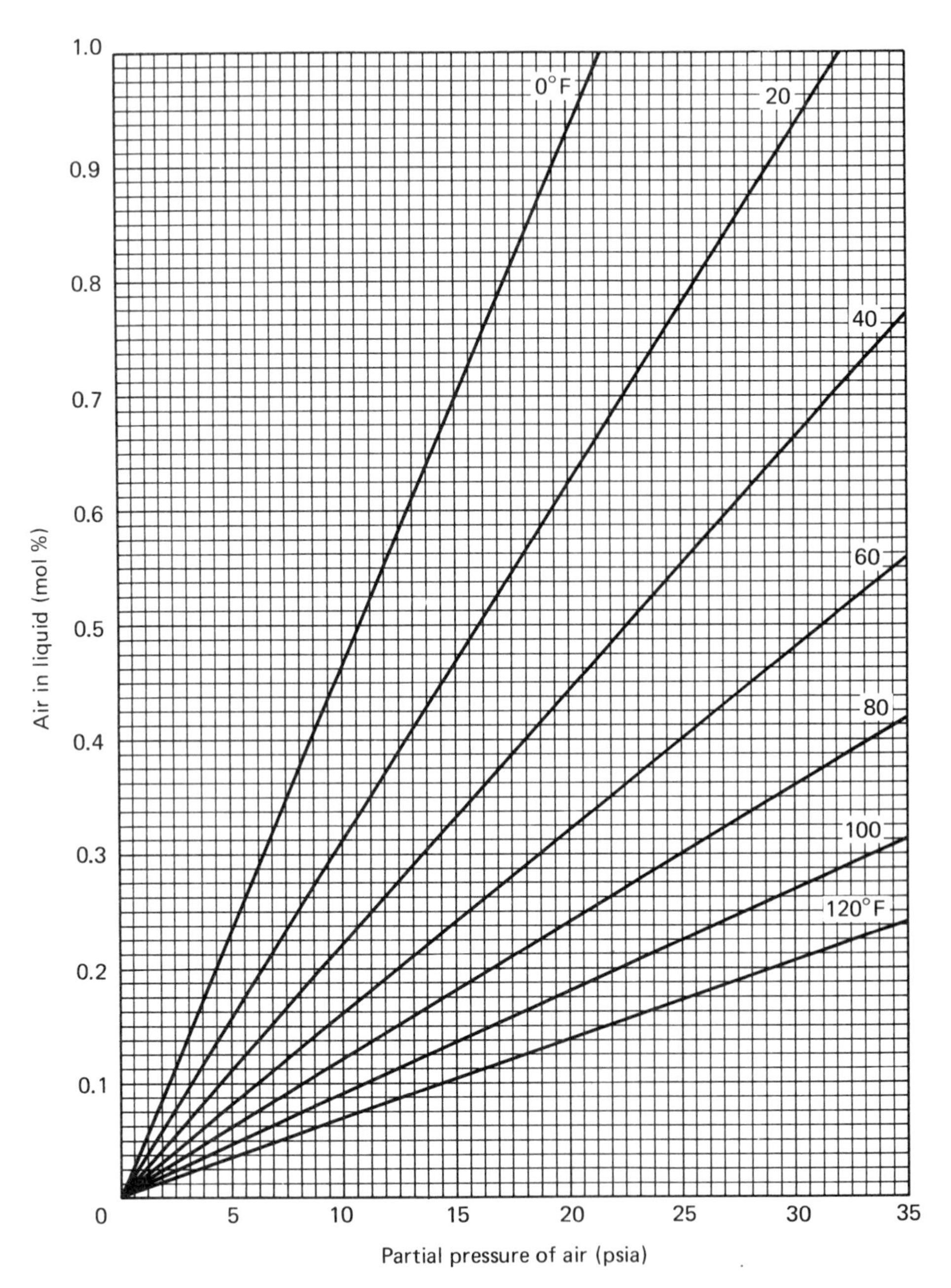

Figure 8–1 Solubility of air in liquid R-11.

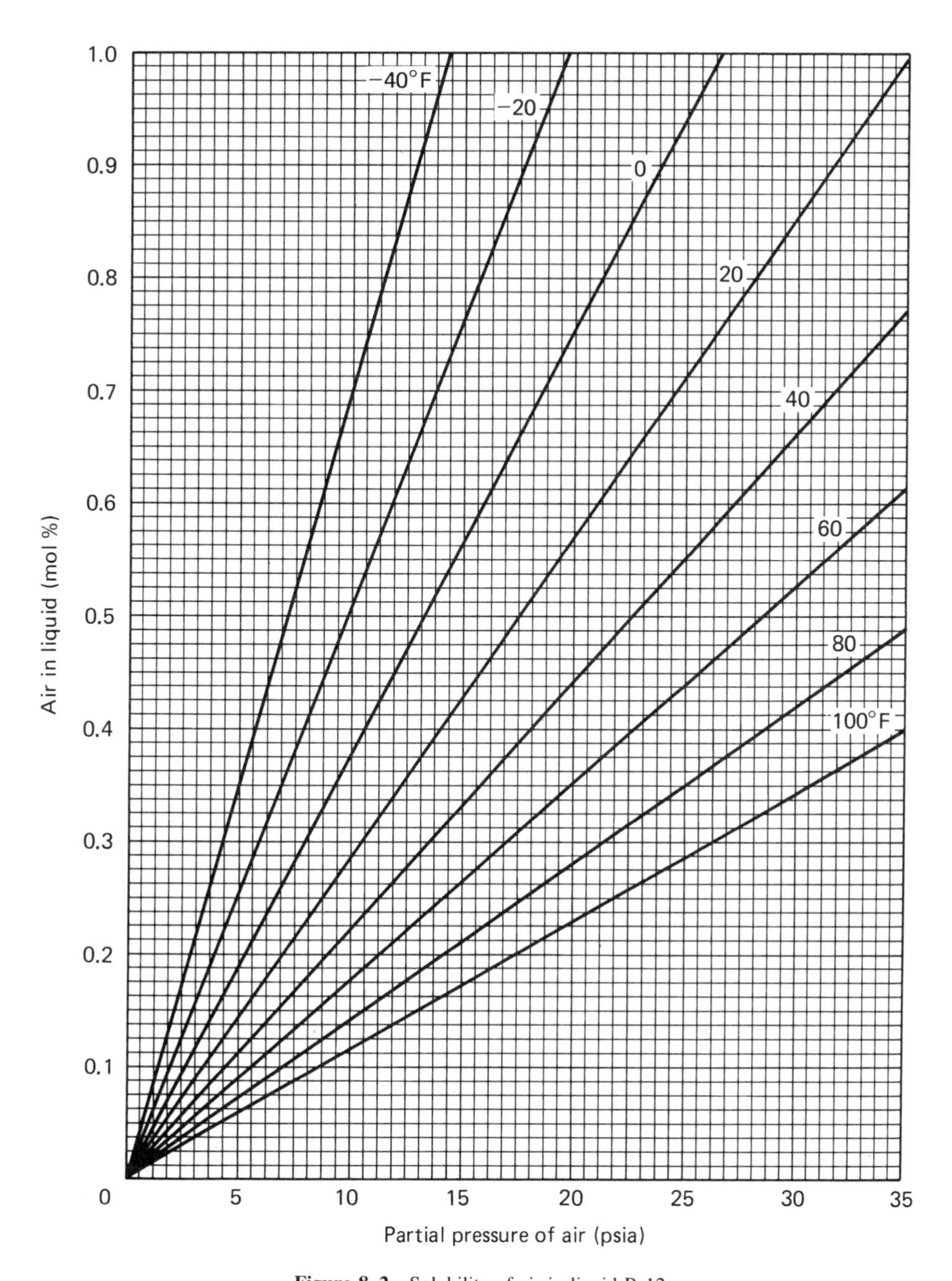

Figure 8–2 Solubility of air in liquid R-12.

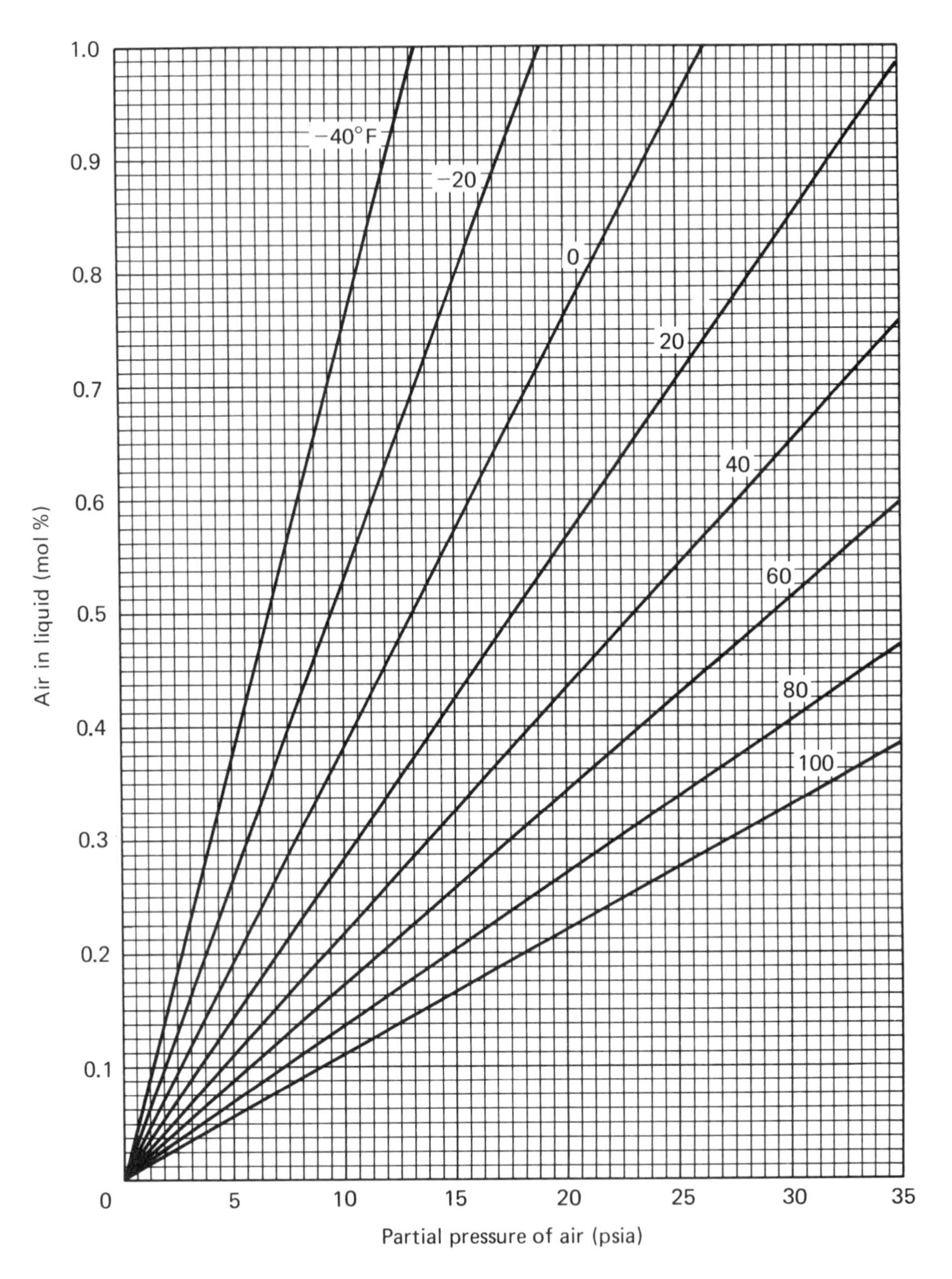

Figure 8–3 Solubility of air in liquid R-22.

$$\frac{\text{J}}{(\text{kg mol})(°\text{K})} \qquad R = 8314.86$$

$$\frac{\text{cal}}{(\text{g mol})(°\text{K})} \qquad R = 1.9873$$

$$\frac{(\text{kPa})(\text{m}^3)}{(\text{kg mol})(°\text{K})} \qquad R = 8.31464$$

$$\frac{(\text{kg/cm}^2)(\text{cm}^3)}{(\text{g mol})(°\text{K})} \qquad R = 84.786$$

For mixtures of ideal gases the compressibility factor is 1 and equation (8–2) becomes $P_2 = Y_2P$ and mole fraction and pressure fraction are numerically the same. At low pressures, both air and refrigerant vapors are nearly ideal and values of mole fraction, volume fraction, and pressure fraction can be used interchangeably. The correction factor, a, can be calculated from tables of thermodynamic properties. Some examples are listed in Table 8–1.

TABLE 8–1 COMPRESSIBILITY FACTORS, PV/nRT

°F	R-11	R-12	R-22	R-113	R-114	R-502
−40	0.9956	0.9684	0.9633		0.9915	0.9533
−30				0.9983		
−20	0.9928	0.9559	0.9453	0.9982	0.9868	0.9351
0	0.9888	0.9406	0.9300	0.9963	0.9800	0.9130
20	0.9836	0.9225	0.9080	0.9943	0.9709	0.8867
40	0.9768	0.9013	0.8823	0.9911	0.9593	0.8560
60	0.9685	0.8770	0.8526	0.9864	0.9452	0.8205
80	0.9585	0.8492	0.8185	0.9808	0.9281	0.7798
100	0.9467	0.8175	0.7794	0.9731	0.9079	0.7330
120	0.9330	0.7815	0.7344	0.9638	0.8845	0.6784

The experimental data shown in Table 8–2 are based on a number of measurements at various concentrations of air at the same temperature.

TABLE 8–2 SOLUBILITY COEFFICIENT, K_2

°F	R-11	R-12	R-22
−40		7.0×10^{-4}	7.4×10^{-4}
0		3.5×10^{-4}	3.9×10^{-4}
75.2		1.5×10^{-4}	1.4×10^{-4}
77	1.25×10^{-4}		
95	0.92×10^{-4}		
122	0.67×10^{-4}		

Change with Temperature

The solubility of air in the liquid decreases with increasing temperature. The experimental measurements listed in Table 8–2 are represented by the following empirical equations:

$$\text{R-11:} \qquad \log K_2 = \frac{1863.8}{T} - 7.3835 \tag{8–3}$$

$$\text{R-12:} \qquad \log K_2 = \frac{1320}{T} - 6.30 \tag{8–4}$$

$$\text{R-22:} \qquad \log K_2 = \frac{1390}{T} - 6.45 \tag{8–5}$$

where $T = °\text{R} = °\text{F} + 459.67$
K_2 = reciprocal pressure, psia

Solubility Ratio

The relationship between air in the liquid and air in the vapor is shown in Figures 8–4, 8–5, and 8–6 for R-11, R-12, and R-22, respectively. The curves at various temperatures are based on

$$X_2 = \frac{Y_2 H_1}{(a/K_2) - Y_2[(a/K_2) - H_1]} \tag{8–6}$$

where X_2 = mole fraction air in the liquid phase
Y_2 = mole fraction air in the vapor phase
H_1 = vapor pressure of the refrigerant at a given temperature, psia
a = compressibility factor (Table 8–1)
K_2 = solubility coefficient = X_2/P_2, where P = partial pressure of air in the vapor

In equation (8–6) it is assumed that Raoult's law is valid and that equilibrium is established. It is also assumed that mole fraction and volume fraction are equivalent—a valid assumption at low pressures. The measurements on which equation (8–6) is based were made at a pressure of about 1 atm.

Since the solubility of air in the liquid phase is slight, the concentration of air in the vapor is much higher than in the liquid. Nevertheless, in a container nearly full of liquid most of the air will be in the liquid. As the liquid level or liquid volume changes, the percentage of air in the liquid also changes as shown in equations (8–7) and (8–8).

$$\text{For R-12:} \qquad \text{AL}_{12} = \frac{100}{1 + 1.78(V_v/V_L)} \tag{8–7}$$

$$\text{For R-22:} \qquad \text{AL}_{22} = \frac{100}{1 + 1.45(V_v/V_L)} \tag{8–8}$$

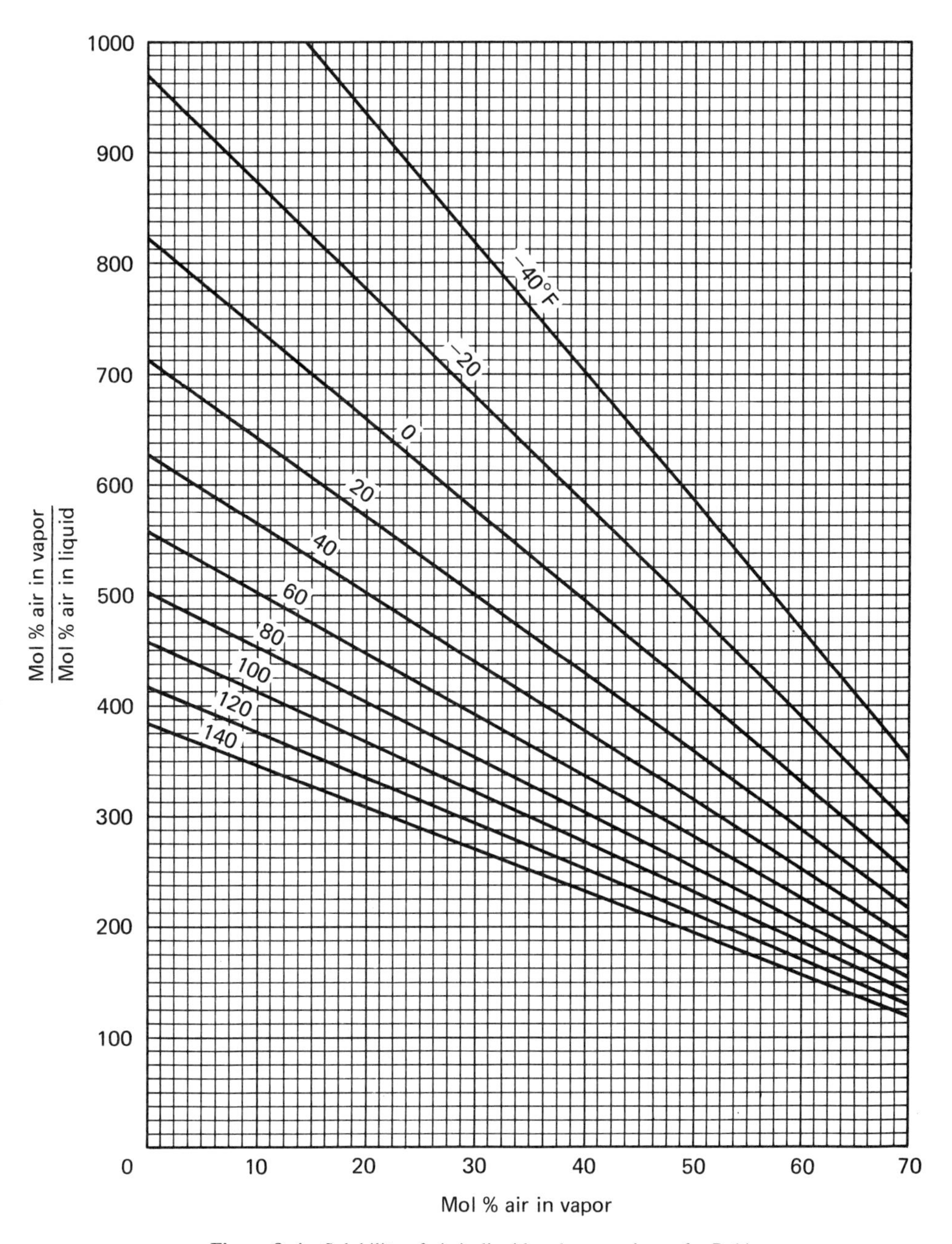

Figure 8–4 Solubility of air in liquid and vapor phases for R-11.

where AL_{12} = percent of total air in container present in the liquid phase of R-12
AL_{22} = percent of total air in container present in the liquid phase of R-22
V_v = volume of the vapor phase
V_L = volume of the liquid phase

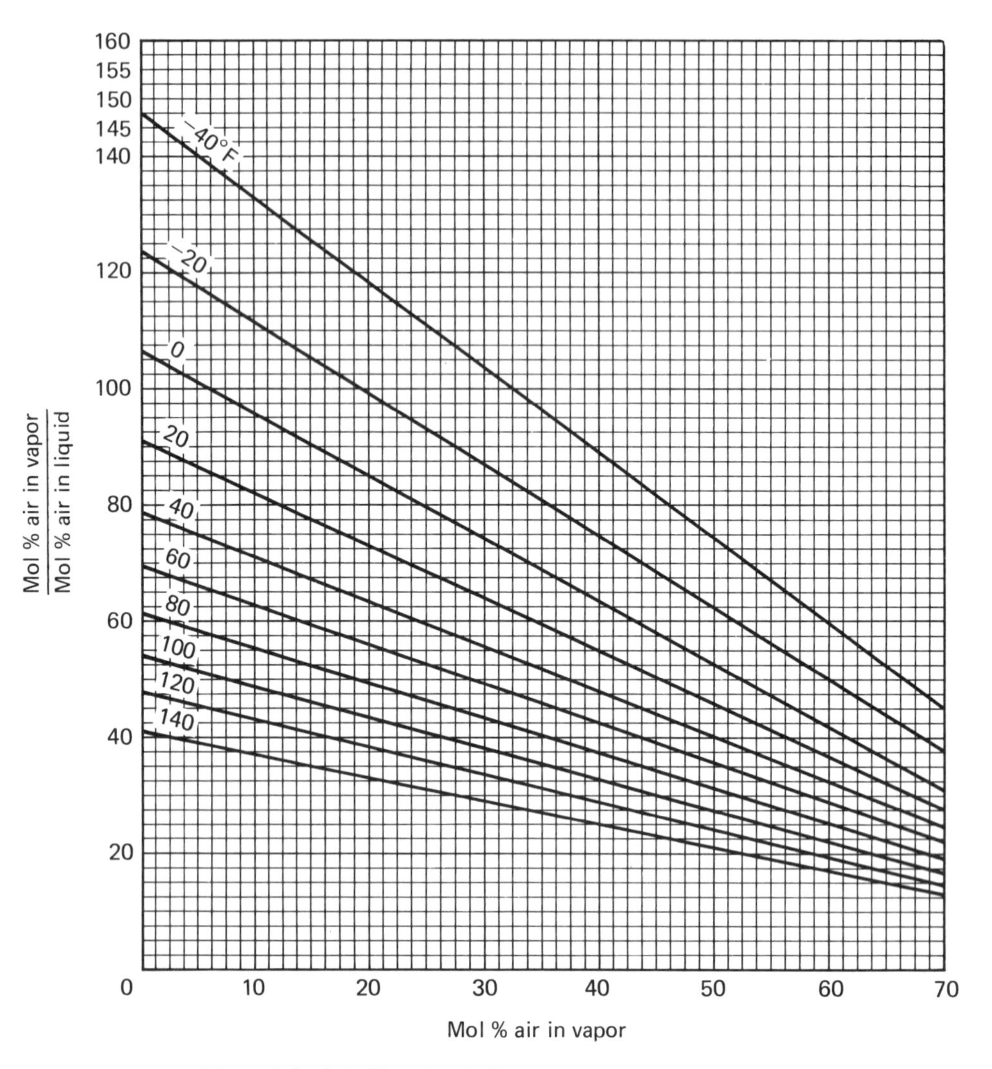

Figure 8–5 Solubility of air in liquid and vapor phases for R-12.

Equilibrium

All relationships between air in the liquid and vapor phases of the refrigerant depend on equilibrium being established. If a cylinder or other container is at rest and is pressurized with air (or other gas), a considerable amount of time is necessary to reach equilibrium. In one series of tests with R-12, about 150 hr was required for equilibrium solubility of air in the liquid without agitation. However, when the cylinder was rotated the concentration of air in the liquid reached a maximum, constant value in about 5 min.

If the amount of air in a cylinder of refrigerant is higher than desired, it can be reduced by venting a small amount of gas from the vapor phase. Between each

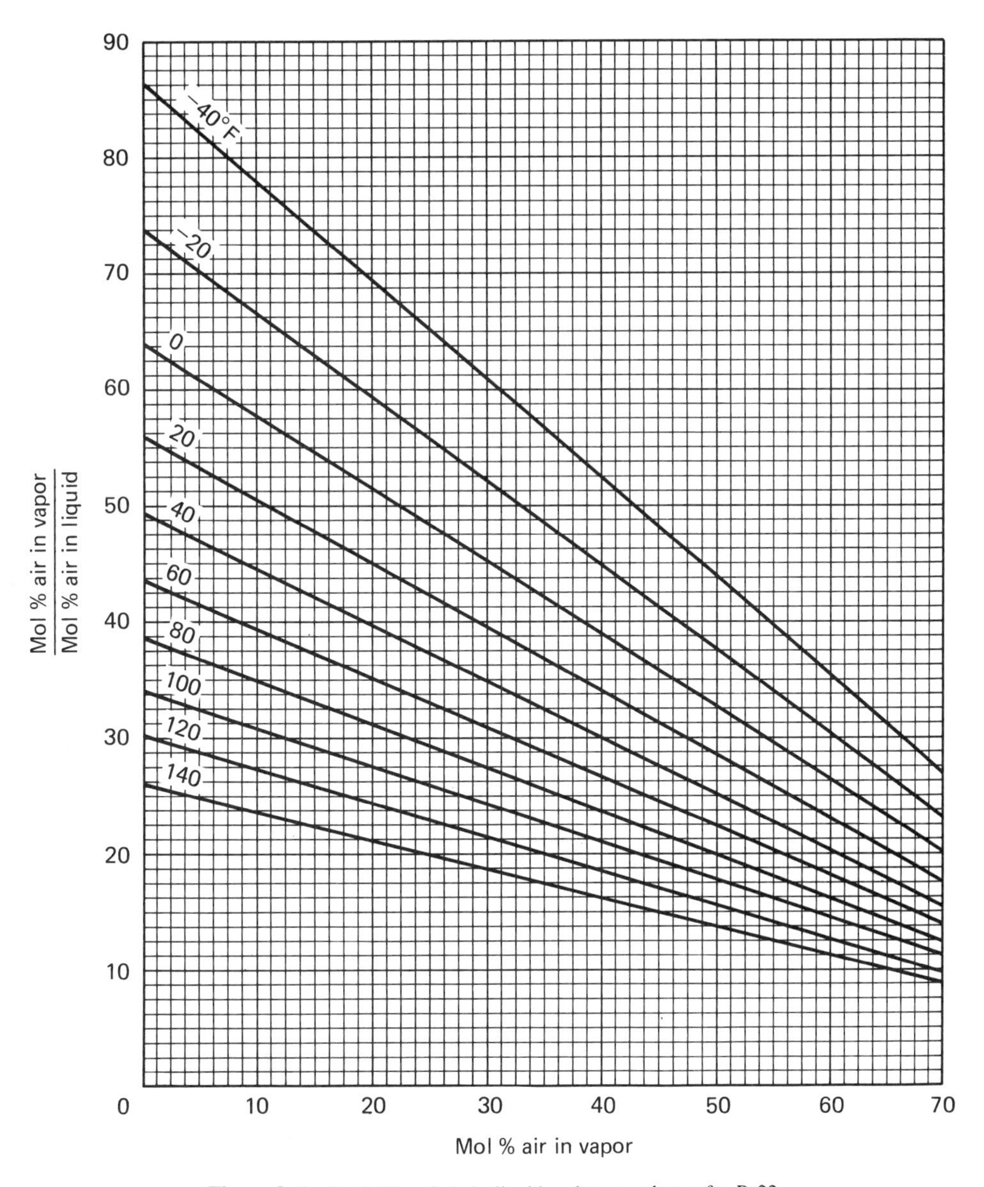

Figure 8–6 Solubility of air in liquid and vapor phases for R-22.

venting the cylinder must be rolled to reestablish equilibrium. In this way the air content can be reduced to a very low concentration after relatively few ventings.

Nearly all of the air in a refrigerant can also be removed by a slow reflux of the refrigerant. A vertical reflux condenser is attached to the cylinder or other container. A pressure gage and valve are installed at the top of the condenser. The temperature of the condenser is kept about 15 to 20° cooler than the liquid refrigerant by either cooling the condenser or warming the refrigerant. As the refluxing continues, air will accumulate at the top of the condenser and the pressure gage will indicate a slight rise in pressure. Air is occasionally vented from the top of the condenser and repeated until there is no further rise in pressure and the observed pressure is equal to the vapor pressure of the refrigerant at the temperature of the liquid.

Specification

The amount of air in new refrigerant is limited, usually to 1.5 vol % in the vapor phase. Vapor-phase analysis is specified because the air concentration is much higher than in the liquid and the analytical result generally more reliable. However, it is now recognized that the result can also be misleading unless the temperature is specified and equilibrium is maintained. The solubility of air in the liquid is affected by the temperature, but the change in the density of the refrigerant vapor is much greater and has more affect on the analysis. For example, suppose that a container of R-12 is in equilibrium at 70°F. A vapor sample is analyzed and found to contain 1 vol %, well below the specification limit. Suppose that the container (cylinder, truck, or tank car) arrives at a customer's plant and at the time of sampling for analysis the refrigerant is at a temperature of 0°F. The analysis now shows an air concentration of more than 3% and the shipment is unacceptable even though the air content has not changed.

The following calculation illustrates the effect of changing refrigerant vapor density on the analysis for air. Volume percent and mole percent are here considered equivalent.

At 70°F, the vapor density of R-12 is 2.0913 lb/ft^3. Dividing by the molecular weight gives the number of moles in 1 ft^3—2.0913/120.93 equals 0.017293 mol. The air content is 1% by volume (also by moles) or 0.000173 mol. The total is 0.017466.

At 0°F the vapor density of R-12 is 0.62156 lb/ft^3 or 0.00514 mol. The air concentration is now 3.26%.

$$\begin{array}{rl} & 0.005140 \text{ mol of R-12} \\ + & \underline{0.000173 \text{ mol of air}} \\ & 0.005313 \text{ mol total} \end{array}$$

The following empirical equation was developed by T. A. Armstrong [8–2] of DuPont and has been used to provide data for the curves in Figure 8–7 for the change in air concentration in the vapor phase with R-12 as the temperature changes. The equation agrees very well with experimental measurements and indicates that changes in the vapor pressure of the refrigerant and the effect of temperature change on air are much greater factors than the change in air solubility in the liquid and the change in liquid level (change in vapor volume)—at least for moderate temperature changes.

$$C_2 = \frac{(C_1)(\text{VP}_1)(T_2)}{(\text{VP}_2)(T_1)}$$

where C_1 = concentration of air in refrigerant vapor at T_1
C_2 = concentration of air at a different temperature
VP_1 = vapor pressure of refrigerant at T_1

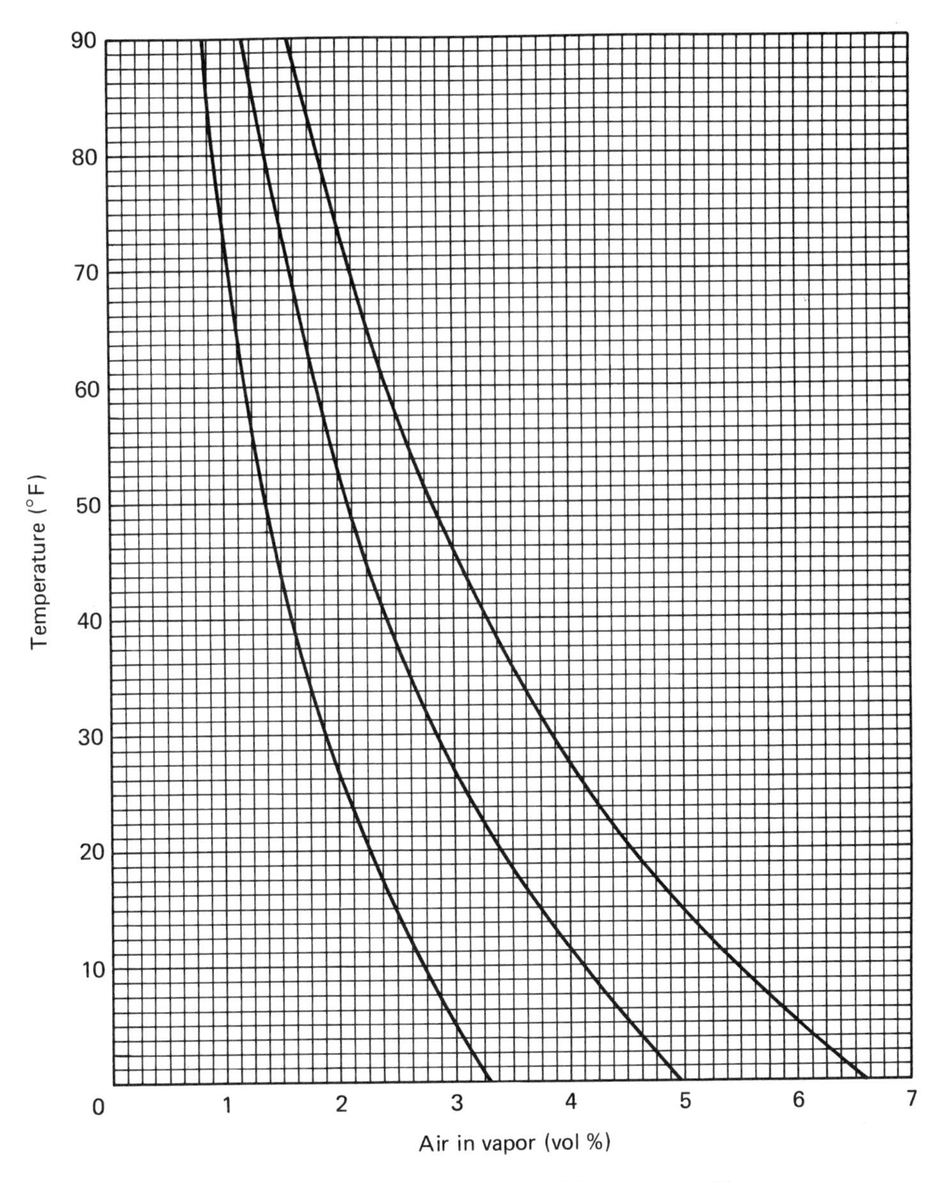

Figure 8–7 Solubility of air in the vapor phase of R-12: change with temperature (three levels of air).

VP_2 = vapor pressure of refrigerant at T_2
T = temperature in °R = °F + 459.67

NITROGEN

The solubility of nitrogen in R-114 was reported by Williams in 1959 [8–3]. Parmelee used the experimental data to develop equations similar to those used in his work with air [8–2].

$$X_2 = K_2 P_2 \quad (8\text{–}9)$$

$$\log K_2 = \frac{1919.6}{T} - 6.8914 \quad (8\text{–}10)$$

where X_2 = mole fraction of nitrogen in liquid R-114
P_2 = partial pressure of nitrogen, psia
T = absolute temperature = °R = °F + 459.67
K_2 = solubility coefficient

The experimental values of K_2 are compared with those from equation (8–10) as follows:

°F	K_2 from equation (8–10)	K_2, experimental
68	5.56×10^{-4}	5.5×10^{-4}
104	3.26×10^{-4}	3.7×10^{-4}
140	2.04×10^{-4}	2.0×10^{-4}

Ratios of the concentration of nitrogen in the vapor to the concentration of nitrogen in the liquid phase were determined. The nitrogen concentration in the vapor phase was less than 1%.

°F	Ratio
68	64
104	56
140	50

OTHER RESULTS AND COMMENTS

The importance of controlling the amounts of air and other gases in the fluorocarbon refrigerants was recognized very early in their development. For many years the standard analytical method involved dissolving the fluorocarbon in a liquid—first kerosene and later perchloroethylene—saturated with air and measuring volumetrically the remaining gas. The impurity was essentially air but became known as *nonabsorbable gas* (NAG). Later, probably as a result of Parmelee's work, it was also sometimes called *noncondensable gas* (NCG). These labels still persist, even though the analytical procedures responsible have long since been replaced, chiefly by the gas chromatograph.

In 1976, Teufel and Quick [8–4] reported a number of measurements of the solubility of NAG in R-12. Their results were reported as percent air in the liquid and vapor phases and seem generally to support Parmelee, although with many scatters and in a number of cases a temperature effect opposite to that found by Parmelee.

TABLE 8–3 SOLUBILITY OF NITROGEN AT 80 ± 5°F

	Pressure of nitrogen and fluorocarbon (psia)	Partial pressure nitrogen (psia)	Concentration of nitrogen in liquid phase (mol %)	Solubility coefficient, K_2
R-114B2	101.7	94.7	1.19	1.26×10^{-4}
($CBrF_2CBrF_2$)	614.7	607.7	7.25	1.19×10^{-4}
	964.7	957.7	11.04	1.15×10^{-4}
	1995	1988	19.48	0.98×10^{-4}
R-114	64.7	31.7	0.73	2.30×10^{-4}
($CClF_2CClF_2$)	114.7	81.7	1.60	1.96×10^{-4}

Gorski [8–5]made some careful and reproducible determinations of the solubility of nitrogen in R-114 and R-114B2, as shown in Table 8–3.

At 80°F the solubility coefficient for nitrogen in R-114B2 reported by Williams was 4.6×10^{-4}, about four times the value found by Gorski. It does seem that the Williams results are high. Comparison with other fluorocarbons offers some confirmation. However, all of the available data show that these gases have very little solubility in the fluorocarbons and the order of magnitude is similar.

There is some evidence that oxygen is a little more soluble than nitrogen, so air should also be slightly more soluble. In a comparison reported by Gorski the solubility of air in R-113 at 77°F is 1.4×10^{-4} and for nitrogen is 0.93×10^{-4}.

The solubility of air in liquid R-12 is about the same as in liquid R-22 at the same temperature at low concentrations (see Table 8–4). However, the ratio of air in the vapor to air in the liquid is less for R-22 than for R-12 and just about inversely proportional to the vapor pressures.

TABLE 8–4 COMPARISON OF VAPOR PRESSURE AND SOLUBILITY

	Vapor pressure (psia)			Solubility ratio[a]		
°F	R-22	R-12	R-22/R-12	R-22	R-12	R-12/R-22
−40	15.222	9.3076	1.635	87	148	1.70
0	38.657	23.849	1.621	64	107	1.67
75	146.91	91.682	1.602	40	63	1.58

[a] $\frac{\text{Air in vapor phase}}{\text{Air in vaporized liquid phase}}$.

Some solubility coefficients for various gases in fluorocarbon liquids are summarized in Table 8–5.

Some additional measurements of the solubility of air in R-12 are recorded in Table 8–6. Air pressure data were not given, so the ratio of solubility in the vapor and liquid phases is shown. Comparisons from Figure 8–5 at the same temperature and vapor composition are also given.

TABLE 8–5 SOLUBILITY OF INERT GASES IN THE LIQUID PHASE

R	Gas	°F	Solubility coefficient, K_2	Reference
12	Air	−40	7.0×10^{-4}	Parmelee [8–1]
	Air	0	3.5×10^{-4}	Parmelee [8–1]
	Air	26	1.6×10^{-4}	Penn [8–2]
	Air	75.2	1.5×10^{-4}	Parmelee [8–1]
	Air	96	0.83×10^{-4}	Penn [8–2]
22	Air	−40	7.4×10^{-4}	Parmelee [8–1]
	Air	0	3.9×10^{-4}	Parmelee [8–1]
	Air	75.2	1.4×10^{-4}	Parmelee [8–1]
114	Nitrogen	68	5.5×10^{-4}	Williams [8–3]
	Nitrogen	104	3.7×10^{-4}	Williams [8–3]
	Nitrogen	140	2.0×10^{-4}	Williams [8–3]
114	Nitrogen	80	2.1×10^{-4}	Gorski [8–5]
114B2	Nitrogen	80	1.2×10^{-4}	Gorski [8–5]
	Helium	80	1.58×10^{-4}	Gorski [8–5]
	Argon	80	2.23×10^{-4}	Gorski [8–5]
11	Air	77	1.25×10^{-4}	Parmelee [8–2]
	Air	95	0.92×10^{-4}	Parmelee [8–2]
	Air	122	0.67×10^{-4}	Parmelee [8–2]
112	Air	77	0.8×10^{-4}	Gorski [8–5]
113	Nitrogen	73	0.93×10^{-4}	Gorski [8–5]
	Air	77	1.4×10^{-4}	Gorski [8–5]
	Helium	80	0.9×10^{-4}	Gorski [8–5]

TABLE 8–6 SOLUBILITY RATIO FOR AIR IN R-12

°F	Air volume fraction $\times 10^4$		Ratio from Fig. 8–5	Other DuPont measurements	Reference
	Liquid	Vapor			
100	1.1	24.5	36.5	27	8–2
	1.7	20	39.5	34	8–2
	2.3	13	42	30	8–2
91	0.12	4.54	35	37.8	Floria [8–2]
	0.32	9.96	32.8	31.1	Floria [8–2]
75	0.12	4.45	38	37.1	Floria [8–2]
	0.31	10.9	35.7	35.2	Floria [8–2]
	1.3	41.5	38	32	8–2
	1.7	67	22	39	8–2
	2.1	103	—	49	8–2
55	1.4	29	52	41	8–2
	1.6	36	46	57	8–2
	2.3	26	47	60	8–2
32	2.0	34	56	67	8–2
	2.1	45	46	95	8–2
−8	0.24	33	76	138	Floria [8–2]
	0.085	16.2	95	191	Floria [8–2]

The calculations in Table 8–7 compare volume percent with mole percent for mixtures of air and R-12 and air and R-22 at a temperature of 70°F and a pressure of 1 atm. It is assumed that the gas volumes are additive, that is, that 98.5 ft^3 of fluorocarbon plus 1.5 ft^3 of air will give 100 ft^3 of the mixture. The refrigerant density data are from standard tables of thermodynamic properties published by ASHRAE.

TABLE 8–7 COMPARISON OF VOLUME PERCENT AND MOLE PERCENT (GAS)

	R-12	Air	R-22	Air
Molecular weight	120.93	28.966	86.476	28.966
Volume (ft^3)	98.5	1.5	98.5	1.5
Specific volume				
cm^3/g		833.329		833.329
ft^3/lb	3.1306		4.4036	
Density				
g/cm^3		0.0012000		0.0012000
lb/ft^3	0.31943	0.074914	0.22709	0.074914
Weight (lb)	31.4638	0.11237	22.3684	0.11237
Moles	0.260182	0.003879	0.258666	0.003879
Mole percent		1.469		1.477

It is sometimes desirable to calculate the volume of the liquid phase and that of the vapor phase in a cylinder or other container of known total volume. The following relationship is handy:

$$\text{volume of liquid} = \frac{(\text{total weight}) - (\text{vapor density})(\text{total volume})}{(\text{density of liquid}) - (\text{density of vapor})}$$

REFERENCES

8–1. H. M. Parmelee, ''Solubility of Air in Freon 12 Dichlorodifluoromethane and Freon 22 Monochlorodifluoromethane,'' *Refrig. Eng.* 59 (June 1951), 573.

8–2. DuPont Company, Freon Products Division, unpublished information.

8–3. V. D. Williams, ''The Solubility of Nitrogen in Freon 114,'' *J. Chem. Eng. Data*, 4 (1959), 92.

8–4. R. J. Teufel and Q. Quick, ''Nonabsorbable Gases in Fluorocarbons,'' *Aerosol Age* (Apr. 1976), 28.

8–5. Robert A. Gorski, DuPont Company, Freon Products Laboratory, unpublished information.

9

OIL RELATIONSHIPS

In refrigeration systems lubricating oil is at once a necessity and the source of considerable trouble. The refrigeration and petroleum industries have for years labored jointly to solve oil–refrigerant relationships and as a result, oil problems are much less serious than they have been in the past. Progress is due to better refrigeration equipment design, significant improvement in the refining of oils, and more knowledge of how oils behave in refrigeration systems.

Oil is needed in refrigeration systems to lubricate compressor bearings, and the properties of the oil must be suitable for this purpose. If the oil stayed in the crankcase where it belongs, many problems in refrigeration would be eliminated. But it does not. Oil leaves the crankcase by slipping past the piston rings in reciprocating compressors, by entrainment with the refrigerant, and by excessive foaming as refrigerant is released from solution in the oil. A small amount of oil circulating with the refrigerant may be beneficial by lubricating valves, controls, and so on. A large amount, however, may cause problems.

OIL PROBLEMS

Oil may be exposed to high temperatures in the compressor—especially at the discharge valves—leading to thermal breakdown and copper plating and varnish deposits on the valve surfaces, thus interfering with their operation. Sludging and acid formation may become serious and greatly limit the life of the system. If too much oil

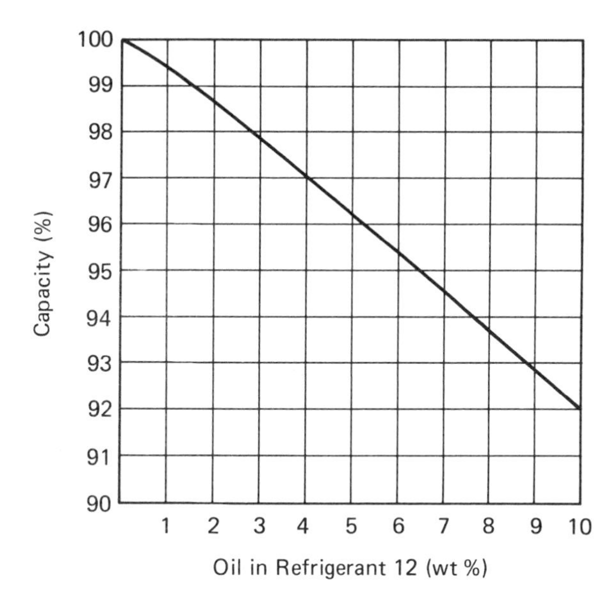

Figure 9–1 Loss of capacity when oil is present in R-12. (From Ref. 9–1.)

leaves the crankcase, there may not be enough left to lubricate the bearings properly and failure may result.

Oil may accumulate in the evaporator and cause a decrease in the capacity of the system. A typical result is illustrated in Figure 9–1. Heat transfer through the walls of the evaporator coil will be impaired.* The effective area of the coil will be less if part of the space is occupied by oil. Both of these factors will require a larger temperature difference between the load and the evaporating refrigerant. The capacity will be further reduced to the extent that the vapor pressure of the refrigerant–oil solution is less than that of the refrigerant alone. A review of the effect of oil in the evaporator by Green led to the following conclusions [9–2]:

1. Heat transfer coefficients are decreased by the presence of oil (as measured on flat plates).
2. At oil concentrations of 0 to 3 wt %, foaming may occur and certain conditions of temperature and heat flux will produce a boiling heat transfer coefficient slightly greater than that of oil-free refrigerant.
3. At oil concentrations of about 0 to 5 wt % in evaporators with geometry similar to a flat plate (outside tubes), heat transfer coefficients are only slightly reduced from those of refrigerant alone.
4. At oil concentrations of 0 to 6 wt % for evaporation inside horizontal tubes, the heat transfer coefficients may be increased over those for oil-free refrigerants. The apparent increase is due to the greater boiling liquid contact area.
5. At oil concentrations greater than 10 wt % in R-12 and R-22, the heat transfer coefficients are below those of refrigerants alone within tubes.

*A very thin coating of oil on the inside of the coil may improve heat transfer by filling in very slight imperfections on the surface. With more oil the heat transfer is decreased.

6. At oil concentrations of 3 to 10 wt % pressure drop in the tubes of an evaporator for R-12 or R-22 is markedly increased.

Experimental work by Heckmatt et al. [9.3] also demonstrated a decrease in the heat transfer coefficient for R-12 when 10 wt % oil was added (see Table 9–1). They found that the reduction was greater with a mineral oil than with a synthetic alkylate-base oil.

TABLE 9–1 EFFECT OF OIL ON HEAT TRANSFER

	Overall heat transfer coefficient with R-22 at −33°F evaporating temperature (Btu/hr-ft^2-°F)		
Heat flux Btu/hr-ft^2	R-22 alone	R-22 and 10 wt % alkylate-base oil	R-22 and 10 wt % mineral oil
400	24.2	22.2	20.9
600	27.5	25.3	23.5
800	30.8	28.3	26.3

The authors point out that the heat transfer coefficients obtained in their work are lower than those usually found in commercial equipment and attribute the difference to the use of a small evaporator with low turbulence. Comparisons in the same equipment, however, seem valid.

Too much oil in the refrigerant may cause restrictions in the expansion valve or capillary tubing and reduce the flow of refrigerant. This possibility is more likely in low-temperature freezers and other applications than at higher evaporating temperatures. When the flow of refrigerant is reduced, temperatures in the evaporator change. Near the expansion device, temperatures are much lower than normal and farther along the evaporator they are higher than normal. These changes are illustrated in Figure 9–2. Temperatures at the inlet and outlet and at various intermediary positions in an evaporator were measured. The solid line shows normal operation at a gage pressure of 2 psig. Liquid refrigerant at the expansion valve inlet was at 10°F. The temperature dropped to −27°F in the valve and remained at that temperature until near the end of the coil. When refrigerant flow was arbitrarily reduced, temperatures illustrated by the dashed line were found. In the expansion valve the temperature was −52°F. From this low point the temperature quickly rose and continued to climb thoughout the length of the coil, and as a result the capacity markedly declined. Changes of this sort would occur if the refrigerant flow is restricted for any reason, such as a shortage of refrigerant or partial plugging of a capillary tube, expansion valve, drier, strainer, sharp bend in the liquid line, and so on.

OIL RETURN [9–5,9–6]

When oil escapes from the crankcase it must be returned to avoid accumulation in some other part of the system. An oil separator is often installed in the compressor discharge line to trap the oil and return it to the crankcase. Although very helpful,

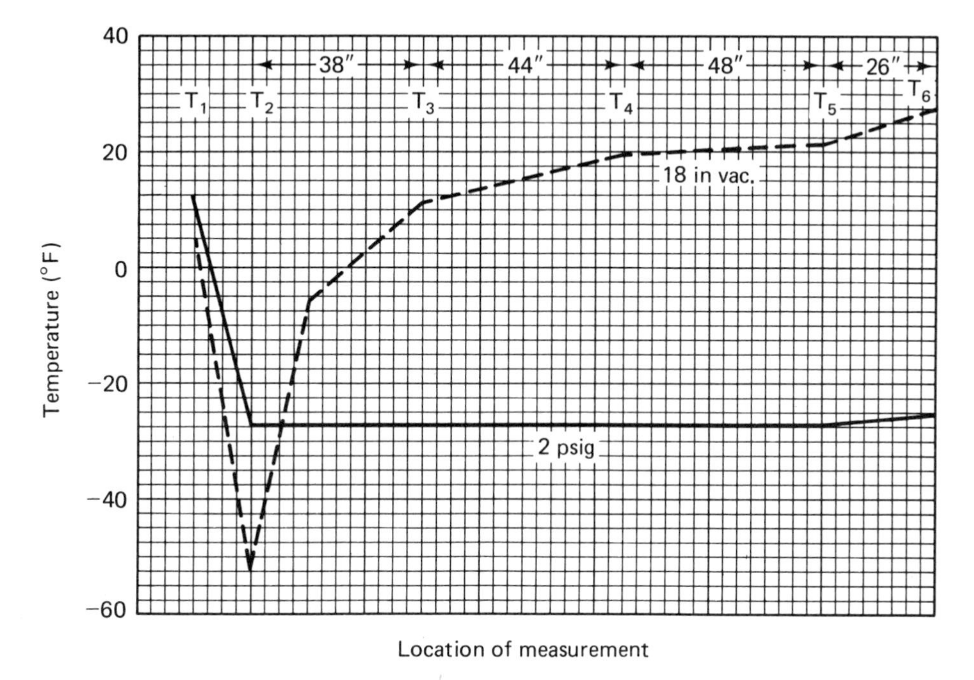

Figure 9–2 R-12 evaporator temperatures with restricted flow. (From Ref. 9–4.)

these devices are not 100% effective and may postpone the problem rather than solve it completely.

In the condenser and liquid line the oil is dissolved in the liquid refrigerant and moves with it. In the compressor discharge line, the evaporator, and the compressor suction line the oil is pushed along by the flowing refrigerant gas. The oil moves in two ways, as a film around the inside of the pipe and as a mist. In either case, how well the oil moves depends on the slope of the line (oil will obviously move better downhill than uphill), on the velocity of the refrigerant, and on the fluidity or viscosity of the oil. The viscosity of the oil depends on:

1. The nature of the oil
2. The temperature
3. The amount of refrigerant dissolved in the oil, which, in turn, depends on:
   a. The nature of the oil
   b. The nature of the refrigerant
   c. The temperature
   d. The pressure

Refrigerant velocity. The recommended refrigerant velocity for good oil movement is at least 750 ft/min in horizontal lines and at least 1500 ft/min in vertical pipes. A maximum velocity of 3000 ft/min is suggested since the pressure drop increases with velocity and too much pressure drop will cause a decrease in capacity. The velocity is affected by the vapor volume of the refrigerant, by the temperature, and by the size of the pipe. See Figures 4–12 to 4–17 for charts useful in estimating refrigerant velocities.

OIL PROPERTIES*

Many properties of lubricating oils have been identified and described, such as viscosity, viscosity index, density, color, refractive index, molecular weight, pour point, floc point, flash point, fire point, aniline point, and others [9–7]. However, in refrigeration applications, the following properties are especially important:

Viscosity	Aniline point
Density	Carbon type
Molecular weight	Molecular composition

Viscosity

The viscosity of the oil is one of the key factors in good oil return. It cannot be so high that the oil becomes viscous at lower temperatures or so low that it does not lubricate properly at higher temperatures. A viscosity around 150 Saybolt universal seconds (SUS) at 100°F seems to be about right for most low and medium temperatures and about 300 SUS at 100°F seems to be about right for higher temperatures such as air conditioning. The dividing lines are not sharp and in most cases either type could be used except for low temperature applications. The following viscosity units are in more or less common use:

$$(\text{mm})^2/\text{s} = \text{cSt}$$

$$\text{cSt} = \frac{\text{cP}}{\text{density}}$$

$$\text{stoke} = (0.0022)(\text{SUS}) - \frac{1.8}{\text{SUS}}$$

where $(\text{mm})^2/\text{s}$ = millimeters per second
cSt = centistoke
cP = centipoise
(stoke)(100) = centistoke
SUS = Saybolt universal seconds
density = grams per cubic centimeter

Centipoise and centistoke are specific properties, related through the density. SUS is an arbitrary measurement of flow through an orifice and the relationship above is not exactly precise but is close enough for most purposes. It tends to be

*Comments about oil properties apply to most oils used in refrigeration service. Where specific data are presented, they generally are for the oils produced by the Sun Oil Company with the trade name Suniso since they are widely used. Other oils, such as those produced by Texaco, have similar properties.

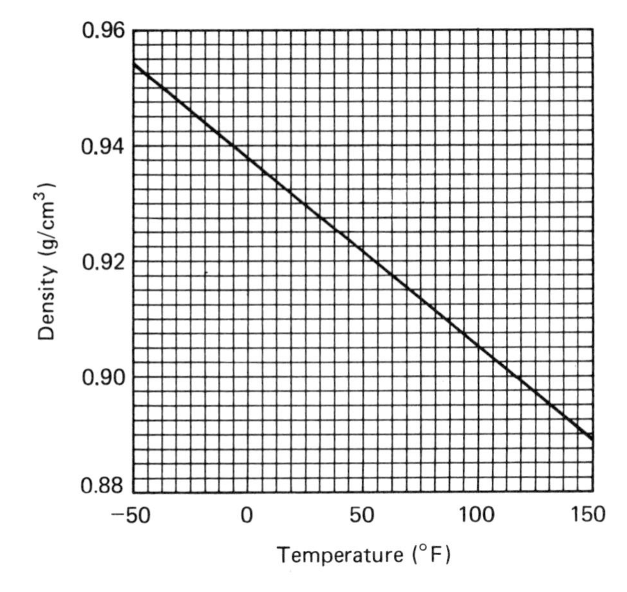

Figure 9–3 Liquid density of Suniso 3GS oil.

Carbon Type

Natural or mineral oil is obtained from underground sources and distilled, refined, and treated until it meets the quite stringent requirements of a refrigeration oil. It is made up of hundreds of different chemical compounds—even after purification and refinement. The character of the oil is determined by the types and amounts of the chemical compounds present and may be quite different from one oil to another. Most of the compounds are hydrocarbons, that is, contain only carbon and hydrogen atoms, although a few may also include nitrogen, oxygen, or sulfur. The hydrocarbons are divided into three general types and the distribution of these types in oil tells much about its nature and usefulness. Each carbon atom can be attached to four other atoms, including another carbon atom. When four different atoms are attached, the compound is called *saturated*. When two bonds are attached to the same carbon atom, the compound is called *unsaturated*. When the carbons in the saturated structure are in a straight line or branched they are called *paraffinic*. When the saturated carbon atoms are in a ring structure they are called *naphthenic*. When the unsaturated carbon atoms are in a ring structure they are called *aromatic*.

paraffinic naphthenic aromatic

Paraffinic compounds are noted for good chemical and thermal stability but poor solubility for fluorocarbon refrigerants, especially polar ones such as R-22. They are also rather poor lubricants. The aromatic compounds are more reactive (less stable), are better solvents for fluorinated refrigerants, and are better lubricants. Naphthenic compounds have properties in between those of the other two classifications. Natural or mineral oils are a mixture of all three types of molecules. Since oils from the same geographical source have about the same general composition, the source is important. It is much easier to refine and purify a base stock with about the right percentage of each carbon type than to alter the composition by addition or subtraction. A good oil for use with fluorocarbon refrigerants would have a fair amount of aromatic carbons and not too many paraffinic atoms. A rather new synthetic oil made by alkylating benzene has a high percentage of paraffinic carbons but also a rather high aromatic content and no naphthenic carbons. It has a good blend of solubility and stability properties. Some typical oil compositions are shown in Table 9–2.

TABLE 9–2 TYPICAL OILS

	Percent carbon atoms		
	C_a	C_n	C_p
Sun oils (Suniso)	14	42	44
Texaco oils (Capella)	8	40	52
Typical paraffinic oil	4	25	71
Alkylated benzene	24	0	76

Molecular Composition

The terms *paraffinic* for oils with a high percentage of paraffinic carbons and *naphthenic* for oils with fewer paraffinic atoms are in general use and identify the areas where the oil might be used. For example, a paraffinic oil might be a good automobile lubricant where good thermal stability is especially important. It could be used in air conditioning, but would be a disaster in low-temperature refrigeration, where low viscosity and high refrigerant solubility are necessary.

Aniline Point

Aniline is an amino derivative of benzene with the chemical formula $C_6H_5—NH_2$. It is a rather poor solvent for paraffinic-type hydrocarbons, a much better solvent for aromatic types, and in between for naphthenic molecules. The aniline point is the temperature where a mixture of aniline and an oil become miscible or change from two liquid phases to one. A high aniline point (above 220°F, for example) indicates a paraffinic-type oil and a lower aniline point (generally below 200°F) would indicate one that was more naphthenic in nature.

SOLUBILITY RELATIONSHIPS

R-12

Most common refrigeration oils are completely miscible with liquid R-12. For example, solutions of R-12 and Suniso 3GS do not separate into two liquid phases at any temperature (at least down to −100°F). The solubility of R-12 as a function of pressure and temperature has been studied with various oils to gain a better understanding of system stability and the role of oil in system operation. A majority of the studies have been with oils produced by the Sun Oil Company with the trade name Suniso and to some extent with the Capella oils from Texaco. Other specialty oils have also been examined.

Before about 1970, the 150 SUS viscosity oil produced by Sun was called Suniso 3G. Since then the trade name has been 3GS. Both oils are apparently from the same base stock, but the 3GS has been further refined to improve the stability. The solubility of R-12, however, is about the same in each oil. Some solubility data for R-12 in Suniso 3G is given in Table 9–3 and in Table 9–4 for Suniso 3GS. In both cases, the original data, principally from Sun and DuPont, have been smoothed and correlated. The solubility of R-12 in Suniso 3GS is also shown in Figure 9–4.

The solubility of R-12 was studied by Bambach [9–11] with an oil having a higher molecular weight, higher aniline point, and greater viscosity than Suniso 3GS. The oil would be expected to be a somewhat poorer solvent for R-12—and is a little, but not enough to be very significant. The two oils are compared in Table 9–5, page 219. The solubility data are in Table 9–6, page 219.

Some measurements of the solubility of R-12 in Suniso 3G were made at the University of Colorado and are shown in Table 9–7, page 221. Smoothed data for the solubility of R-12 in Suniso 3GS oil are shown in Figure 9–4, page 218, at regular values of temperature. The solubility coefficient, K, as a function of temperature is given in Figure 9–5, page 221.

$$K = \frac{\text{mole fraction refrigerant}}{\text{pressure (psia)}}$$

Average values of K from the tabular data at each temperature are used here.

In Figure 9–6, page 222, the solubility data are plotted at refrigerant temperature versus oil temperature at several different compositions. The pressure is constant for any given refrigerant temperature. For example, if the refrigerant in the evaporator is at 60°F and the oil is at a temperature of 120°F, the oil will contain about 18 wt % R-12 at equilibrium, assuming that the pressure has equalized between the evaporator and the compressor crankcase. If the oil has cooled down to 100°F, it will contain 25% R-12.

Spauschus has studied the thermodynamic properties of solutions of R-12 and oil based on the vapor pressure data of Bambach [9–13]. He found that R-12-oil solutions (Bambach's oil) exhibit positive deviations from Raoult's law over the

TABLE 9–3 SOLUBILITY OF R-12 IN SUNISO 3G OIL

Pressure (psia)	Refrigerant		
	wt %	mol %	mol %/psia
		60°F	
24.7	11.5	26.17	1.060
34.7	18.5	38.25	1.102
44.7	27.4	50.73	1.135
54.7	38.8	63.37	1.159
		Av.	1.114
		80°F	
24.7	8.5	20.22	0.819
34.7	12.7	28.41	0.819
44.7	17.5	36.66	0.820
54.7	23.5	45.60	0.834
64.7	30.5	54.49	0.842
74.7	40.5	65.00	0.870
		Av.	0.834
		100°F	
24.7	6.4	15.72	0.636
34.7	8.8	20.84	0.601
44.7	12.0	27.12	0.607
54.7	15.7	33.69	0.616
74.7	24.8	47.36	0.634
94.7	38.2	62.78	0.663
		Av.	0.626
		120°F	
34.7	7.3	17.69	0.510
54.7	12.3	27.68	0.506
74.7	17.7	36.98	0.495
94.7	24.2	46.55	0.492
114.7	32.2	56.44	0.492
		Av.	0.499
		140°F	
34.7	5.6	13.93	0.401
54.7	9.2	21.66	0.396
74.7	13.3	29.51	0.395
94.7	17.8	37.14	0.392
114.7	22.8	44.62	0.389
134.7	29.0	52.71	0.391
154.7	36.7	61.27	0.396
		Av.	0.394
		160°F	
34.7	4.4	11.16	0.322
54.7	7.8	18.75	0.343
74.7	11.2	25.60	0.343
94.7	14.5	31.63	0.334
114.7	18.4	38.09	0.332
134.7	22.8	44.62	0.331
154.7	27.5	50.86	0.329
		Av.	0.333
		175°F	
54.7	6.0	14.83	0.271
94.7	12	27.12	0.286
134.7	18.4	38.09	0.283
174.7	26.2	49.20	0.282
214.7	35.7	60.24	0.281
254.7	47.8	71.41	0.280
		Av.	0.281
		180°F	
34.7	3.1	8.03	0.231
54.7	5.3	13.25	0.242
74.7	7.8	18.75	0.251
94.7	10.4	24.05	0.254
114.7	13.2	29.33	0.256
134.7	16.8	35.52	0.264
		Av.	0.250
		200°F	
54.7	5.0	12.56	0.230
94.7	8.5	20.22	0.214
134.7	13.8	30.40	0.226
174.7	19.2	39.33	0.225
214.7	26.0	48.95	0.228
254.7	33.3	57.67	0.226
		Av.	0.225
		250°F	
54.7	2.5	6.54	0.120
94.7	4.5	11.39	0.120
134.7	7.5	18.12	0.135
174.7	10.8	24.83	0.142
214.7	12.4	27.86	0.130
254.7	18.8	38.72	0.152
294.7	24.0	46.28	0.157
		Av.	0.137

Source: Ref. 9–9.

TABLE 9–4 SOLUBILITY OF R-12 IN SUNISO 3GS OIL

Pressure (psia)	Refrigerant		
	wt %	mol %	mol %/psia
		75°F	
25	10	23.26	0.930
40	20	40.55	1.014
52	30	53.90	1.037
62	40	64.53	1.041
70	50	73.18	1.045
75	60	80.36	1.071
83	80	91.61	1.104
			Av. 1.035
		100°F	
37.5	10		0.620
65	20		0.624
87	30		0.620
101	40		0.639
112	50		0.653
120	60		0.670
134	80		0.684
			Av. 0.644
		125°F	
47	10		0.495
88	20		0.461
117	30		0.461
143	40		0.451
156	50		0.469
165	60		0.487
180	80		0.509
			Av. 0.476
		150°F	
67.5	10	23.26	0.345
115	20	40.55	0.353
155	30	53.90	0.348
185	40	64.53	0.349
203	50	73.18	0.360
217	60	80.36	0.370
235	80	91.61	0.390
			Av. 0.359
		175°F	
77	10		0.302
137	20		0.296
184	30		0.293
217	40		0.297
245	50		0.299
267	60		0.301
305	80		0.300
			Av. 0.298
		200°F	
90	10		0.258
160	20		0.253
213	30		0.253
259	40		0.249
295	50		0.248
330	60		0.244
			Av. 0.251
		225°F	
100	10		0.233
180	20		0.225
258	30		0.209
330	40		0.196
			Av. 0.216
		250°F	
145	10		0.160
247	20		0.164
345	30		0.156
			Av. 0.160

Source: Ref. 9–10.

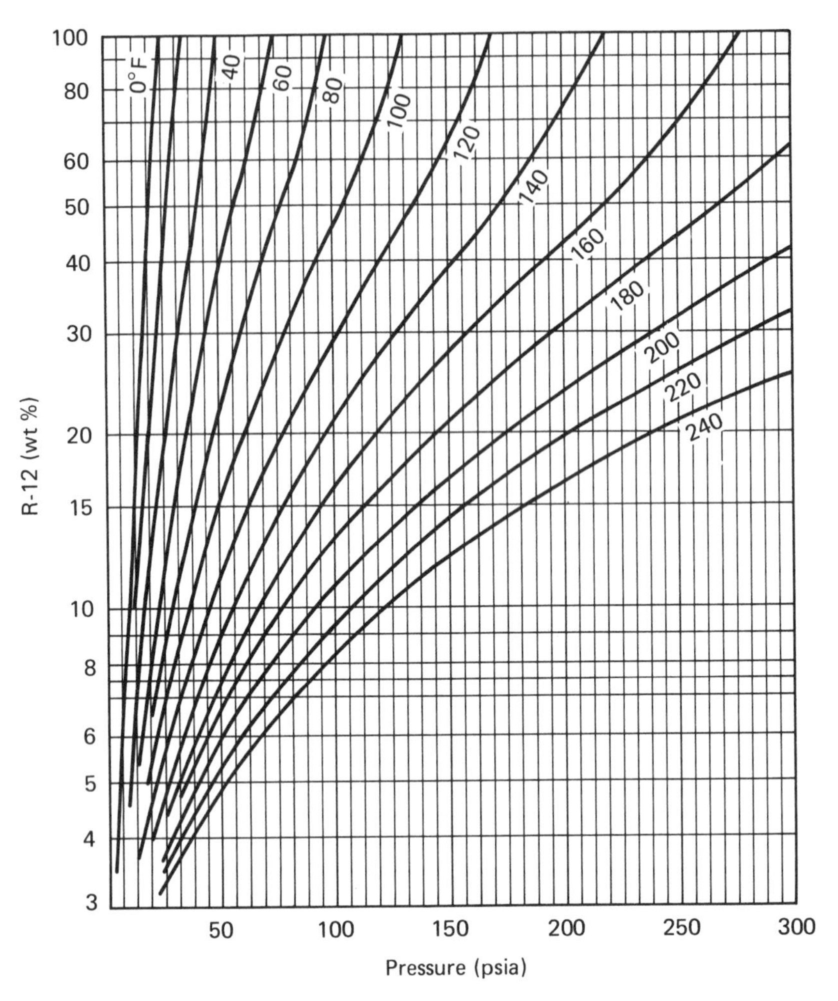

Figure 9–4 Solubility of R-12 in Suniso 3GS oil as a function of pressure.

temperature range −30 to 90°C. The excess free energy is positive and almost symmetrical with respect to composition. The excess entropy, enthalpy, and excess volume are complex temperature- and concentration-dependent functions. Qualitatively, the excess thermodynamic properties can be explained in terms of molecular interactions produced by the mixing process. At low temperatures the excess thermodynamic properties resemble those of simple fluorocarbon–hydrocarbon mixtures, but at higher temperatures (above 0°C) the analogy is no longer valid.

R-22

R-22 has more polarity than R-12 and is less soluble in lubricating oil. A number of measurements of the solubility of R-22 in oils have been made and some of the

TABLE 9–5 DESCRIPTION OF OILS

	Bambach	3GS
Density at 68°F (20°C) (g/ml)	0.91105	0.917
Flash point (open cup) (°F)	421	
Aniline point (°F)	184.1	160
Average molecular weight	398	330
Refractive index at 68°F (D line)	1.5024	1.5015
Viscosity		
68°F		
Centistoke	129.5	
Centipoise	118.0	
SUS	590	
100°F		
Centistoke		33.1
SUS		155
122°F		
Centistoke	32.9	
Centipoise	29.3	
SUS	155	
Carbon type		
C_a		14
C_n		42
C_p		44

TABLE 9–6 SOLUBILITY OF R-12 IN BAMBACH'S OIL

Pressure (psia)	Refrigerant			Pressure (psia)	Refrigerant		
	wt %	mol %	mol %/psia		wt %	mol %	mol %/psia
	40°F				80°F		
10.209	5		1.446	15.543	5		0.950
18.814	10		1.423	31.210	10		0.858
29.025	20		1.555	51.484	20		0.877
35.196	30		1.663	64.544	30		0.907
39.489	40		1.740	73.944	40		0.929
42.712	50		1.796	81.084	50		0.946
45.246	60		1.838	86.642	60		0.960
47.299	70		1.871	90.986	70		0.972
48.994	80		1.897	94.358	80		0.985
50.415	90		1.919	96.891	90		0.998
		Av.	1.7148			Av.	0.9382
	60°F				100°F		
12.709	5	14.764	1.162	18.716	5		0.789
24.485	10	26.777	1.094	39.063	10		0.685
39.102	20	45.139	1.154	66.433	20		0.679
48.234	30	58.515	1.213	84.569	30		0.692
54.702	40	68.692	1.256	97.820	40		0.702
59.586	50	76.696	1.287	107.95	50		0.710
63.405	60	83.156	1.312	115.80	60		0.718
66.441	70	88.478	1.332	121.86	70		0.720
68.868	80	92.940	1.350	126.40	80		0.735
70.795	90	96.734	1.366	129.63	90		0.746
		Av.	1.2526			Av.	0.7176

TABLE 9–6 (Continued)

Pressure (psia)	Refrigerant wt %	Refrigerant mol %	Refrigerant mol %/psia	Pressure (psia)	Refrigerant wt %	Refrigerant mol %	Refrigerant mol %/psia
	120°F				200°F		
22.226	5	14.764	0.664	39.436	5		0.374
48.096	10	26.777	0.557	96.674	10		0.277
84.182	20	45.139	0.536	187.15	20		0.240
108.74	30	58.515	0.538	254.61	30		0.230
126.93	40	68.692	0.541	307.16	40		0.224
140.91	50	76.696	0.544	348.39	50		0.220
151.73	60	83.156	0.548	380.02	60		0.219
159.95	70	88.478	0.553	402.95	70		0.220
165.94	80	92.940	0.560	417.81	80		0.222
169.95	90	96.734	0.569	425.13	90		0.228
		Av.	0.561			Av.	0.2455
	140°F				220°F		
26.065	5		0.566	44.460	5		0.332
58.353	10		0.459	111.97	10		0.239
104.95	20		0.430	221.67	20		0.204
137.44	30		0.426	305.10	30		0.192
161.86	40		0.424	370.82	40		0.185
180.72	50		0.424	422.65	50		0.181
195.27	60		0.426	462.33	60		0.180
206.21	70		0.429	490.79	70		0.180
213.93	80		0.434	508.70	80		0.182
218.78	90		0.442	516.66	90		0.187
		Av.	0.446			Av.	0.2063
	160°F				240°F		
30.222	5		0.489	49.739	5		0.297
69.859	10		0.383	128.53	10		0.200
128.92	20		0.350	259.93	20		0.174
171.10	30		0.342	361.82	30		0.162
203.18	40		0.338	442.95	40		0.155
228.09	50		0.336	507.22	50		0.151
247.28	60		0.336	556.35	60		0.149
261.52	70		0.338	591.22	70		0.150
271.29	80		0.343	612.51	80		0.152
276.99	90		0.349	620.92	90		0.156
		Av.	0.3604			Av.	0.175
	180°F						
34.684	5		0.426				
82.633	10		0.324				
156.28	20		0.289				
210.03	30		0.279				
251.43	40		0.273				
283.76	50		0.270				
308.60	60		0.269				
326.80	70		0.271				
338.96	89		0.274				
345.51	90		0.280				
		Av.	0.2955				

Source: Ref. 9–11.

TABLE 9–7 SOLUBILITY OF R-12 IN SUNISO 3G OIL FROM THE UNIVERSITY OF COLORADO

Oil (wt %)	°F	Pressure (psia)	F-12 (mol %)	mol %/psia
40.2	2.12	22.70	80.23	3.534
	34.34	41.71		1.924
	56.3	60.34		1.330
20.2	−0.04	22.71	91.51	4.030
	31.28	41.80		2.189
	52.34	60.40		1.515

Source: Ref. 9–12.

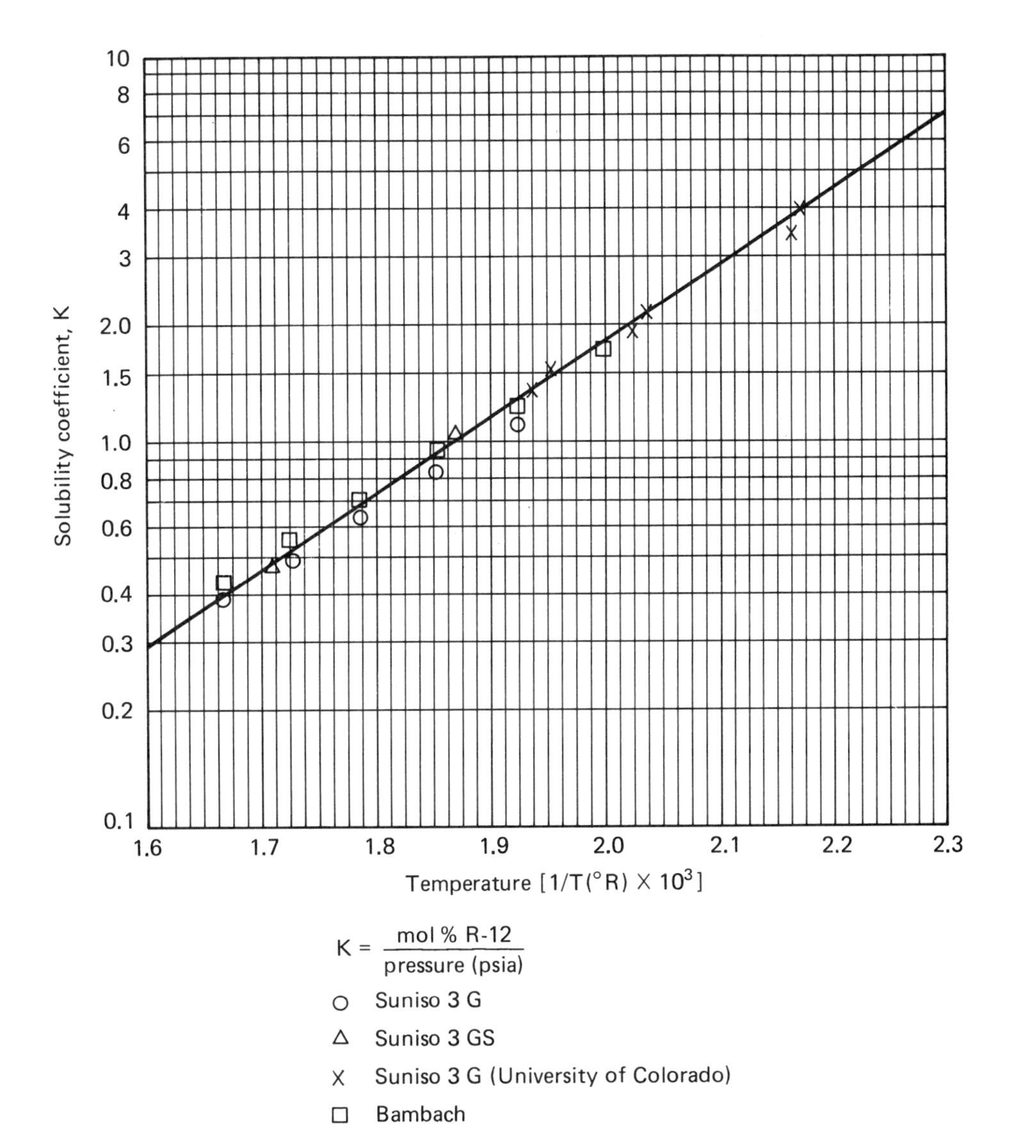

Figure 9–5 Solubility coefficients for R-12 and various oils as a function of temperature.

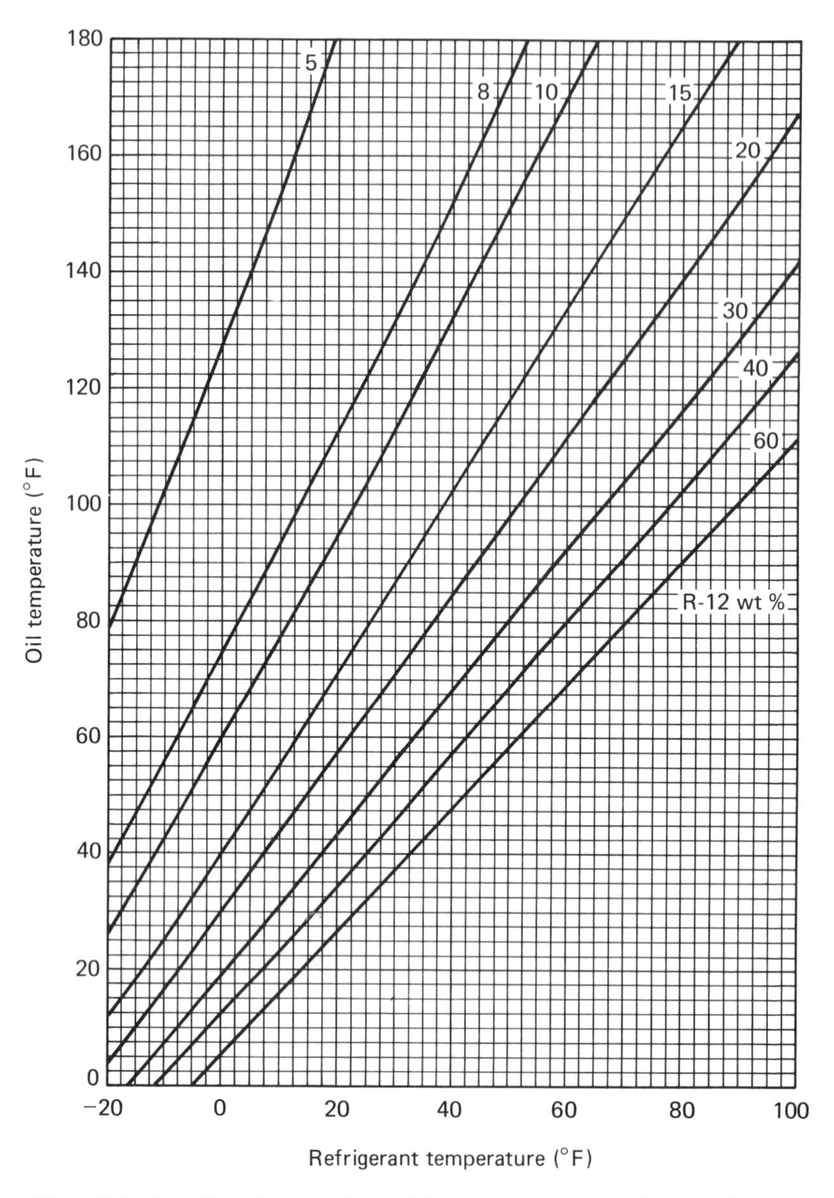

Figure 9–6 Solubility of R-12 in Suniso 3GS oil as a function of oil and refrigerant temperatures.

results are included here. They do not always agree very well, but perfect agreement is not especially important for refrigeration purposes.

As mentioned earlier, the Sun oils, 3G and 3GS, are from the same stock—the 3GS being more refined. Most investigators have found the solubility of R-22 about the same in either oil. Some early Sun data [9–9], however, show the pressure of 3G solutions about 15 psi higher than 3GS solutions—a possible confusion between gage and absolute pressures. In most cases, solutions of refrigerants and oils show higher pressure (less solubility) with paraffinic oils than with naphthenic types. However, in one reference (Albright and Mandelbaum [9–15]) the reverse was reported.

The solubility data for solutions of R-22 and oils are shown in the following tables (Tables 9–8 to 9–17) and charts (Figures 9–7 to 9–12, pages 232–236). In most cases the reported data have been smoothed and correlated to give a reasonable and consistent result.

Table 9–8, page 224. Solubility of R-22 in Suniso 3G oil. Pressures were reported in psig and 14.7 has been added to give psia. Conversions from percent by weight to mole percent were made as shown above.

Table 9–9, page 225. Solubility of R-22 in Suniso 3GS oil. Comparison with Table 9–8 at similar temperatures and concentrations, shows consistent differences of about 15 psi (see Figure 9–7).

Table 9–10, page 226. Solubility of R-22 in Suniso 3G oil. Measurements at DuPont with Suniso 3G oil seem to agree better with the Sun Oil data for Suniso 3GS oil than with 3G (see Figure 9–8).

Table 9–11, page 227. Solubility of R-22 in Suniso 3G oil. Some measurements at lower temperatures than much of the other data but fit in well in general.

Table 9–12, page 227. Solubility of R-22 in other oils. Suniso 351 was a specialty oil not widely used with somewhat higher viscosity than Suniso 3GS. Capella D is a Texaco oil with a viscosity of about 300 SUS at 100°F. The solubility of R-22 is about the same in these two oils and in both cases somewhat less than in the 150 SUS oils.

Table 9–13, page 228. Solubility of R-22 in a paraffinic oil. The measurements by Albright and Mandelbaum are reported to be with a "150 SUS paraffinic oil." The solubility of R-22 in this oil is less than in Suniso 3GS, as would be expected with a paraffinic oil compared with a naphthenic oil.

TABLE 9–8 SOLUBILITY OF R-22 IN SUNISO 3G OIL

Pressure (psia)	Refrigerant wt %	Refrigerant mol %	Refrigerant mol %/psia
	32°F		
34.7	10.2	30.23	0.871
54.7	19.5	48.03	0.878
64.7	28.4	60.21	0.931
		Av.	0.893
	40°F[a]		
24.7	7.5	23.63	0.957
34.7	9.7	29.07	0.838
44.7	12.3	34.86	0.780
54.7	16	42.09	0.769
64.7	21.3	50.80	0.785
74.7	30.8	62.94	0.843
		Av.	0.828
	60°F		
54.7	11.2	32.49	0.594
64.7	13.2	36.72	0.568
74.7	16	42.09	0.563
84.7	20.4	49.44	0.584
94.7	27.5	59.14	0.624
104.7	40	71.78	0.686
		Av.	0.603
	70°F[a]		
15.7	2.0	7.02	0.447
19.7	2.3	8.01	0.407
24.7	3.5	11.83	0.479
39.7	5.8	18.56	0.468
54.7	8.4	25.34	0.463
69.7	11.4	32.25	0.463
84.7	16.4	42.06	0.497
		Av.	0.461
	80°F		
64.7	10.2	30.23	0.467
74.7	11.4	32.93	0.441
84.7	13.0	36.31	0.429
94.7	15.6	41.36	0.437
104.7	18.8	46.90	0.448
114.7	23.2	53.54	0.467
124.7	29.7	61.71	0.495
134.7	38.7	70.66	0.525
144.7	54.0	81.75	0.565
		Av.	0.475
	100°F		
94.7	11.2	32.49	0.343
114.7	14.0	38.31	0.334
134.7	18.2	45.91	0.341
154.7	24.8	55.72	0.360
174.7	37.0	69.14	0.396
		Av.	0.355
	120°F		
114.7	10.7	31.37	0.273
134.7	13.2	36.72	0.273
154.7	16.7	43.34	0.280
174.7	21.3	50.80	0.291
194.7	27.2	58.77	0.302
214.7	35.0	67.26	0.313
234.7	46.2	76.62	0.326
		Av.	0.294
	140°F		
154.7	11.5	33.15	0.214
174.7	14.3	38.90	0.223
194.7	17.2	44.21	0.227
214.7	20.6	49.75	0.232
234.7	24.0	54.64	0.233
254.7	28.3	60.10	0.236
274.7	33.7	65.97	0.240
294.7	42.5	73.82	0.250
		Av.	0.232
	160°F		
174.7	10.4	30.69	0.176
194.7	12.5	35.28	0.181
214.7	15.0	40.24	0.187
234.7	17.4	44.56	0.190
254.7	20.3	49.29	0.194
274.7	23.2	53.54	0.195
294.7	26.3	57.65	0.196
		Av.	0.188

[a]DuPont data [9–14].

Source: Ref. 9–9.

TABLE 9–9 SOLUBILITY OF R-22 IN SUNISO 3GS OIL

Pressure (psia)	Refrigerant wt %	Refrigerant mol %	Refrigerant mol %/psia
	50°F		
25	5	16.72	0.669
40	10	29.77	0.774
60	20	48.82	0.814
72	30	62.05	0.862
76	40	71.78	0.944
80	50	79.23	0.990
82	60	85.13	1.038
86	70	89.90	1.045
90	80	93.85	1.043
93	90	97.17	1.080
		Av.	0.923
	75°F		
30	5		0.557
55	10		0.541
80	20		0.610
98	30		0.633
109	40		0.659
120	50		0.660
127	60		0.670
130	70		0.692
133	80		0.706
137	90		0.709
		Av.	0.644
	100°F		
45	5		0.372
77	10		0.387
117	20		0.417
142	30		0.437
167	40		0.430
178	50		0.445
190	60		0.448
200	70		0.450
205	80		0.458
208	90		0.467
		Av.	0.431
	125°F		
55	5		0.304
100	10		0.298
170	20		0.287
216	30		0.287
242	40		0.297
263	50		0.301
276	60		0.308
280	70		0.321
284	80		0.330
290	90		0.335
		Av.	0.307
	150°F		
60	5	16.72	0.279
125	10	29.77	0.238
208	20	48.82	0.235
268	30	62.05	0.232
310	40	71.78	0.232
338	50	79.23	0.234
360	60	85.13	0.236
373	70	89.90	0.241
383	80	93.85	0.245
390	90	97.17	0.249
		Av.	0.242
	175°F		
100	5		0.167
152	10		0.196
250	20		0.195
330	30		0.188
392	40		0.183
		Av.	0.186
	200°F		
140	5		0.119
203	10		0.147
310	20		0.157
388	30		0.160
		Av.	0.146
	225°F		
190	5		0.088
260	10		0.115
380	20		0.128
		Av.	0.110
	250°F		
250	5		0.067
323	10		0.092
		Av.	0.080

Source: Ref. 9–10.

TABLE 9–10 SOLUBILITY OF R-22 IN SUNISO 3G OIL

Pressure (psia)	Refrigerant wt %	Refrigerant mol %	Refrigerant mol %/psia
	40°F (RT-31)		
24.7	7.5	23.63	0.957
34.7	9.7	29.07	0.838
44.7	12.3	34.86	0.780
54.7	16.0	42.09	0.769
64.7	21.3	50.80	0.785
74.7	30.8	62.94	0.843
		Av.	0.828
	85°F (D-34D)		
19.7	3.09	8.4	0.426
34.7	3.74	12.6	0.362
49.7	7.23	22.5	0.453
64.7	8.05	24.5	0.379
79.7	11.44	32.2	0.404
94.7	11.31	32.0	0.337
109.7	13.20	35.8	0.326
124.7	43.43	50.0	0.401
		Av.	0.386
	85°F (KLLB-255, pp. 77–81)		
25	2.96	10.0	0.400
35	4.87	16.1	0.460
50	5.2	17.0	0.340
65	8.51	26.19	0.403
90	12.95	36.21	0.402
		Av.	0.401
	150°F		
19.7	1.28	4.71	0.239
34.7	2.36	8.44	0.243
49.7	3.36	11.71	0.236
64.7	4.00	13.72	0.212
79.7	5.46	18.06	0.227
94.7	6.46	20.85	0.220
109.7	7.47	23.55	0.215
124.7	8.40	25.92	0.208
		Av.	0.225
	200°F		
19.7	0.84	1.81	0.134
34.7	1.57	5.74	0.165
49.7	2.20	7.90	0.159
64.7	2.91	10.26	0.159
79.7	3.67	12.69	0.159
94.7	4.31	14.66	0.155
109.7	5.09	16.99	0.155
124.7	5.66	18.63	0.149
		Av.	0.154
	70°F (KSS-5321)		
15.7	2.0	7.22	0.460
19.7	2.3	8.24	0.418
24.7	3.5	12.16	0.492
39.7	5.8	19.02	0.479
54.7	8.4	25.92	0.474
69.7	11.4	32.93	0.472
84.7	16.4	42.81	0.505
		Av.	0.471

Source: Ref. 9–14.

Table 9–14, page 229. Solubility of R-22 in a General Electric oil. The GE oil is reported to be a highly refined ''water white'' oil with a viscosity of 150 SUS at 100°F. The measurements by Spauschus in general fit well with data for other oils (Figure 9–10).

Table 9–15, page 230. Solubility of R-22 in a General Electric oil. The measurements are in a lower temperature range than those of Spauschus and where overlap occurs have a somewhat lower solubility coefficient.

Table 9–16, page 231. Comparison of three oils. The properties of a naphthenic oil, Suniso 3GS, and two paraffinic oils, the General Electric oil and the oil used by Albright and Mandelbaum, are compared.

TABLE 9–11 SOLUBILITY OF R-22 IN SUNISO 3G OIL

Pressure (psia)	Refrigerant wt %	mol %	mol %/psia
		5.9°F	
42.81	64	87.15	2.036
		33.3°F	
72.88	64	87.15	1.196
		46.2°F	
91.09	64	87.15	0.957
		14.54°F	
52.16	79	93.49	1.792
		31.82°F	
71.94	79	93.49	1.300
		45.5°F	
91.09	79	93.49	1.026

Source: Ref. 9–12.

TABLE 9–12 SOLUBILITY OF R-22 IN OTHER OILS

Pressure (psia)	Refrigerant wt %	mol %	mol %/psia	Pressure (psia)	Refrigerant wt %	mol %	mol %/psia
	Capella D				*Suniso 351*		
	100°F				100°F		
64.7	6.2	21.1	0.326	64.7	5.9	20.3	0.314
114.7	12.7	37.0	0.323	114.7	12.3	36.2	0.316
164.7	23.6	55.6	0.338	164.7	21.7	52.2	0.317
		Av.	0.329			Av.	0.316
	175°F				175°F		
64.7	3.2	11.6	0.179	64.7	2.9	10.7	0.165
114.7	4.7	16.7	0.146	114.7	5.6	19.4	0.169
164.7	9.1	28.8	0.175	164.7	8.2	26.6	0.162
264.1	15.0	41.5	0.157	264.7	15.2	42.9	0.162
		Av.	0.164			Av.	0.164
	250°F				250°F		
64.7	1.6	6.1	0.094	64.7	1.8	7.5	0.116
114.7	3.3	12.2	0.106	114.7	3.5	12.8	0.112
164.7	5.2	18.5	0.112	164.7	5.2	18.1	0.110
264.7	8.8	29.0	0.110	264.7	8.2	29.6	0.112
		Av.	0.106			Av.	0.113

Source: Ref. 9–14.

TABLE 9–13 SOLUBILITY OF R-22 IN A PARAFFINIC OIL

Pressure (psia)	Refrigerant wt %	Refrigerant mol %	Refrigerant mol %/psia	Pressure (psia)	Refrigerant wt %	Refrigerant mol %	Refrigerant mol %/psia
	50°F				125°F		
8	1.12	5	0.625	20		5	0.250
16	2.35	10	0.625	40		10	0.250
30	5.13	20	0.667	78		20	0.256
36	6.72	25	0.694	97		25	0.258
42	8.48	30	0.714	119		30	0.252
55	12.60	40	0.727	155		40	0.258
69	17.78	50	0.725	194		50	0.258
		Av.	0.682			Av.	0.255
	75°F				150°F		
10		5	0.500	28		5	0.179
21		10	0.476	53		10	0.189
40		20	0.500	101		20	0.198
51		25	0.490	130		25	0.192
60		30	0.500	157	8.48	30	0.191
80		40	0.500			Av.	0.190
100		50	0.500				
		Av.	0.495				
	100°F						
14		5	0.357				
30.5		10	0.328				
57		20	0.351				
71		25	0.352				
85		30	0.353				
112		40	0.357				
140		50	0.357				
		Av.	0.351				

Source: Ref. 9–15.

Table 9–17, page 231. Phase separation with Suniso 3G oil and R-22. With polar refrigerants such as R-22 and/or highly fluorinated refrigerants such as R-115, solutions with oils may separate into two liquid phases. The upper, oil layer will be rich in oil and the lower refrigerant layer will have a higher concentration of refrigerant. The temperatures separating the single-phase and two-liquid-phase conditions for several different compositions are given in Table 9–17.

R-502

R-502 does not become completely miscible with oils at any temperature. Normally, the refrigerant-rich liquid layer is on the bottom and the oil-rich layer on top. R-115, one of the components of R-502, is not very soluble in lubricating oils, while R-22, the other component, is completely miscible at temperatures above about 35°F (Suniso 3GS). Mixtures of R-502 and oils are essentially a liquid R-115 layer on the bottom and an oil layer on top with R-22 distributed between them. The

TABLE 9–14 SOLUBILITY OF R-22 IN A GENERAL ELECTRIC OIL

Pressure (psia)	Refrigerant wt %	Refrigerant mol %	Refrigerant mol %/psia	Pressure (psia)	Refrigerant wt %	Refrigerant mol %	Refrigerant mol %/psia
	4°F (−20°C)				68°F (20°C)		
22.9	5	16.04	0.700	40.2	5		0.399
28.4	10	28.52	1.004	68.9	10		0.415
33.9	20	47.58	1.404	97.3	20		0.489
35.0	30	60.88	1.739	114.0	30		0.534
35.1	40	70.77	2.016	124.2	40		0.570
35.2	50	78.41	2.228	127	50		0.617
35.3	60	84.49	2.393	128	60		0.660
35.3	70	89.44	2.534	128	70		0.699
35.4	80	93.56	2.643	129	80		0.725
35.5	90	97.03	2.733	129	90		0.752
		Av.	1.939			Av.	0.586
	14°F (−10°C)				86°F (30°C)		
25.8	5		0.622	47.0	5		0.341
34.9	10		0.817	84.8	10		0.336
43.9	20		1.084	122.6	20		0.388
49.2	30		1.237	145	30		0.420
50.0	40		1.415	160	40		0.442
50.0	50		1.568	163	50		0.481
50.1	60		1.686	165	60		0.512
50.3	70		1.778	167	70		0.536
50.5	80		1.853	168	80		0.557
51.0	90		1.903	169	90		0.574
		Av.	1.396			Av.	0.459
	32°F (0°C)				104°F (40°C)		
28.7	5		0.559	54.5	5		0.294
43.2	10		0.660	103.8	10		0.275
57.8	20		0.823	153	20		0.311
66.4	30		0.917	182	30		0.335
68.0	40		1.041	202	40		0.350
69.2	50		1.133	207	50		0.379
70.0	60		1.207	212	60		0.399
70.8	70		1.263	215	70		0.416
71.4	80		1.310	217	80		0.431
72.0	90		1.348	218	90		0.445
		Av.	1.026			Av.	0.364
	50°F (10°C)				122°F (50°C)		
33.6	5		0.477	65.9	5		0.243
54.6	10		0.522	122.2	10		0.233
75.7	20		0.629	178	20		0.267
88.0	30		0.692	220	30		0.277
92.4	40		0.766	248	40		0.285
94.8	50		0.827	263	50		0.298
96.0	60		0.880	271	60		0.312
96.5	70		0.927	276	70		0.324
97.0	80		0.965	278	80		0.337
97.5	90		0.995	280	90		0.347
		Av.	0.768			Av.	0.292

TABLE 9–14 (Continued)

Pressure (psia)	Refrigerant wt %	Refrigerant mol %	Refrigerant mol %/psia	Pressure (psia)	Refrigerant wt %	Refrigerant mol %	Refrigerant mol %/psia
	140°F (60°C)				158°F (70°C)		
77.7	5		0.206	90.2	5		0.178
144	10		0.198	166	10		0.172
210	20		0.227	241	20		0.197
272	30		0.224	330	30		0.184
306	40		0.231	366	40		0.193
327	50		0.240	389	50		0.202
339	60		0.249	402	60		0.210
345	70		0.259	411	70		0.218
347	80		0.270	418	80		0.224
350	90		0.277	425	90		0.228
		Av.	0.238			Av.	0.201

Source: Ref. 9–16.

partition of R-22 between the two liquids will depend on its solubility in each. As the temperature increases, the concentration of R-22 in the liquid refrigerant layer will probably decline. As the critical temperature of the refrigerant mixture is approached, the position of the two layers is reversed and the refrigerant-rich layer is on top. Refrigerant density decreases faster with temperature than that of oil. When the critical temperature is reached, the refrigerant layer vanishes in a manner typical of liquids at the critical point. The temperature at which this phenomenon occurs is about 180°F. The measured critical temperature for pure R-502 is 179.9°F, so perhaps the composition of the refrigerant layer is about that of R-502 at the critical point—and the presence of oil may have little effect. The critical temperature of R-115 is 175.9°F and of R-22, 204.8°F.

TABLE 9–15 SOLUBILITY OF R-22 IN A GENERAL ELECTRIC OIL

Pressure (psia)	Refrigerant wt %	Refrigerant mol %	Refrigerant mol %/psia	Pressure (psia)	Refrigerant wt %	Refrigerant mol %	Refrigerant mol %/psia
	−40°F				50°F		
15.7	16.3	41.42	2.638	15.7	2.55	8.68	0.553
	−22°F				68°F		
15.7	10.6	30.10	1.917	15.7	1.88	6.50	0.414
	−4°F				86°F		
15.7	7.05	21.59	1.375	15.7	1.43	5.00	0.318
	14°F				104°F		
15.7	4.87	15.67	0.998	15.7	1.1	3.88	0.247
	32°F						
15.7	3.9	12.84	0.818				

Source: Ref. 9–14.

TABLE 9–16 COMPARISON OF THREE OILS

	Naphthenic Suniso 3GS	Paraffinic	
		GE	Albright
Viscosity			
SUS at 100°F	155	192	149.5
cSt at 100°F	33.1	108 (68°F)	32.0
Molecular weight	330	314	400
Density at 68°F	0.917	0.8823	—
Gravity (API degrees)	22.8	—	31.6
Refractive index	1.5015	1.4780	—
Carbon type (%)			
C_p		45	
C_n		55	
C_a	38	0	
R-22 Solubility at 15 wt%			
50°F	52	68	62
125°F	138	155 (122°F)	174

TABLE 9–17 PHASE SEPARATION WITH SUNISO 3G OIL AND R-22

Oil conc. (wt %)	Minimum temperature for single liquid phase (°F)	Reference	Oil conc. (wt%)	Minimum temperature for single liquid phase (°F)	Reference
0.7	−63	9–14	18.54	33.6	9–17
0.99	−59.8	9–17	22.29	31.8	9–17
1	9	9–14	22.36	31.6	9–17
1.06	−59.8	9–17	26.15	29.1	9–17
2.68	−25.8	9–17	29.83	27.3	9–17
2.93	−21.1	9–17	35.66	22.6	9–17
3.5	−10	9–14	36.54	21.7	9–17
6.4	8.8	9–17	40.5	20	9–14
7.0	16	9–14	40.85	18.9	9–17
7.34	13.5	9–17	44.91	14	9–17
10	23	9–14	64.22	−25.4	9–17
12.16	27.5	9–17	65.16	−27.9	9–17
13.43	29.8	9–17	68.6	−38	9–17
14.5	30	9–14	73.2	−81	9–14
18.17	33.6	9–17			

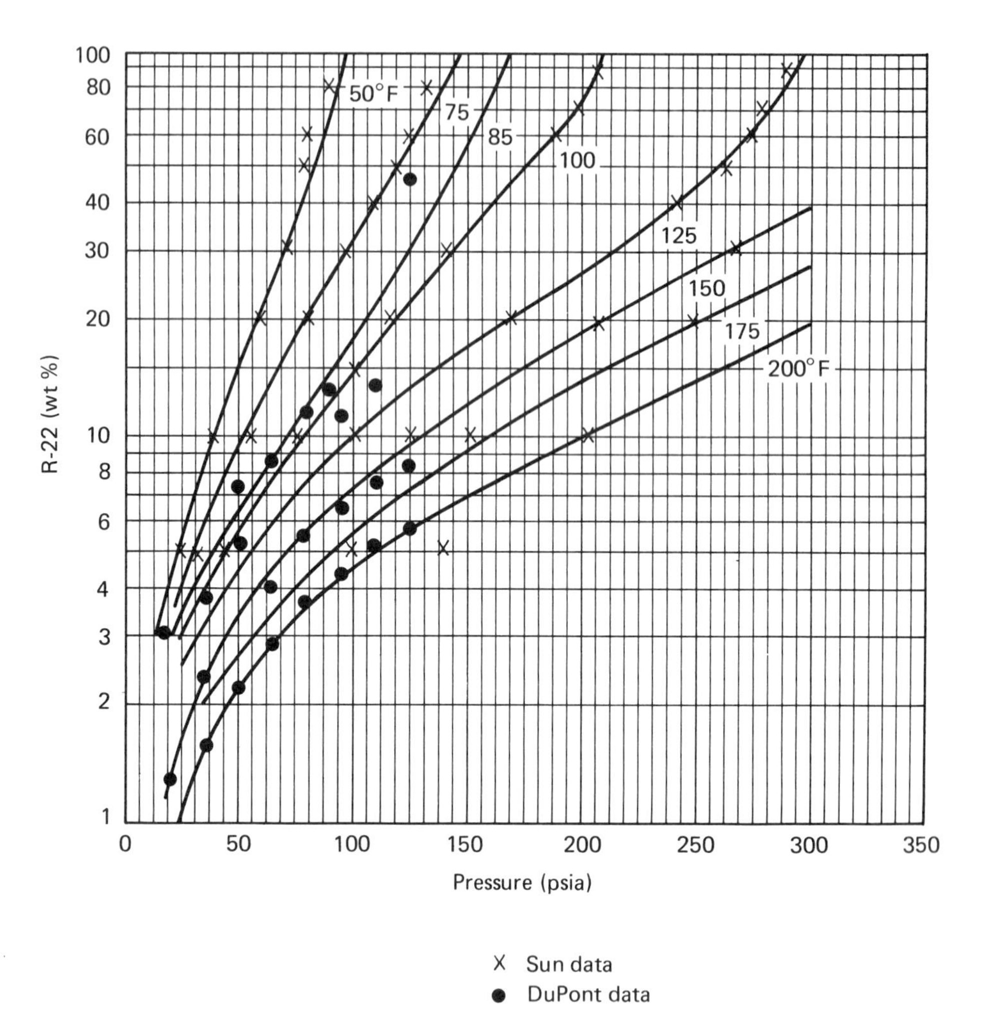

Figure 9–7 Solubility of R-22 in Suniso 3G and 3GS oils.

The solubility of R-502 in Suniso 3G and Suniso 3GS oils is about the same and is shown here in Tables 9–18 and 9–19, pages 238 and 239. The solubility data in Table 9–18 were measured by DuPont; the data in Table 9–19 were presented in chart form in a bulletin from the Sun Oil Company. The solubility data in these two tables were smoothed by plotting pressure versus log refrigerant concentration at various temperatures. The central portions of these isothermal curves were fairly straight, but at both low and high refrigerant concentrations there was considerable curvature. Data from this chart were replotted as log pressure versus reciprocal absolute temperature. Straight lines were drawn at regular refrigerant concentrations, producing solubility points used in preparing Figure 9–13, page 240. The data are shown as pressure versus R-502 concentration at regular intervals of temperature. The curves at the left of the chart are for the solubility of R-502 in oil and at the right for the solubility of oil in liquid R-502. In between, two liquid phases exist

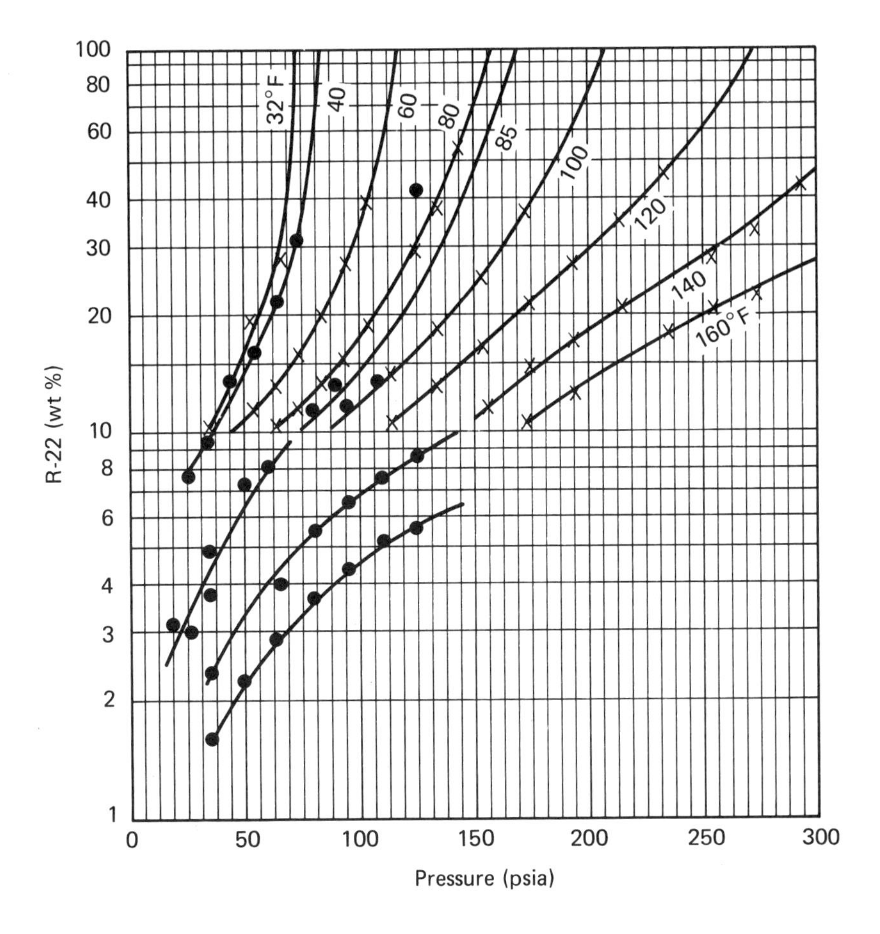

Figure 9–8 Solubility of R-22 in Suniso 3G oil.

and the pressure remains constant. The R-502 scale at the bottom shows the total, overall composition. The curves *AB* and *CD* show conditions at the boundary lines between single and two-liquid phases. In the two-phase region, the composition of each phase is given by the intersection with the boundary lines. For example, the total composition at *E* is 30% R-502, the pressure is 100 psia, and the temperature is about 50+°F. The composition of the large, upper, oil layer at *F* is about 25.5% R-502 and the small refrigerant layer is about 95.5% R-502. If more R-502 were added, the total composition would change but the composition of each layer would stay the same. At the same time, the size of the refrigerant layer would increase and that of the oil layer decrease. For additional solubility data see Figures 9–14, 9–15, and 9–16, pages 245, 246, and 247.

Suniso 4G and 4GS are similar to 3G and 3GS in nature but have viscosities of about 300 SUS rather than 150 SUS. The solubility of R-502 in both series of

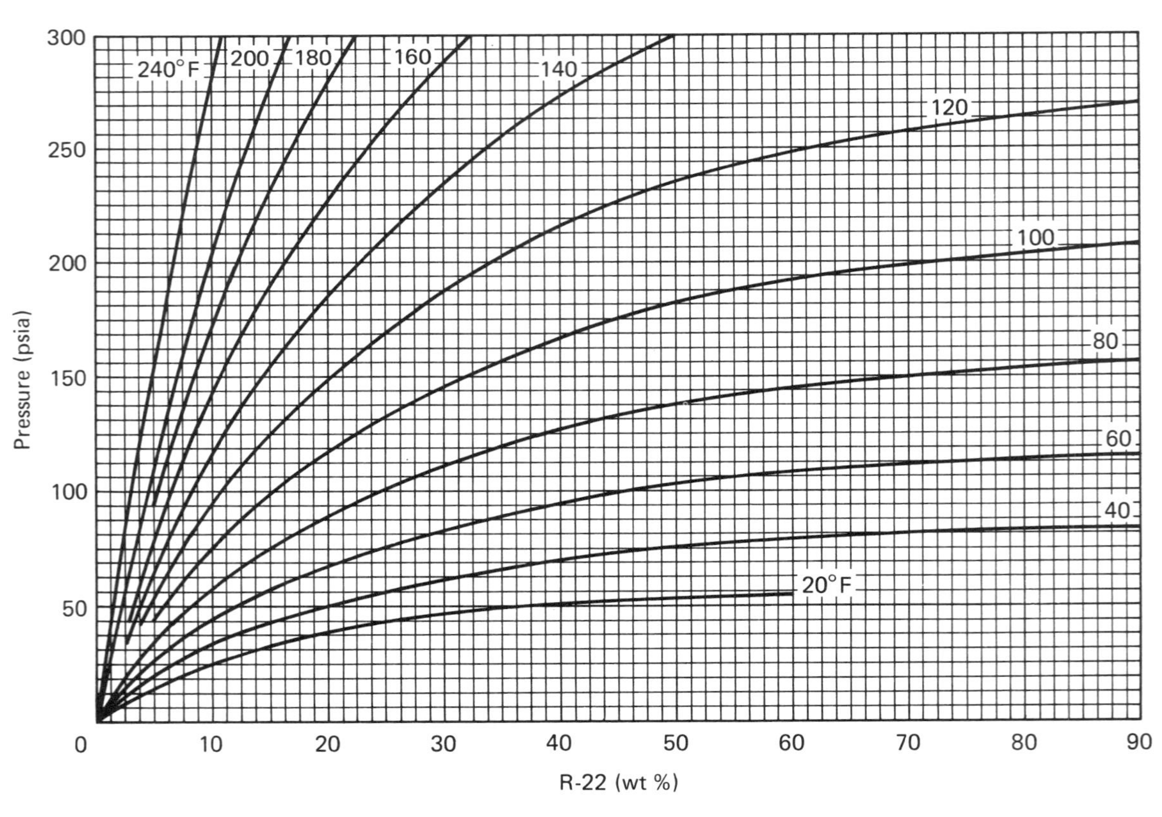

Figure 9–9 Solubility of R-22 in Suniso 3GS oil.

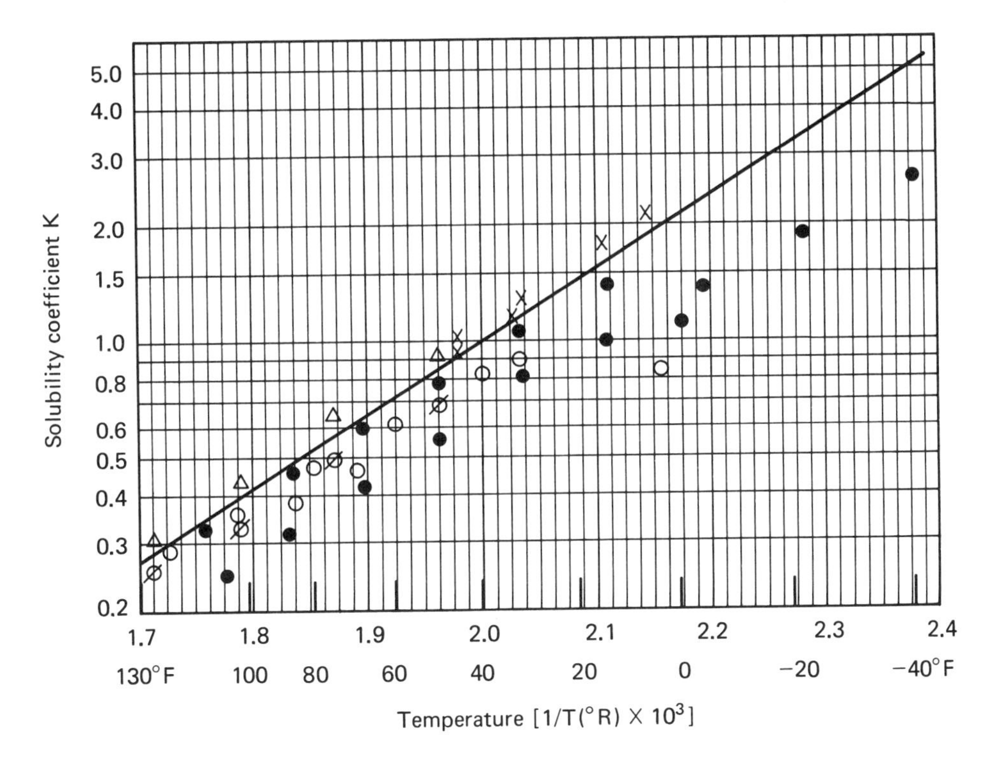

Figure 9–10 Solubility of R-22 in oils.

oils is about the same. Some DuPont measurements are given in Table 9–20, page 241, and Sun data in Table 9–21, page 242.

Solubility data in Capella D (Texaco) and Zerice S41 (Exxon) are given in Table 9–22, page 243. Zerice oils were formerly called "Alaska Bleu" and are reported to be mixtures of alkylated benzenes. Solubility relationships of R-502 and small amounts of oil are shown in Table 9–23, page 244.

R-13 and R-503 [9–14]

The solubility of R-13 and of R-503 in two oils is shown in Figure 9–18, page 250. The solid line is for a 150 SUS naphthenic oil such as Suniso 3GS and the dashed line is for a synthetic, alkylated benzene oil. With both refrigerants the

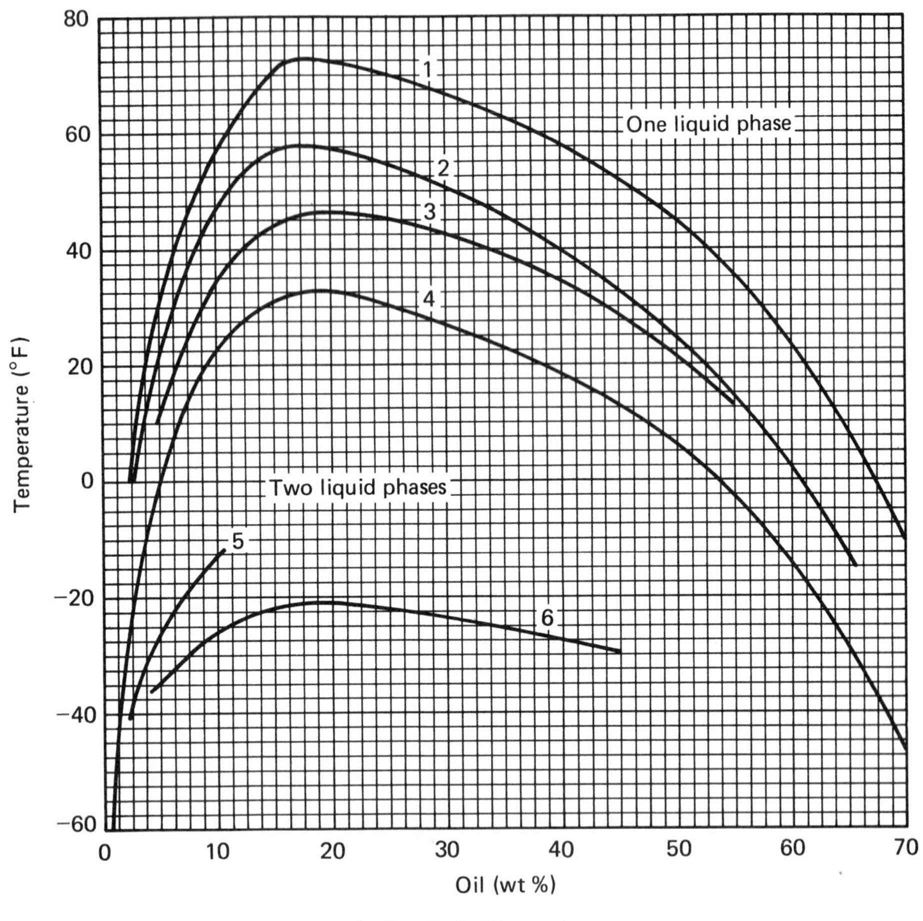

1 Capella D (Texaco)
2 Suniso 4G (Sun)
3 Capella B (Texaco)
4 Suniso 3G (Sun)
5 Suniso 3GS with R-22/R-12 (90/10 by wt)
6 Suniso 3G with R-22/R-12 (85/15 by wt)

Figure 9–11 Solubility relationships of R-22 and oils.

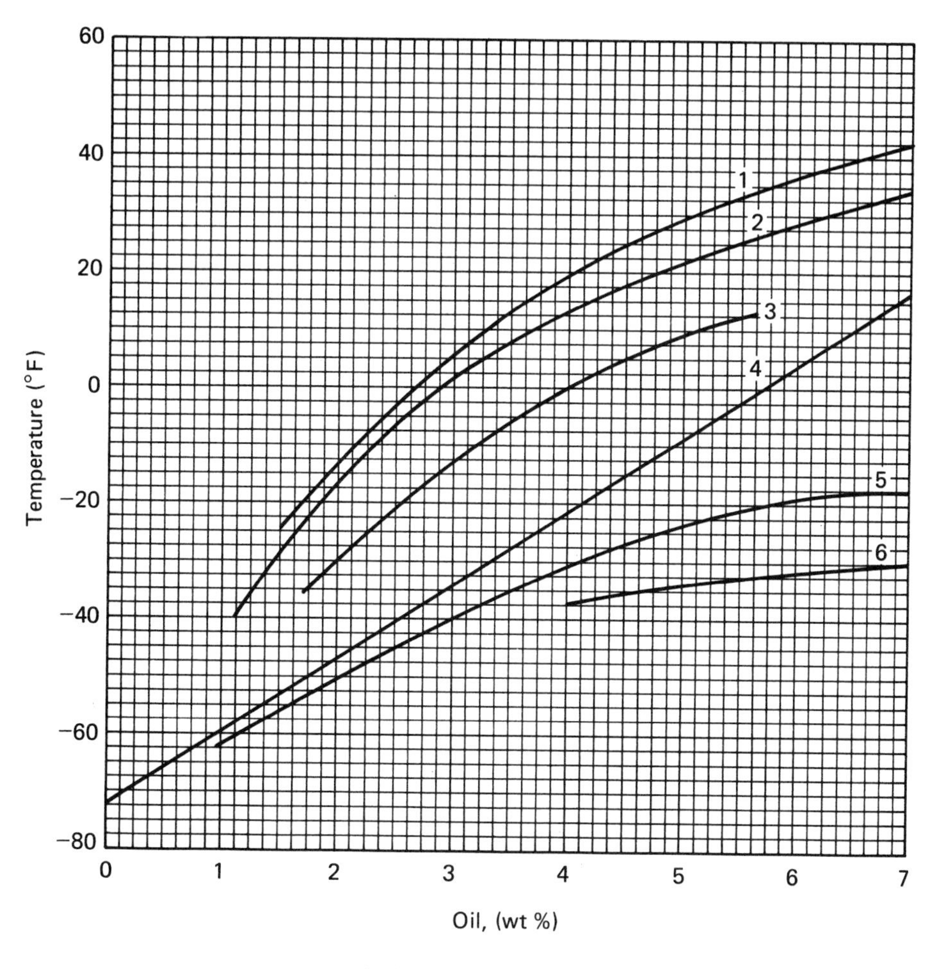

1 Capella D (Texaco)
2 Suniso 4G (Sun)
3 Capella B (Texaco)
4 Suniso 3G (Sun)
5 Suniso 3GS with R-22/R-12 (90/10 by wt)
6 Suniso 3G with R-22/R-12 (85/15 by wt)

Figure 9–12 Solubility of oil in liquid R-22.

TABLE 9–18 SOLUBILITY OF R-502 IN SUNISO 3G OIL

Pressure (psia)	Refrigerant wt %	Refrigerant mol %	Refrigerant mol %/psia
	−32°F		
19.7	15.8	35.68	1.811
	−20°F		
19.7	13.3	31.20	1.584
24.7	17.3	38.21	1.547
		Av.	1.565
	−10°F		
19.7	9.1	22.83	1.159
24.7	12.6	29.88	1.210
34.7	18.9	40.79	1.176
		Av.	1.182
	0°F		
19.7	9.2	23.05	1.170
44.7	22.0	45.47	1.017
		Av.	1.094
	10°F		
19.7	5.2	13.95	0.708
24.7	6.4	16.81	0.681
34.7	11.5	27.75	0.800
44.7	18.6	40.31	0.902
		Av.	0.773
	25°F		
19.7	5.8	15.40	0.782
24.7	5.6	14.92	0.604
34.7	8.5	21.54	0.621
44.7	12.5	29.69	0.664
59.7	19.2	41.26	0.691
		Av.	0.672
	40°F		
34.7	6.7	17.51	0.505
49.7	9.9	24.52	0.493
74.7	20.3	42.95	0.575
89.7	28.1	53.60	0.598
		Av.	0.543
	50°F		
44.7	7.8	20.00	0.447
59.7	10.7	26.15	0.438
74.7	14.7	33.75	0.452
89.7	19.9	42.34	0.472
		Av.	0.452
	60°F		
19.7			
59.7	9.1	22.83	0.382
89.7	15.9	35.85	0.400
		Av.	0.391

Pressure (psia)	Refrigerant wt %	Refrigerant mol %	Refrigerant mol %/psia
	70°F		
18.7	2.4	6.78	0.363
23.7	2.9	8.11	0.342
33.7	4.1	11.22	0.333
48.7	6.0	15.87	0.326
74.7	9.7	24.10	0.323
77.7	9.5	23.68	0.305
88.7	13.5	31.57	0.356
112.7	14.9	34.10	0.303
		Av.	0.331
	85°F		
19.7	2.2	6.23	0.316
24.7	2.0	5.69	0.230
34.7	2.9	8.11	0.234
49.7	5.2	13.95	0.281
64.7	6.2	16.34	0.253
74.7	7.9	20.23	0.271
79.7	10.0	24.72	0.310
94.7	10.3	25.34	0.268
109.7	11.2	27.16	0.248
124.7	13.2	31.01	0.249
		Av.	0.266
	150°F		
19.7	1.4	4.03	0.204
24.7	1.5	4.31	0.174
34.7	1.6	4.59	0.132
39.7	1.8	5.14	0.129
49.7	2.9	8.11	0.163
59.7	3.6	9.94	0.167
64.7	3.6	9.94	0.154
79.7	4.4	11.98	0.150
89.7	4.7	12.72	0.142
94.7	5.0	13.46	0.142
109.7	6.2	16.34	0.149
124.7	6.8	17.74	0.142
		Av.	0.154
	200°F		
19.7	0.8	2.33	0.118
34.7	1.3	3.75	0.108
49.7	2.0	5.69	0.115
64.7	2.9	8.11	0.125
74.7	3.5	9.68	0.130
94.7	3.8	10.46	0.110
109.7	4.2	11.47	0.105
124.7	5.2	13.95	0.112
		Av.	0.115

Source: Ref. 9–14.

TABLE 9–19 SOLUBILITY OF R-502 IN SUNISO 3GS OIL

Pressure (psia)	Refrigerant wt %	Refrigerant mol %	Refrigerant mol %/psia
	50°F		
63	10	24.72	0.392
88	20	42.50	0.483
99.5	30	55.89	0.562
106	40	66.34	0.626
		Av.	0.516
	75°F		
83	10		0.298
126	20		0.337
143	30		0.391
155	40		0.428
		Av.	0.364
	100°F		
114	10		0.217
172	20		0.247
202	30		0.277
215	40		0.309
		Av.	0.263
	125°F		
133	10		0.186
217	20		0.196
263	30		0.213
288	40		0.230
		Av.	0.206
	150°F		
163	10		0.151
275	20		0.155
340	30		0.164
387	40		0.171
		Av.	0.160
	175°F		
213	10		0.116
327	20		0.130
		Av.	0.123
	200°F		
270	10		0.092
395	20		0.108
		Av.	0.095

Source: Ref. 9–10.

solubility in the alkylated benzene oil is significantly higher than in the mineral oil. The region to the left of the curves is for one liquid solution and at the right for two liquid phases. The top ends of the curves are at the critical temperatures of the refrigerants.

R-500

R-500 is an azeotropic mixture containing 73.8 wt % R-12 and 26.2 wt % R-152a. The properties of R-500 are influenced considerably by the large amount of R-12 present and it seems reasonable that the liquid phase is completely miscible with oil. The solubility of R-500 in Suniso 3GS oil at pressures below saturation has been estimated assuming that the solubility factor, K, is about the same for a given refrigerant on a mole basis and that the pressure of a refrigerant–oil solution is related to the boiling point of the refrigerant. Solubility factors for R-12, R-22, and R-502 at various temperatures are listed in Table 9–24. From a chart of K versus the boiling point of the refrigerant, similar factors for R-500 were estimated and converted to solubility as weight percent, as in Table 9–25. A molecular weight of 325 was used for Suniso 3GS and of 99.31 for R-500. The estimated solubility at different pressures is shown in Table 9–26.

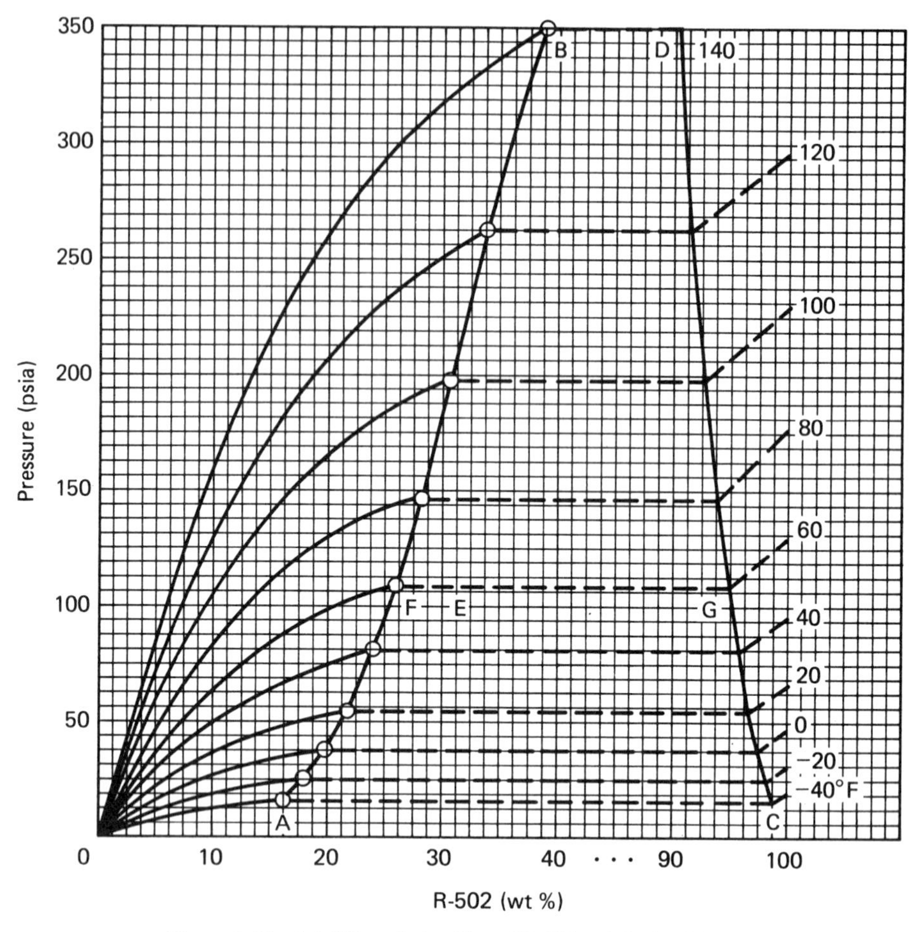

Figure 9–13 Solubility relationships of R-502 and Suniso 3G oils.

REFRIGERANT MIGRATION

When refrigeration and air-conditioning systems are at rest, pressures throughout the system tend to reach equilibrium. Pressure is primarily governed by the vapor pressure of the liquid refrigerant. If liquid is present at two or more locations, the refrigerant will move to the coolest location. Oil is also involved to the extent that refrigerant dissolves in it. The vapor pressure of the oil–refrigerant solution is always less than that of the pure refrigerant at the same temperature. As pressures equalize, refrigerant will dissolve in the oil. The temperature of the liquid refrigerant will always be higher than that of the oil—if not, all of the refrigerant will migrate to the oil. Oil temperatures, amount of refrigerant in the oil, and refrigerant temperatures at equilibrium—that is, all at the same pressure—are given in Figure 9–6, page 222, for R-12 and Figure 9–17, page 249, for R-22.

The migration of refrigerant into the oil in the crankcase can be a problem. If too much refrigerant dissolves in the oil, the viscosity and lubricity may be

TABLE 9–20 SOLUBILITY OF R-502 IN SUNISO 4G OIL

Pressure (psia)	Refrigerant wt %	Refrigerant mol %	Refrigerant mol %/psia
	60°F		
19.7	0.9	2.65	0.1347
24.7	3.0	8.49	0.3438
34.7	3.1	8.76	0.2524
44.7	5.3	14.38	0.3217
59.7	7.2	18.88	0.3163
74.7	11.6	28.25	0.3782
89.7	18.5	40.52	0.4517
			Av. 0.314
	75°F		
29.7	2.1	6.05	0.2036
44.7	4.3	11.88	0.2658
59.7	5.5	14.87	0.2490
74.7	7.7	20.02	0.2680
89.7	10.2	25.42	0.2834
104.7	11.3	27.66	0.2641
119.7	14.8	34.26	0.2863
134.7	18.9	41.15	0.3055
149.7	21.4	44.96	0.3004
			Av. 0.270
	100°F		
29.7	1.1	3.23	0.1087
44.7	2.9	8.22	0.1840
59.7	4.2	11.63	0.1947
74.7	5.0	13.64	0.1826
89.7	7.5	19.57	0.2182
104.7	8.5	21.80	0.2082
119.7	10.0	25.00	0.2089
134.7	12.2	29.43	0.2185
149.7	12.5	30.00	0.2004
164.7	16.6	37.39	0.2270
			Av. 0.195
	135°F		
44.7	2.0	5.77	0.1291
59.7	3.0	8.49	0.1423
74.7	3.9	10.86	0.1453
89.7	4.9	13.39	0.1493
104.7	5.6	15.11	0.1443
119.7	6.3	16.79	0.1403
134.7	6.9	18.19	0.1351
149.7	8.9	22.67	0.1514
164.7	8.9	22.67	0.1376
			Av. 0.142
	150°F		
44.7	1.7	4.93	0.1104
59.7	2.7	7.69	0.1288
74.7	2.8	7.96	0.1065
89.7	4.2	11.63	0.1296
104.7	4.5	12.39	0.1183
119.7	5.6	15.11	0.1262
134.7	5.7	15.35	0.1140
149.7	6.9	18.19	0.1215
164.7	8.6	22.02	0.1337
			Av. 0.121
	175°F		
29.7	0.7	2.07	0.0697
44.7	1.7	4.93	0.1104
59.7	3.1	8.76	0.1467
74.7	2.5	7.14	0.0956
89.7	3.9	10.86	0.1210
104.7	3.7	10.34	0.0987
119.7	4.7	12.89	0.1077
134.7	5.4	14.62	0.1086
149.7	6.0	16.07	0.1074
164.7	8.5	21.80	0.1324
			Av. 0.110
	200°F		
29.7	0.4	1.19	0.0401
44.7	0.8	2.36	0.0529
59.7	1.6	4.65	0.0779
74.7	2.2	6.32	0.0846
89.7	2.7	7.69	0.0857
104.7	3.1	8.76	0.0836
119.7	3.8	10.60	0.0885
134.7	4.1	11.37	0.0844
149.7	5.3	14.38	0.0961
164.7	5.5	14.87	0.0903
			Av. 0.078

Source: Ref. 9–14.

TABLE 9–21 SOLUBILITY OF R-502 IN SUNISO 4GS OIL

Pressure (psia)	Refrigerant wt %	Refrigerant mol %	Refrigerant mol %/psia
	50°F		
75	10	25.00	0.3334
105	20	42.86	0.4082
110	30	56.26	0.5114
"	40	66.67	0.6061
"	50	75.00	0.6819
"	60	81.82	0.7438
"	80	92.31	0.8392
112	100		
			Av. 0.589
	75°F		
90	10	25.00	0.2778
125	20	42.86	0.3429
142	30	56.26	0.3962
152	40	66.67	0.4386
160	60	81.82	0.5114
164	100		
			Av. 0.393
	100°F		
105	10	25.00	0.2381
177	20	42.86	0.2421
205	30	56.26	0.2744
215	40	66.67	0.3101
225	50	75.00	0.3333
230	60	81.82	0.3557
231	100		
			Av. 0.292
	125°F		
138	10	25.00	0.1812
220	20	42.86	0.1948
260	30	56.26	0.2164
282	40	66.67	0.2364
290	50	75.00	0.2586
303	60	81.82	0.2700
316	100		
			Av. 0.226
	150°F		
170	10	25.00	0.1471
260	20	42.86	0.1648
310	30	56.26	0.1815
355	40	66.67	0.1878
387	50	75.00	0.1938
423	100		
			Av. 0.175
	175°F		
218	10	25.00	0.1147
325	20	42.86	0.1319
388	30	56.26	0.1450
			Av. 0.131
	200°F		
215	5	13.64	0.0634
300	10	25.00	0.0833
400	20	42.86	0.1072
			Av. 0.085

Source: Ref. 9–10.

affected and bearings not properly lubricated. Furthermore, when the compressor starts, excess refrigerant will be quickly drawn out of the oil, causing foaming and possible loss of oil from the crankcase, leading to further lubrication problems.

A study of the rate of migration of R-22 to oil by Spauschus [9–18] illustrates the more general phenomenon. Quantitative measurements were made to show the effect of temperature difference, oil volume, and oil surface area on migration rate.

In Figure 9–19, page 250, migration rates are shown for three conditions. In one case the refrigerant temperature was held at 95°F and the oil at 77°F. Transfer of refrigerant into the oil was rapid. When the refrigerant and oil were at the same temperature (77°F) so that the temperature difference was zero, the rate was much slower. In the third case, the refrigerant was held at 77°F and the oil at 95°F. The rate was slower still with these conditions, although there was an appreciable difference in pressure:

TABLE 9–22 SOLUBILITY OF R-502 IN OTHER OILS

Oil	Temperature (°F)	Pressure (psia)	Refrigerant wt %	Refrigerant mol %	Refrigerant mol %/psia
Capella D[a]	100	64.7	4.5	12.9	0.199
		114.7	10.7	27.3	0.238
		164.7	17.9	40.6	0.247
					Av. 0.175
	175	64.7	2.3	6.9	0.106
		164.7	7.9	21.2	0.129
		264.7	12.6	31.1	0.118
					Av. 0.118
Alaska Bleu[b]	100	64.7	9.5	24.8	0.383
		114.7	15.8	37.0	0.323
		164.7	33.1	60.8	0.369
					Av. 0.358
	175	64.7	5.7	15.9	0.246
		164.7	13.8	33.4	0.203
					Av. 0.225
Alaska Bleu No. 3[c]	−20	19.7	29.3	54.3	2.70
	−10	19.7	16.9	36.8	1.88
	0	19.7	11.8	27.7	1.42
	20	19.7	6.5	16.6	0.838
		24.7	9.4	22.9	0.942
					Av. 0.890
	30	19.7	5.7	14.8	0.680
		24.7	6.4	16.4	0.660
					Av. 0.670
	50	19.7	4.5	11.9	0.508
		24.7	4.8	12.6	0.506
					Av. 0.507

[a]Molecular weight assumed to be 350.
[b]Now called Zerice S41. Molecular weight assumed to be 350.
[c]Molecular weight assumed to be 320.
Source: Ref. 9–14.

vapor pressure of R-22 at 77°F = 151 psia

vapor pressure of the oil solution at 95°F and 21% R-22 after 20 hr = 112 psia

Migration of R-22 out of oil solutions also occurs, and here again, the rate is increased by an increase in the temperature difference between oil and refrigerant. This condition is illustrated in Figure 9–20, page 250. The R-22 migrating out of

TABLE 9–23 SOLUBILITY OF OILS IN R-502

Oil	Concentration of oil		Lowest temperature for miscibility (°F)	Pressure (psia)
	vol %	wt %		
Suniso 3G	1	0.72	−40.9	
	1	0.72	−40	
	2	1.44	−23.8	
	2	1.44	−27.4	25.2
	3	2.16	−9.5	
	3	2.16	−12.6	19.5
	4	2.88	−8.6	
	4	2.88	5	49.7
	5	3.60	25.7	
	5	3.60	19.8	65.2
	6	4.32	43.7	
	6	4.32	46.4	
	7	5.04	60.8	
	7	5.04	60.8	
Suniso 3GS	0.25	0.18	−61.6	
	1.00	0.72	−41.8	
Capella B	0.25	0.18	−37.4	
	1	0.68	−24.5	
	2	1.36	−6.0	
	3	2.04	16.5	
	4	2.72	36.5	
	5	3.40	71.5	
Zerice S41		3.3	<−76	
		9.8	−9.8	
		13.2	17.6	
		16.7	31.8	
		22.2	17.6	
		27.3	−25.2	
		27.4	−9.2	
		29.7	−14.4	
		31.5	55.8	
		33.3	47.8	
		40.0	31.3	
		47.2	2.1	
		56.6	−9.2	
		75.5	−14.8	

Source: Ref. 9–14.

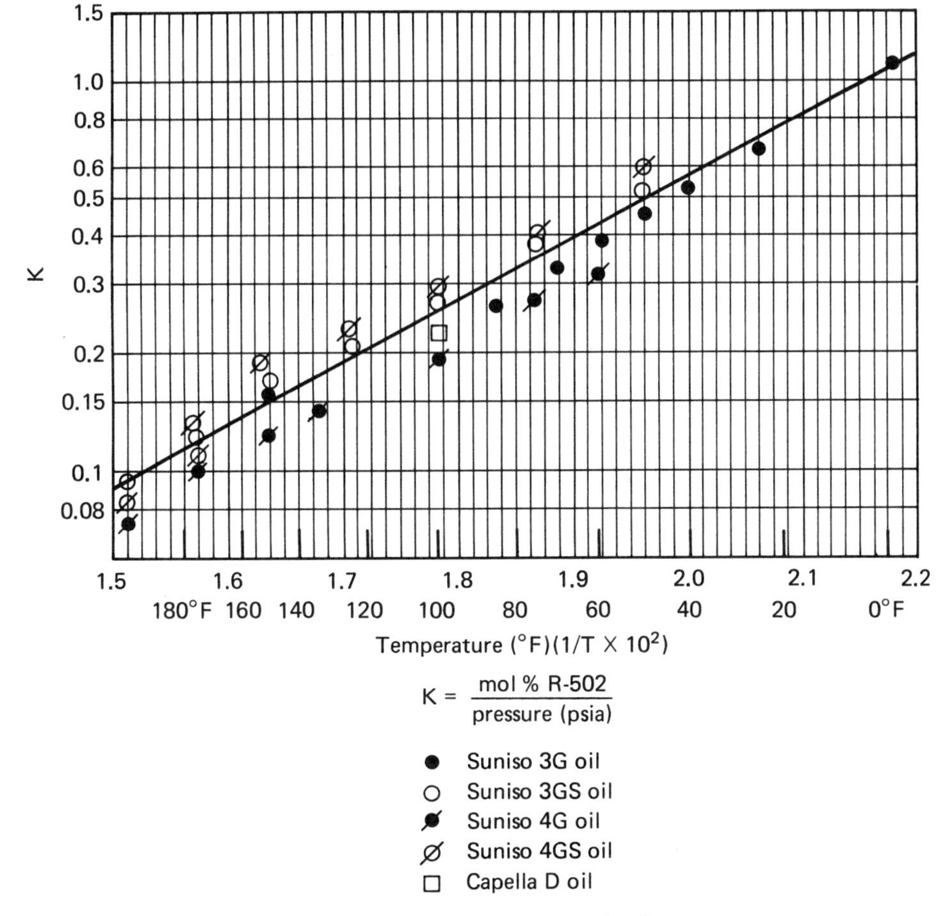

Figure 9–14 Solubility of R-502 in oils.

the oil was condensed at room temperature (about 77°F). The oil solution was maintained at temperatures from 82 to 100°F.

The relative surface area of the oil apparently controls the initial migration rate, but after a short period of time, the volume of the oil becomes more important. These effects are illustrated in Figure 9–21, page 251. The numbered curves represent the following conditions:

Curve	Oil: Area (in.^{2})	Oil: Weight (g)
1	5.7	28.1
2	5.7	9.3
3	2.9	10.4

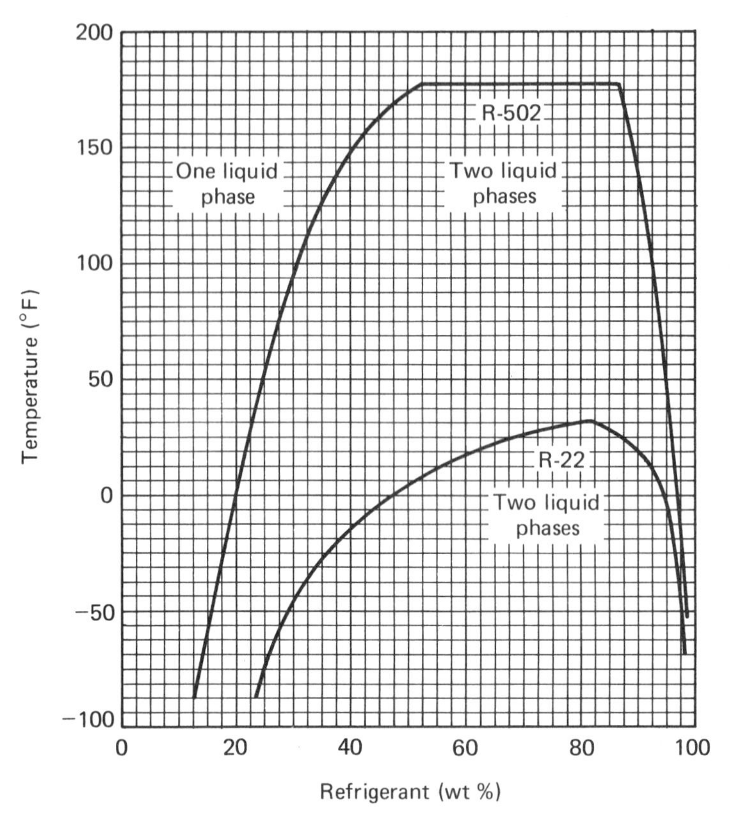

Figure 9–15 Phase diagram for R-502 and R-22 with Suniso 3GS oil.

TABLE 9–24 SOLUBILITY FACTORS FOR R-12, R-22, AND R-502 FROM FIGURES 9–5, 9–10, AND 9–14

Temperature (°F)	K^a		
	R-12	R-22	R-502
−20	6.4	4.0	1.8
0	3.75	2.4	1.2
20	2.4	1.6	0.77
40	1.75	1.0	0.56
60	1.25	0.7	0.41
100	0.64	0.4	0.26
140	0.39	0.26	0.18
160	0.32	0.21	0.15
200	0.23	0.16	0.11

$^aK = \dfrac{\text{mole fraction}}{\text{psia}} \times 10^2.$

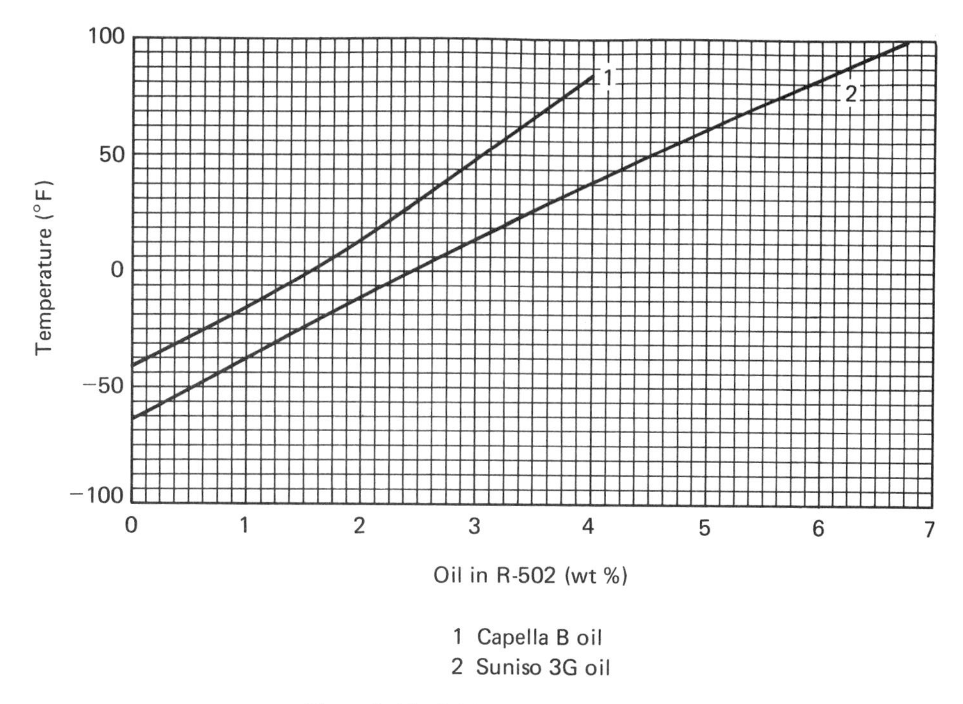

Figure 9–16 Solubility of oil in R-502.

For about the first 100 min, the migration rates, represented by curves 1 and 2 with the same surface area, are about the same. From then on the transfer of refrigerant is much faster for curve 1, where about three times as much oil is present as for curve 2. With the conditions of curves 2 and 3 where the total amount of oil present is about the same, the rate becomes almost the same even though the surface area for curve 3 is about half that of curve 2. The same data are used in Figure 9–22, page 251, showing the change in composition of the oil solution instead of the actual weight of refrigerant added.

VISCOSITY

In refrigerating and air-conditioning equipment the refrigerant and oil are intimately associated throughout the cycle and a knowledge of the properties of solutions of these materials is important for good operation. The chief problem with oil is getting

TABLE 9–25 ESTIMATED SOLUBILITY FACTORS FOR R-500

°F	*K*	mol % oil	wt % 500/psia	°F	*K*	mol % oil	wt % 500/psia
−20	5.8	94.2	1.845	100	0.565	99.435	0.173
0	3.48	96.52	1.090	120	0.44	99.56	0.135
20	2.28	97.72	0.708	140	0.352	99.648	0.108
40	1.52	98.48	0.469	160	0.287	99.713	0.088
60	1.08	98.92	0.333	180	0.243	99.757	0.074
80	0.75	99.25	0.230	200	0.211	99.789	0.065

TABLE 9–26 ESTIMATED SOLUBILITY OF R-500 IN SUNISO 3GS OIL (WT %)

Temperature (°F)	psia										
	10	20	30	50	75	100	120	140	160	180	200
0	10.9	21.8									
20	7.1	14.2									
40	4.7	9.4	14.1								
60	3.3	6.7	10.0	16.7							
80	2.3	4.6	6.9	11.5	17.3						
100	1.7	3.5	5.2	8.7	13.0						
120	1.4	2.7	4.1	6.8	10.1	13.5					
140	1.1	2.2	3.2	5.4	8.1	10.8	13.0				
160	0.88	1.8	2.6	4.4	6.6	8.8	10.6	12.3			
180	0.74	1.5	2.2	3.7	5.6	7.4	8.9	10.4	11.8		
200	0.65	1.3	2.0	3.3	4.9	6.5	7.8	9.1	10.4	11.7	13.0

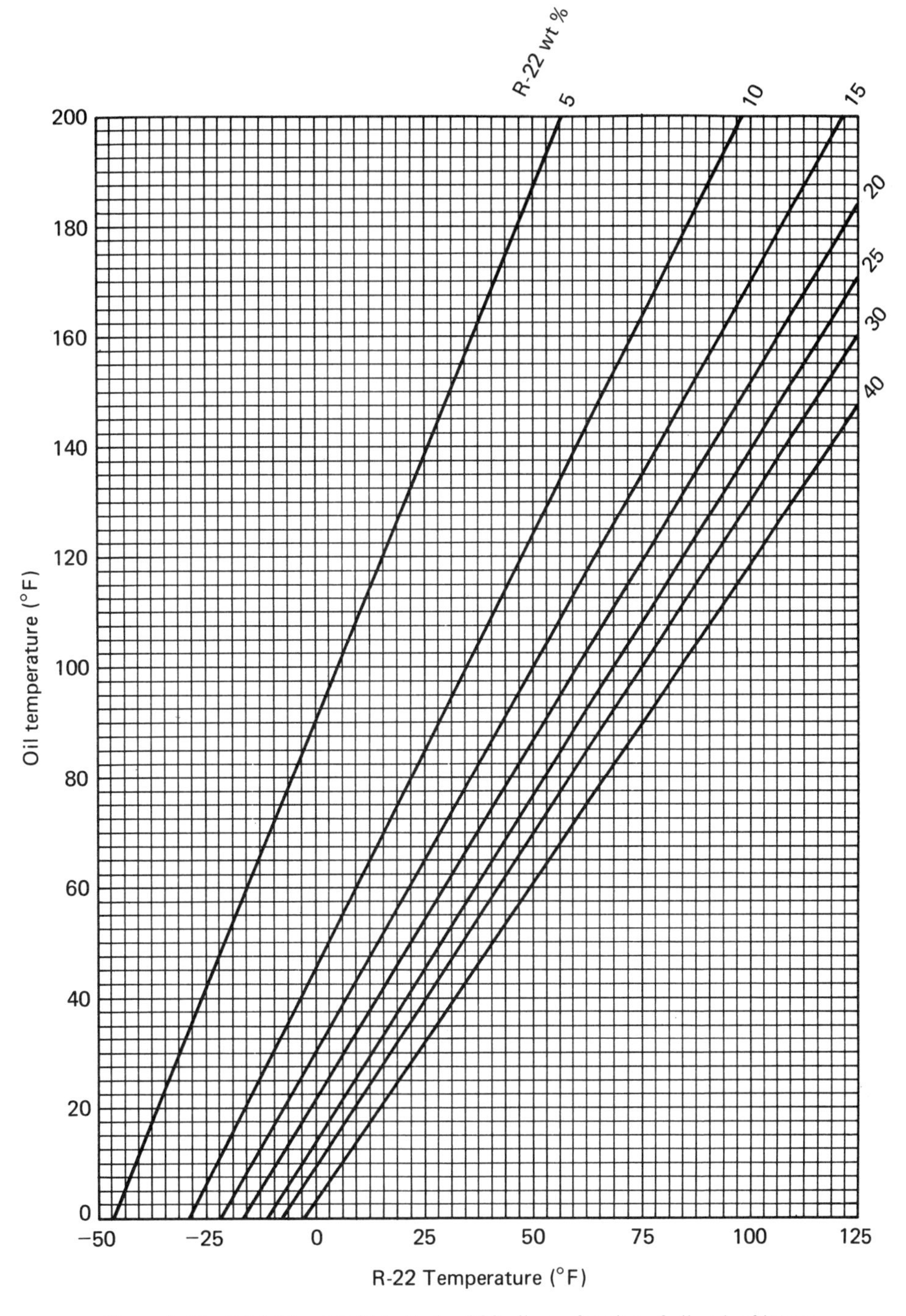

Figure 9–17 Solubility of R-22 in Suniso 3GS oil as a function of oil and refrigerant temperatures.

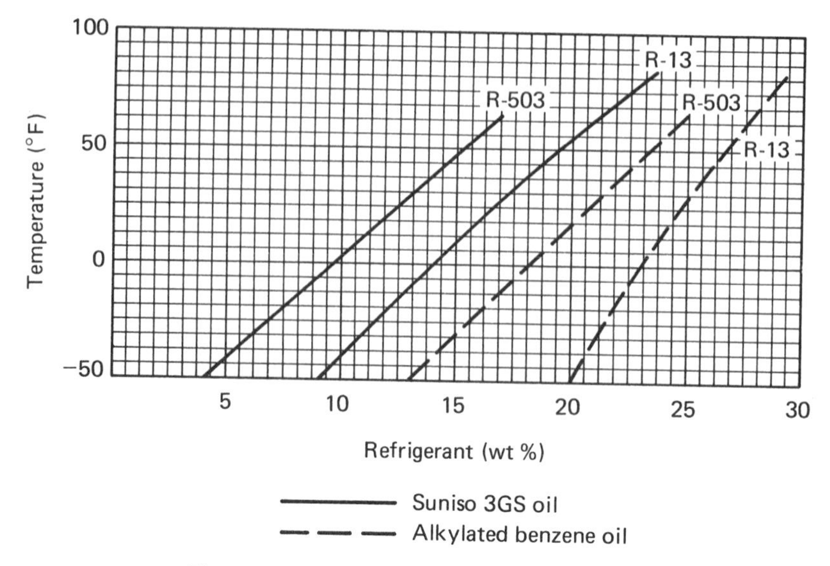

Figure 9–18 Solubility of R-13 and R-503 in oils.

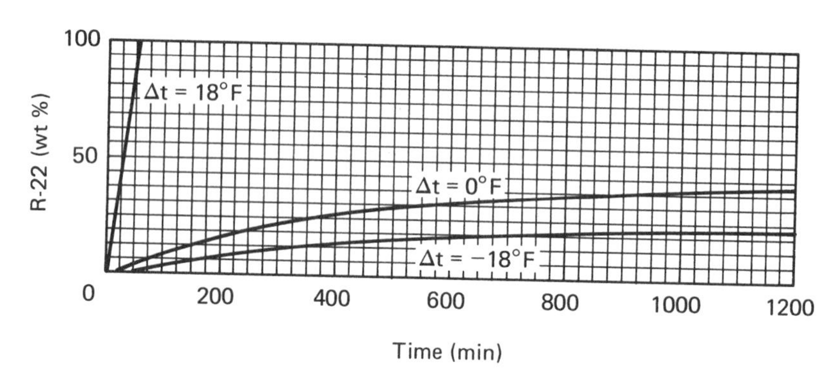

Figure 9–19 Migration rate (percentage) of R-22 into oil.

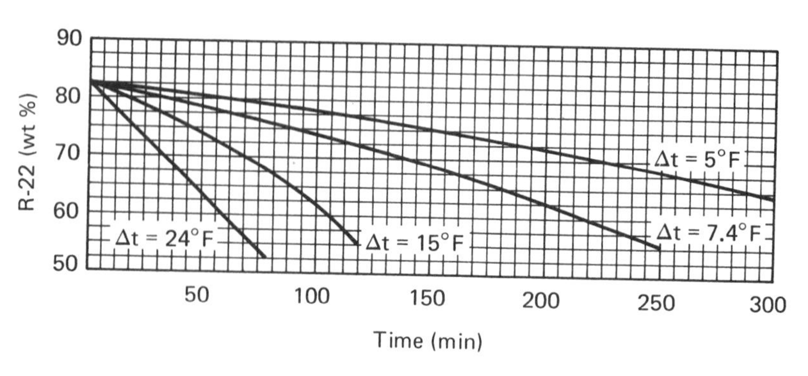

Figure 9–20 Migration rate (percentage) of R-22 from oil.

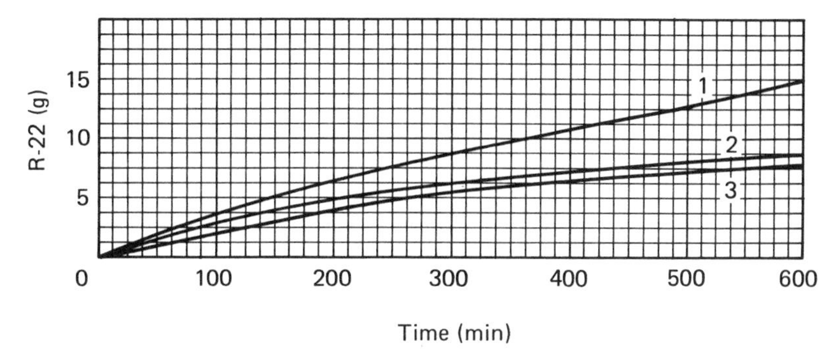

Figure 9–21 Migration rate (grams) of R-22 into oil.

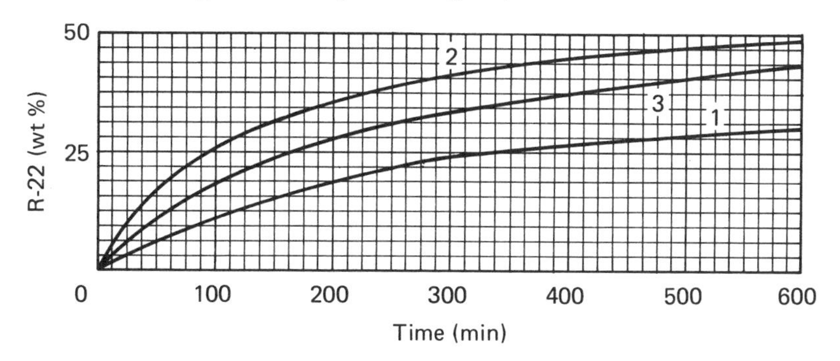

Figure 9–22 Migration rate (grams) of R-22 into oil.

it back to the compressor after it gets out. The viscosity of oil with dissolved refrigerant is the key factor in oil return since the oil is moved by the force of the flowing refrigerant and thin oil moves better than thick or viscous oil. Temperature is a big factor, but the nature of the refrigerant is also very important. For example, oil is a good solvent for R-12 but only moderately good for R-22 and less so for R-502. The latter two refrigerants may have oil-return problems at low temperatures (see Figures 9–23 to 9–26, pages 252–255). The viscosity is a measure of the mobility of the oil–refrigerant solution and can give a good idea of when to expect oil–return problems.

Several measurements of the viscosity of oil–refrigerant solutions have been reported and have aided a great deal in understanding this important property. In 1964, Parmelee published a major contribution, including viscosity data for a number of refrigerants dissolved in a naphthenic-based oil [9–19]. The implications of this work on the return of oil from the evaporator to the compressor were later discussed by Cooper [9–20] and others. The measurements were made over a temperature range at constant pressure to simulate conditions that might be found in the low-pressure area of the refrigerant cycle.

One of the most important discoveries by Parmelee was that the viscosity of the oil solution reaches a maximum at a temperature somewhat higher than the temperature of the evaporating refrigerant. The coldest part of a refrigeration system is in the evaporator, but oil is less viscous here than at warmer temperatures in the

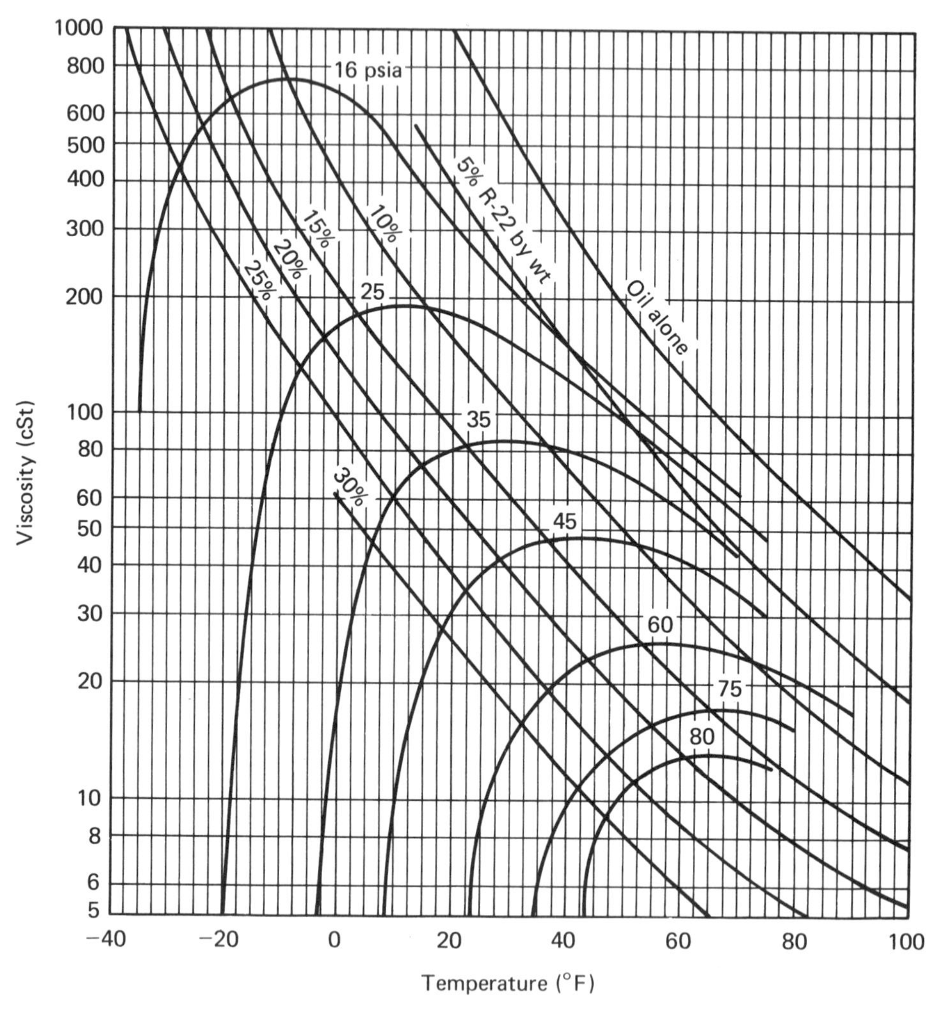

Figure 9–23 Viscosity and solubility relationships of R-22/Suniso 3GS oil solutions (centistoke).

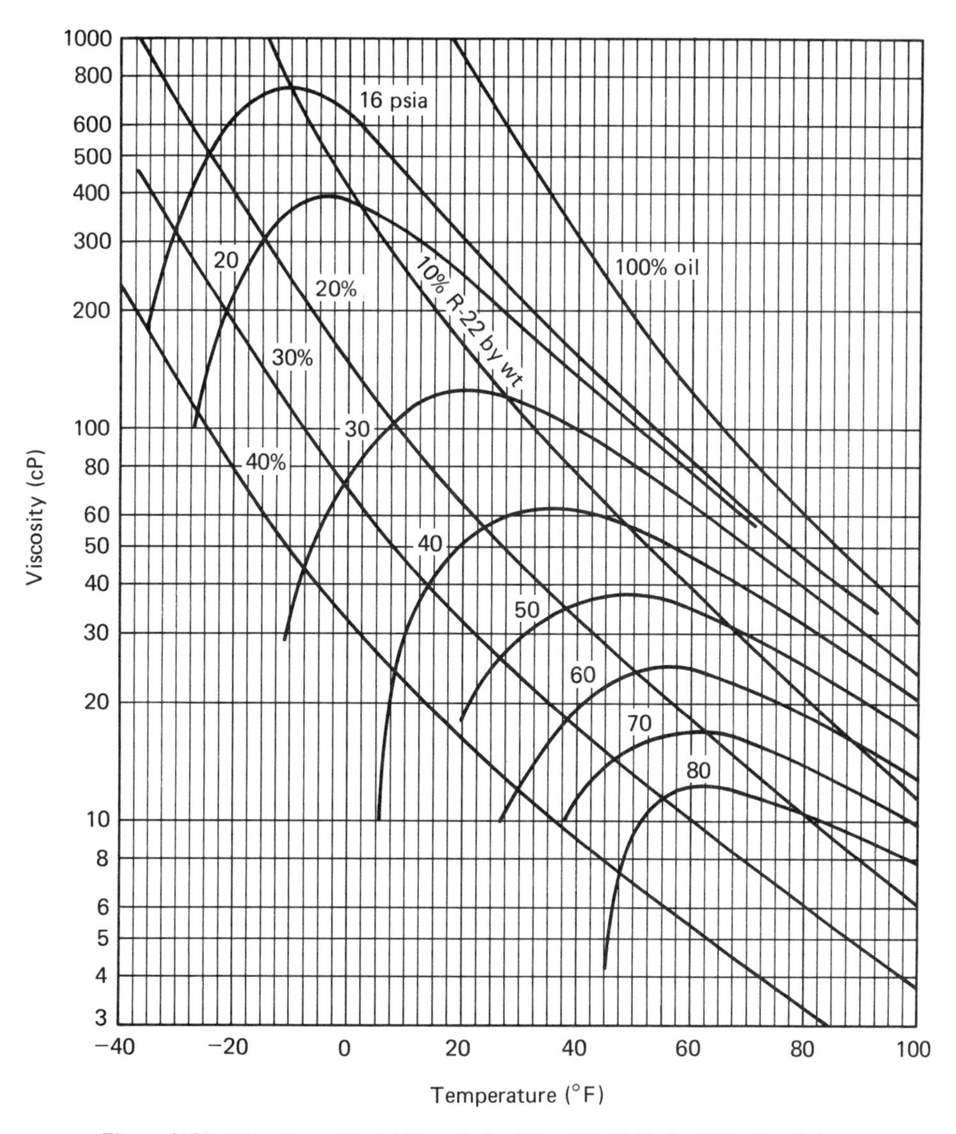

Figure 9–24 Viscosity and solubility relationships of R-22/Suniso 3GS oil solutions (centipoise).

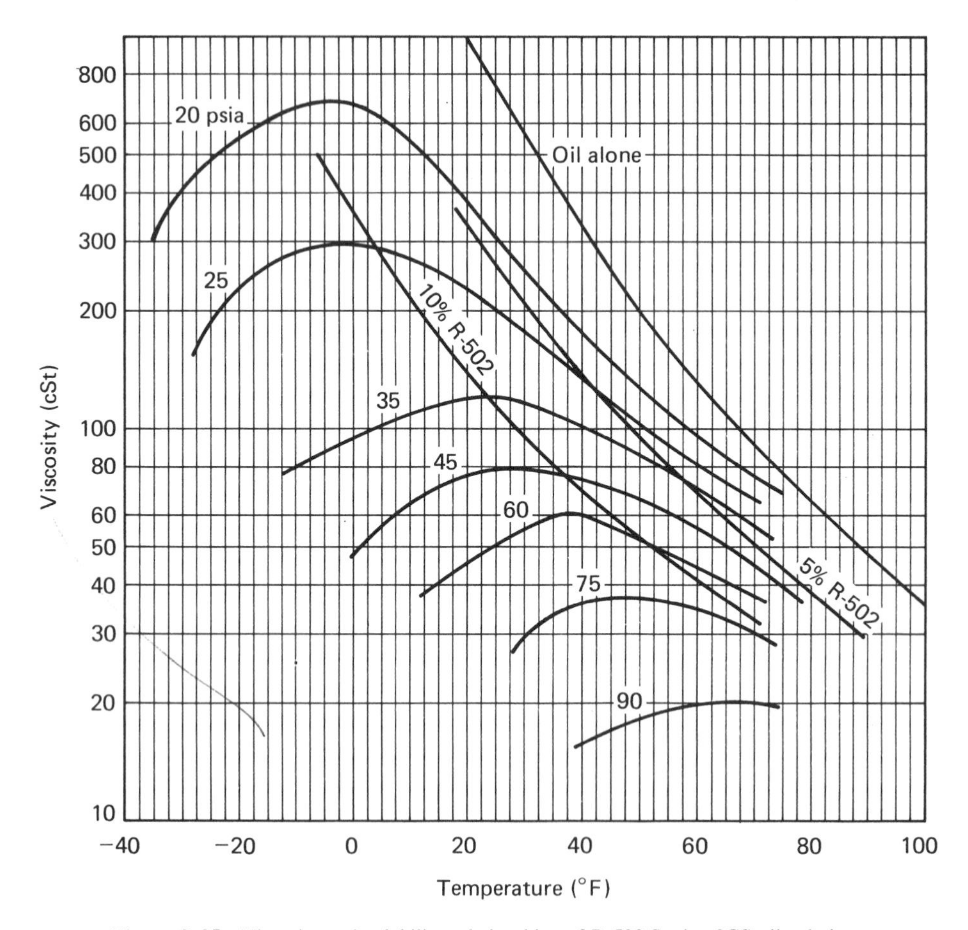

Figure 9–25 Viscosity and solubility relationships of R-502/Suniso 3GS oil solutions (centistoke).

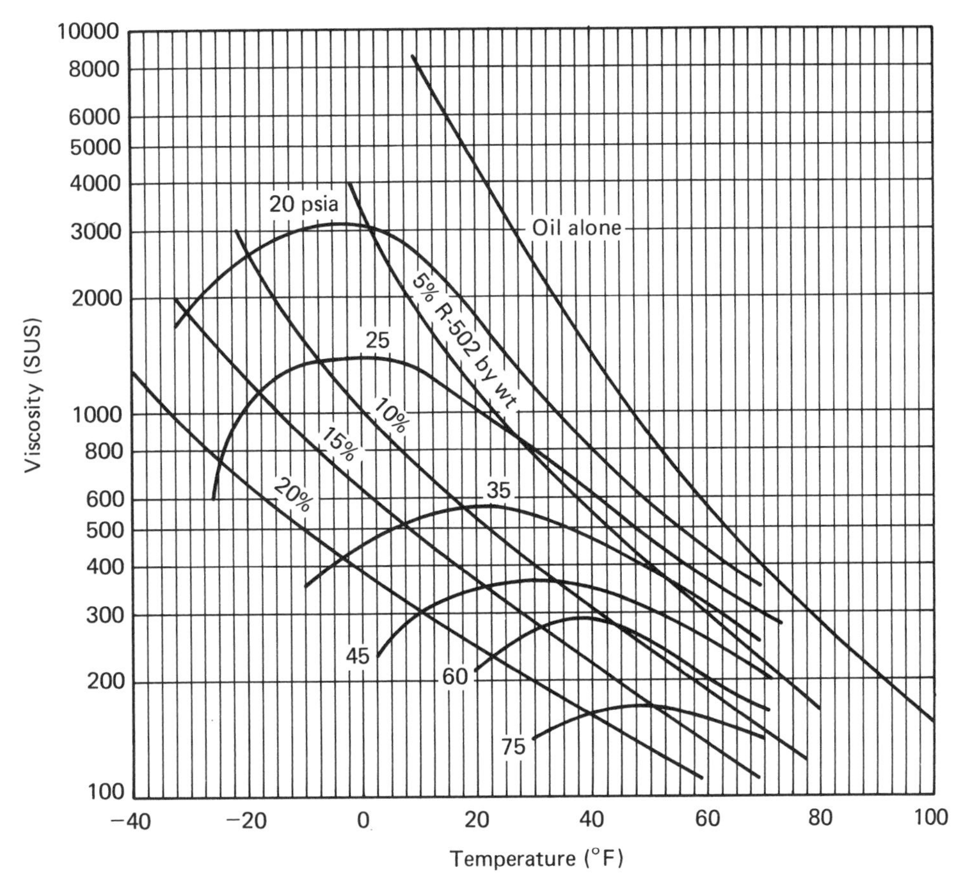

Figure 9–26 Viscosity and solubility relationships of R-502/Suniso 3GS oil solutions (SUS).

TABLE 9–27 CONDITIONS FOR MAXIMUM VISCOSITY OF SOLUTIONS OF REFRIGERANTS IN SUNISO 3G OIL

Refrigerant evaporating temperature (°F)	Pressure (psia)	Oil temperature at same pressure (°F)	Viscosity of oil solution (cSt)	Concentration of refrigerant (wt %)	Superheat (°F)
		R-22			
−40	15.2	−17	730	13	23
−35	17.3	−9	515	13.5	26
−30	19.6	−2	360	12.5	28
−25	22.1	4	258	12.5	29
−20	24.8	10	188	13	30
−15	27.9	16	140	13.5	31
−10	31.2	23	105	13.5	33
−5	34.8	29	81	14	34
0	38.7	34	63	15	34
5	42.9	40	49.5	15	35
10	47.5	45	40	15	35
15	52.4	50	32.5	16	35
20	57.7	55	26.5	17	35
25	63.5	60	22	17.5	35
30	69.6	65	18.5	18	35
		R-502			
−40	18.8	−12	640	8	28
−35	21.2	−5	445	8	30
−30	23.9	1	325	8	31
−25	26.8	7	250	8	32
−20	30.0	13	195	8.5	33
−15	33.5	18.5	155	9	33.5
−10	37.3	24	126	9	34
−5	41.3	29	105	9.5	34
0	45.8	33	87	10	33
5	50.6	38	74	10	33
10	55.7	42	63	11.5	32
15	61.2	47	54	12	32
20	67.2	51	47	13	31
25	73.5	55	41	13.5	30
30	80.3	59	36.5	14	29
		R-12			
−30	12.0				
−25	13.6				
−20	15.3	23	115	11	43
−15	17.1	31	87	11.5	46
−10	19.2	38	67	11	48
−5	21.4	44	53	12	49
0	23.8	50	42.5	12.5	50
5	26.5	56	35	13	51
10	29.3	61	29	14	51
15	32.4	67	24	14.5	52
20	35.7	72	20.5	15	52
25	39.3	77	17.5	15	52
30	43.1	81	15	17	51

Source: Ref. 9–23.

return line. This unexpected finding is easily explained. Oil is dissolved in the liquid refrigerant as it enters the evaporator. When the refrigerant evaporates, a liquid oil phase remains, containing an amount of dissolved refrigerant consistent with the solubility relationships of the two materials at the temperature and pressure in the evaporator. The pressure is assumed to be the vapor pressure of the pure refrigerant at the evaporating temperature. At this point, the residual oil is fairly fluid because of the dissolved refrigerant. When the oil leaves the evaporator area, the temperature increases and the viscosity is affected in two ways:

1. The viscosity increases because refrigerant is being driven out of solution.
2. The viscosity decreases due to the effect of temperature on the oil.

As the temperature of the oil rises in the suction line, the first effect predominates. The viscosity of the oil increases faster due to escaping refrigerant than it decreases due to the effect of temperature. As the temperature continues to rise, the temperature effect becomes the major factor and the viscosity of the oil decreases. The location of the point of maximum oil viscosity in the suction line and corresponding conditions of temperature and pressure are of some interest in the study of refrigeration cycles. These points are shown for various evaporating temperatures in Table 9–27. For example, with R-22 evaporating at −40°F with a pressure of about 15 psia, the maximum viscosity of oil would be at −17°F and the oil would contain 13 wt % R-22. At 0°F evaporation, the maximum point would be at 34°F. In these studies, a 150 SUS naphthenic-base oil was used (Suniso 3G).

A few tests were made with R-13 and Suniso 3GS oil, but viscosities were too high to be of much value. For example, at −40°F and a pressure of 90 psia the viscosity was 6180 cSt or 28,400 SUS. At −50°F and a pressure of 70 psia, the viscosity was 20,300 cSt or 93,400 SUS. With R-13B1, viscosities were also quite high but not as bad as R-13, as shown in Table 9–28.

Parmelee's work was reported in centistoke and converted to SUS with the following approximate relationship:

$$\text{stoke} = (0.0022)(\text{SUS}) - \frac{1.8}{\text{SUS}}$$

TABLE 9–28 MAXIMUM VISCOSITIES OF R-13B1 AND NAPHTHENIC OIL SOLUTIONS

Pressure (psia)	Temperature (°F)	Viscosity	
		cSt	SUS
16	−35	8,700	40,000
25	−25	1,780	8,190
35	−10	551	2,540
60	18	110	—
65	75	70	323
75	40	48	222

TABLE 9–29 VISCOSITY OF R-22/SUNISO 3G OIL

Temperature (°F)	Pressure (psia)	Centistoke	SUS	Pressure (psia)	Centistoke	SUS	Pressure (psia	Centistoke	SUS
−30	16	155	718						
−25	16	410	1900						
−20	16	650	3013						
−15	16	750	3476	25	50	234			
−10	16	750	3476	25	100	464			
−5	16	700	3245	25	140	649			
0	16	680	3152	25	165	765	35	15	78
5	16	600	2781	25	180	834	35	40	186
10	16	460	2132	25	200	927	35	58	270
15	16	380	1761	25	185	857	35	73	338
20	16	320	1483	25	175	811	35	81	374
25	16	275	1275	25	170	788	35	82	380
30	16	220	1020	25	155	718	35	82.5	382
35	16	180	834	25	140	649			
40	16	150	695	25	125	579	35	80	370
45	16	120	556				35	75	348
50				25	95	440	35	70	324
55							35	65	302
60	16	85	394	25	78	362	35	55	256
70	16	65	301	25	55	256	35	44	204
10	45	10	60						
15	45	20	98						
20	45	30	140						
25	45	37.5	174	60	9	56			
30	45	40	186	60	13.5	72			
35	45	46	214	60	17.5	88	75	5.8	44
40	45	48	223	60	20	98	75	9.2	56
45	45	47	219	60	23.5	112	75	12.3	68
50	45	46	214	60	25	118	75	13.5	72
55	45	43	199	60	25.5	120	75	15.5	80
60	45	40	186	60	25	118	75	16.5	83
65							75	17.2	87
70	45	35	162	60	23.5	112	75	17.6	88
45	85	8	52						
50	85	10	60						
55	85	12	66						
60	85	12.5	68						
65	85	13	70						
70	85	13.5	72						

Source: Ref. 9–24.

TABLE 9–30 VISCOSITY OF R-12/SUNISO 3G OIL SOLUTIONS

Pressure (psia)	Temp. (°F)	Oil alone SUS	Solution viscosity		Pressure (psia)	Temp. (°F)	Solution viscosity	
			cSt	SUS			cSt	SUS
16	−10		9.5	58	35	25	7.5	51
	−5		25	118		30	10.5	62
	0		48	223		35	12.5	68
	5		65	302		40	14.5	76
	10	7000	80	370		45	17.2	87
	15	6300	95	440		50	19.5	98
	20	4600	102	472		60	22.5	108
	25	3500	108	500		70	21.5	103
	30	2600	105	486				
	35	1900	100	462				
	40	1550	95	440				
	50	950	80	370				
	55	770						
	60	620	63	292				
	70	430	49	228				
	85	225						

Source: Ref. 9–24.

The viscosity has also been converted to centipoise:

$$\text{centipoise} = (\text{centistoke})(\text{density})$$

The density of the oil–refrigerant solutions was calculated from the densities of the pure components as explained earlier. Data for R-22 and Suniso 3GS oil obtained from Parmelee's published charts are given in Table 9–29, in Table 9–30 for R-12, and in Table 9–31 for R-502.

Some measurements by Parmelee at higher temperatures are shown in Table 9–32, page 261. The viscosity of solutions of R-115 and oil is significantly higher than that of the other refrigerants. Since R-502 is about half R-115 and half R-22, the viscosity with R-502 is in between that of R-22 and R-115.

In Table 9–33, page 262, are listed viscosities of R-22 and R-502 in special oils—special in the sense that they are not as yet widely used. Solutions of R-22 with a typical paraffinic base oil and with the GE highly refined oil are shown. Viscosities are about the same with both oils. For example, at a pressure of 16 psia and a temperature of 5°F, the viscosity with the paraffinic oil is about 1000 cSt and with the GE oil about 800 cSt. Both oils are used for air-conditioning applications and are not recommended for use at low temperatures. At the foregoing conditions the viscosity with a typical naphthenic base oil is about 600 cSt. At lower temperatures the difference in viscosity is greater—and more important. At −10°F the viscosity of a solution of R-22 in the GE oil is about 1650 cSt and about 750 cSt in a naphthenic oil.

TABLE 9–31 VISCOSITY OF R-502/SUNISO 3G OIL SOLUTIONS

Temp. (°F)	Centi-stoke	SUS	Centi-stoke	SUS	Centi-stoke	SUS	Centi-stoke	SUS
	At 20 psia							
−35	300	1390						
−30	410	1900	At 25 psia					
−25	480	2225	160	742				
−20	550	2549	220	1020				
−15	600	2781	270	1251	At 35 psia			
−10	650	3013	285	1321	80	370		
−5	680	3152	290	1344	90	417	At 45 psia	
0	670	3105	295	1367	95	440	47	219
5	620	2874	285	1321	100	462	57	265
10	605	2804	275	1275	108	500	65	302
15	450	2086	250	1159	115	532	70	325
20	380	1761	220	1020	118	546	73	338
25	305	1414	200	927	120	556	76	352
30	270	1251	175	811	115	532	78	360
35	205	950	155	718	108	500	76	352
40	175	811	135	625	100	462	72	333
50	125	579	100	462	85	392	65	302
60	95	440	80	370	70	324	55	255
70	75	348	65	301	55	253	44	204
	At 60 psia							
15	40	185						
20	46	214						
25	50	233	At 75 psia					
30	55	255	29	135				
35	60	278	33	154	At 90 psia			
40	65	302	37	172	15.5	80		
45			37.5	172	17	86		
50	50	233	37	172	18	90		
55					18.5	92		
60	43	200	34	158	19.5	96		
70	37	172	30	140	20	98		

Source: Ref. 9–24.

TABLE 9–32 VISCOSITY OF SOLUTIONS OF REFRIGERANTS IN SUNISO 3G OIL AT HIGHER TEMPERATURES

Temperature (°F)	Vapor pressure[a] (psia)	R-12 viscosity		R-22 viscosity		R-502 viscosity		R-115 viscosity	
		cSt[b]	SUS[c]	cSt	SUS	cSt	SUS	cSt	SUS
85	0	53.8	249.4	53.8	249.4	53.8	249.5	53.8	249.5
	25	21.6	104.5	36.1	168.5	43.1	200.1	48.0	222.7
	40	16.0	81.2	27.6	130.5	35.8	171.6	43.7	202.9
	65	6.5	47.1	18.1	98.4	23.8	135.9	37.6	175.2
	90	2.6	34.7	12.0	66.0	22.0	106.3	31.7	148.6
150	0	11.8	65.5	11.8	65.5	11.8	65.5	11.8	65.5
	25	9.0	55.7	9.8	58.4	10.4	60.5	11.1	63.0
	40	7.9	51.9	9.0	55.7	10.0	59.1	10.7	61.6
	65	6.6	47.6	7.8	51.6	9.0	55.7	10.0	59.1
	90	5.2	43.1	7.0	48.9	8.3	53.3	9.6	57.9
200	0	5.6	44.6	5.6	44.6	5.6	44.6		
	25	5.0	42.6	5.2	43.3	5.5	44.3		
	40	4.7	41.7	5.0	42.6	5.2	43.3		
	65	4.3	40.3	4.6	41.3	4.9	42.3		
	90	4.0	39.4	4.4	40.6	4.5	41.0		

[a]Vapor pressure of the refrigerant at a temperature adjusted to produce the desired pressure.
[b]Centistoke measured by the rolling ball method.
[c]Saybolt universal seconds calculated from centistoke as in ASTM Special Technical Bulletin No. 43B.
Source: Ref. 9–14.

In recent years an alkylated benzene oil has been used very successfully with R-22, R-502, R-13, and R-503 for low-temperature applications [9–21, 9–22]. Its lubricating properties are very similar to other oils but it is a better solvent for the fluorinated refrigerants and the solutions are lower in viscosity (Figures 9–27 to 9–31, pages 263–267). Some properties are compared with a mineral oil below.

	Alkyl benzene	Mineral oil
Viscosity (SUS)		
100°F	150	155
210°F	41.1	40.8
Molecular weight	320	325
Pour point (°F)	−50	−45
Floc point (°F)	<−115	−68
Aniline point (°F)	125	160
Specific gravity (60/60)	0.872	0.917
Dielectric strength (kV)	>34	>34

TABLE 9–33 VISCOSITY OF REFRIGERANT/OIL SOLUTIONS WITH SPECIAL OILS

R-22 at 16 psia			R-502 with an alkylated benzene oil at 20 psia		
Temperature (°F)	Viscosity cSt	Viscosity SUS	Temperature (°F)	Viscosity cSt	Viscosity SUS
150 SUS paraffinic base oil			−30	33	153
5	1000	4635	−25	90	317
10	350	1622	−20	160	742
15	290	1344	−15	230	1066
20	240	1112	−10	300	1390
30	170	788	−5	350	1622
40	125	579	0	400	1854
50	95	440	5	380	1761
GE oil			10	360	1669
−10	1650	7648	15	330	1530
−5	1300	6026	20	290	1344
0	1000	4635	25	250	1159
5	800	3708	30	220	1020
10	650	3013	35	185	857
15	520	2410	40	160	742
20	430	1993	50	112	519
30	290	1344			
40	210	973			
50	150	695			

Source: Ref. 9–24.

The alkylated benzene oil is especially useful with refrigerants where other oils may be difficult to return to the compressor. For example, R-502 dissolved in alkylated benzene at −10°F and a pressure of 20 psia forms a solution with a viscosity of about 300 cSt, about half that with a naphthenic oil such as Suniso 3G.

The viscosity of R-22/Suniso 3GS oil solutions (Table 9–34, page 268) was made available as a chart of temperature versus viscosity in centipoise at curves of constant composition [9–23]. From this chart the data in Table 9–34 were obtained and converted to other units as outlined earlier.

Measurements of solubility and viscosity for solutions of oils and the fluorinated refrigerants at different places and sometimes by different methods do not always agree. To arrive at some consistency, the available data were plotted and replotted using parameters of pressure, temperature, concentration, viscosity, and so on. Errant points were ignored. Examples of the results of this smoothing can be found in Table 9–35, page 269, and in the charts that follow. They may appear to be more accurate than they really are but should be useful unless extremely precise data are essential.

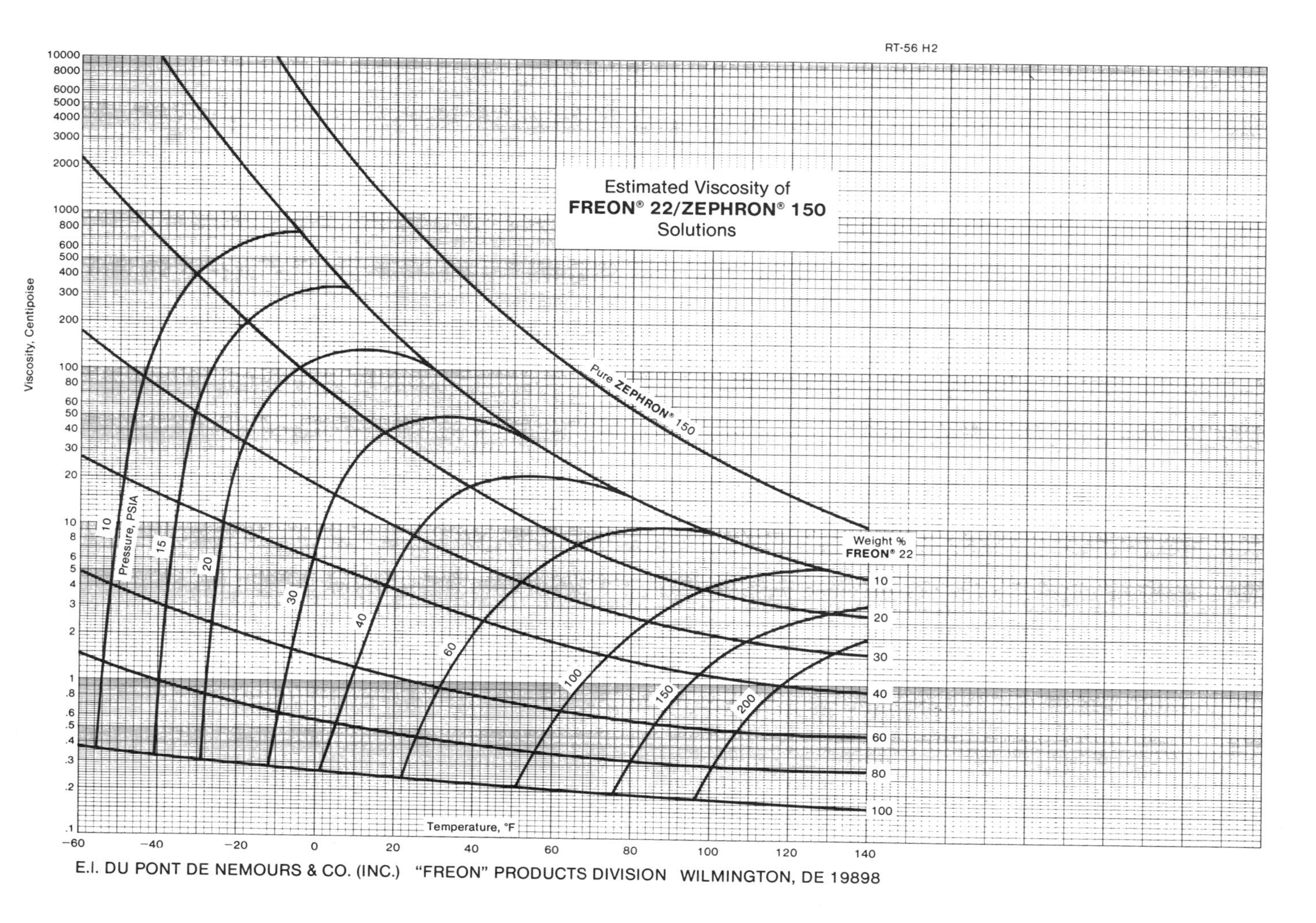

Figure 9–27 Viscosity and solubility relationships of R-22/alkyl benzene oil solutions (centipoise). (Reprinted by permission of the DuPont Company.)

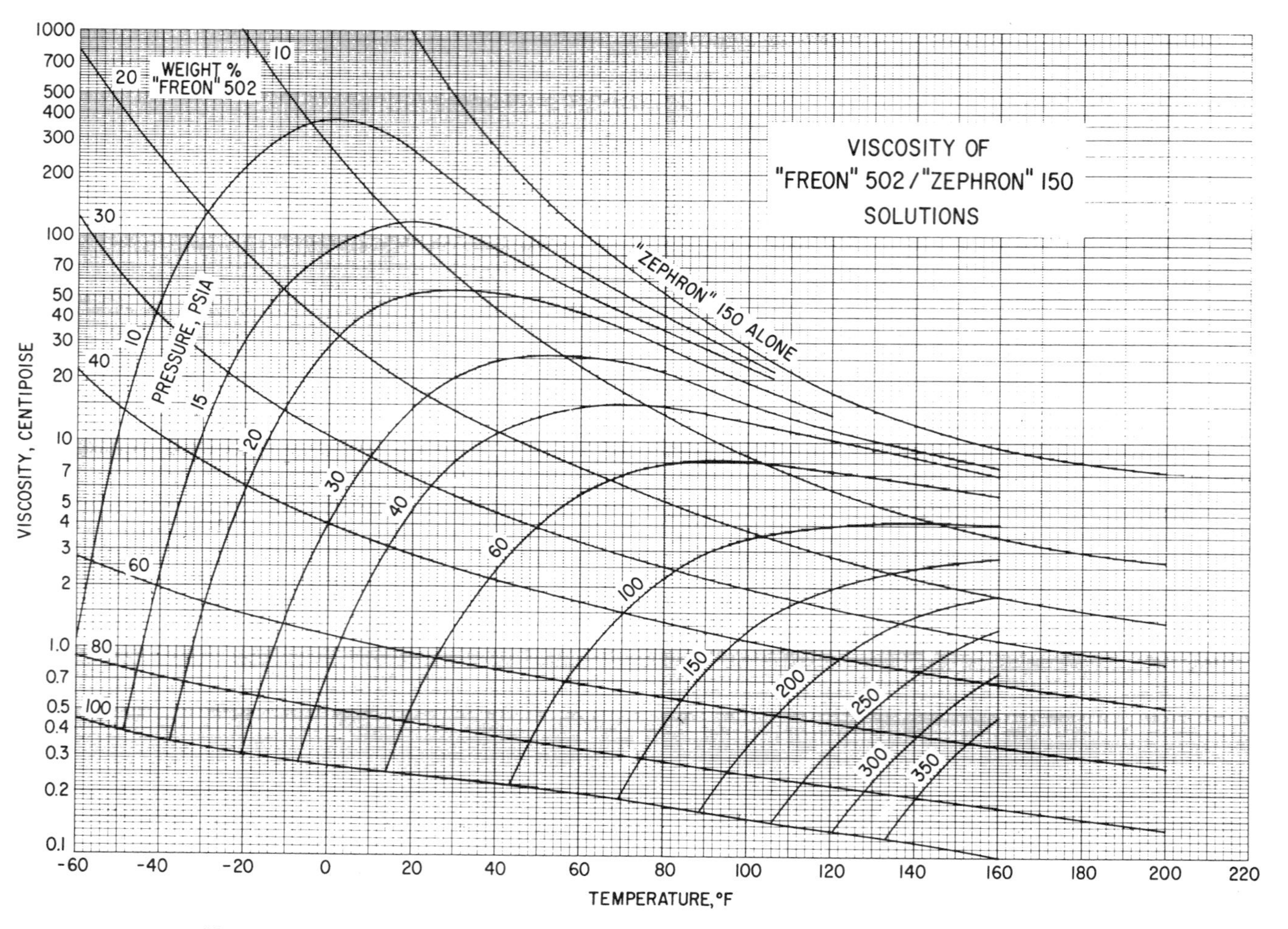

Figure 9–28 Viscosity and solubility relationships of R-22/alkyl benzene oil solutions (SUS). (Reprinted by permission of the DuPont Company.)

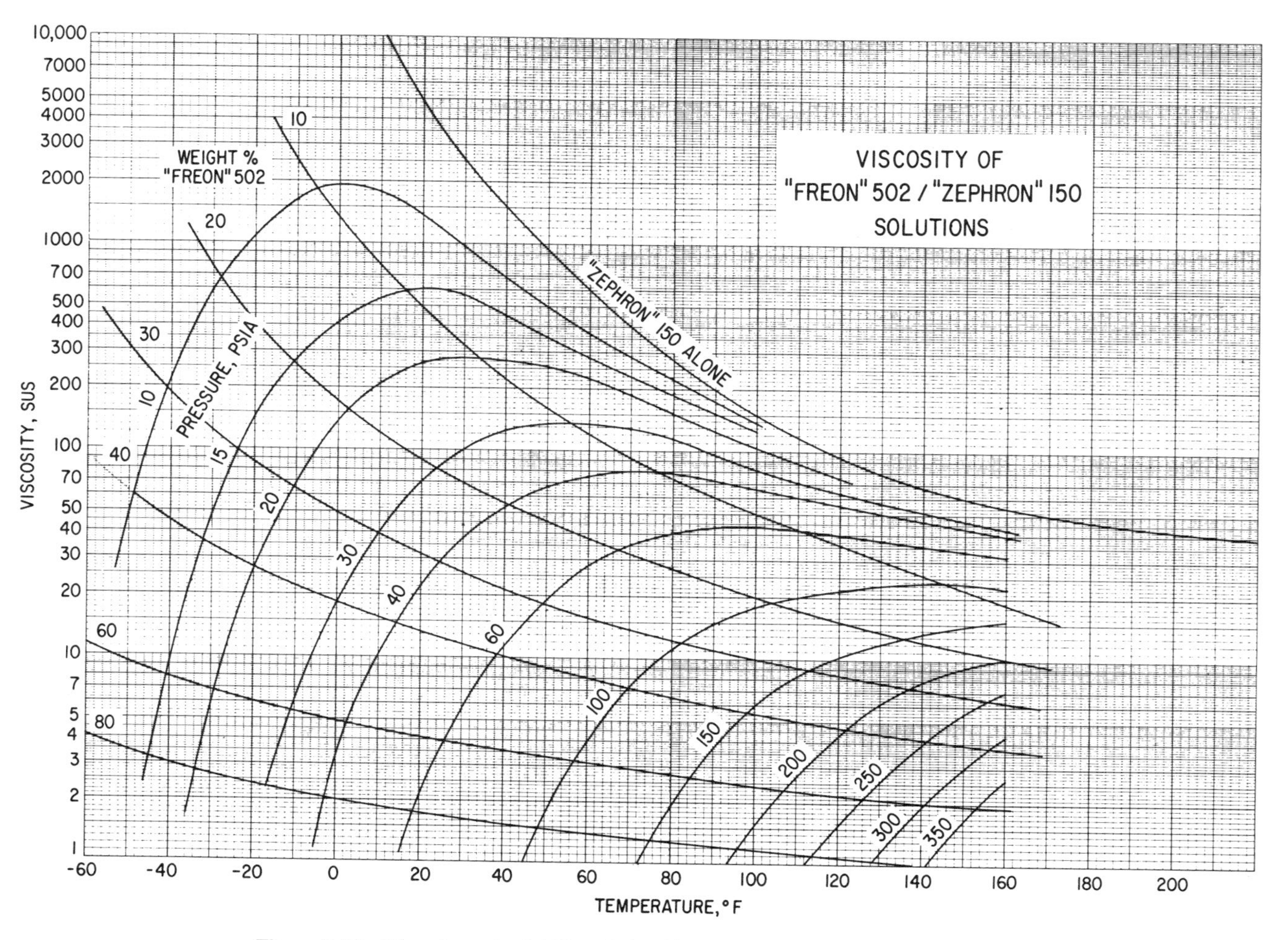

Figure 9–29 Viscosity and solubility relationships of R-502/alkyl benzene oil solutions (centipoise). (Reprinted by permission of the DuPont Company.)

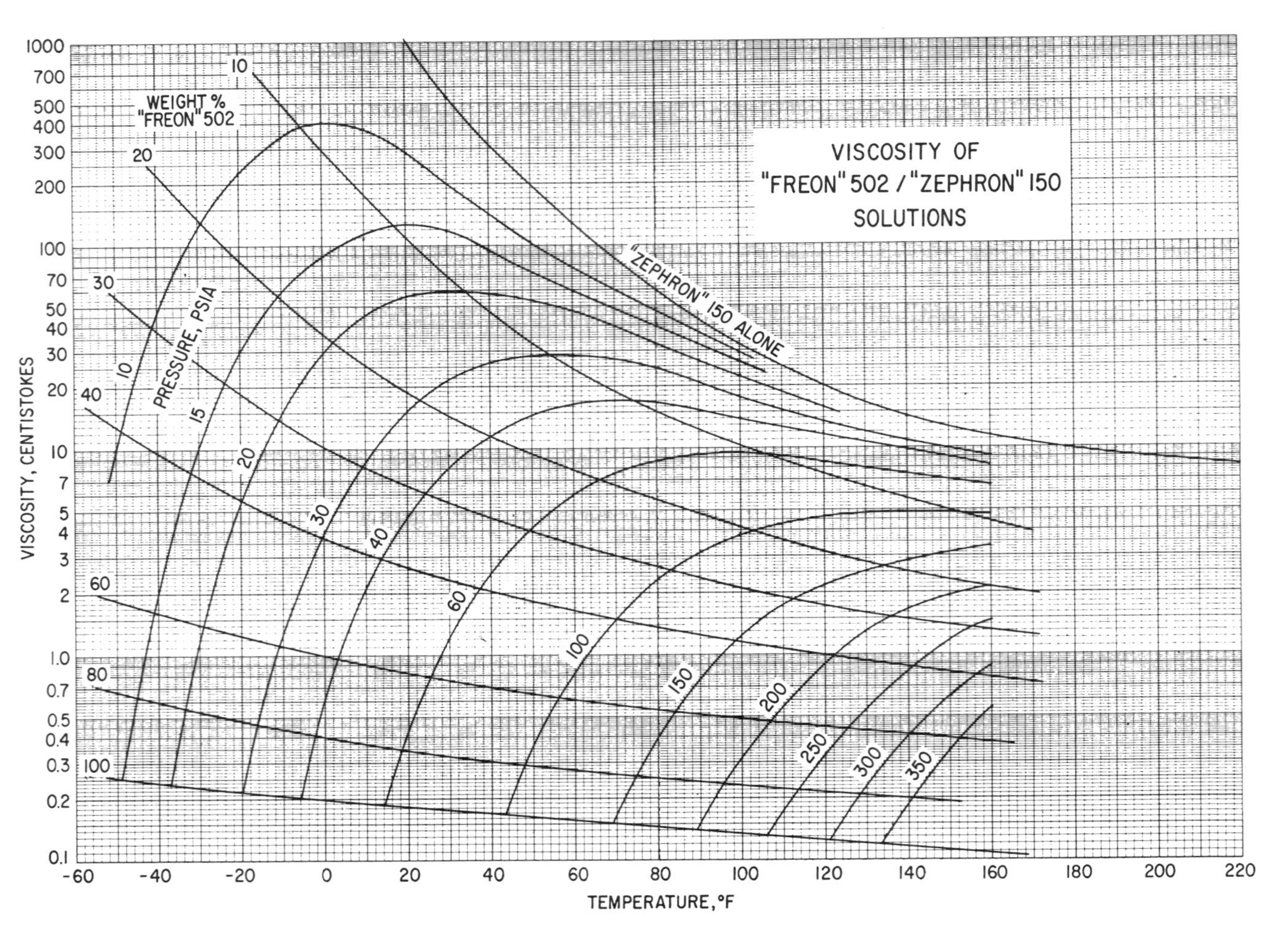

Figure 9–30 Viscosity and solubility relationships of R-502/alkyl benzene oil solutions (centistoke). (Reprinted by permission of the DuPont Company.)

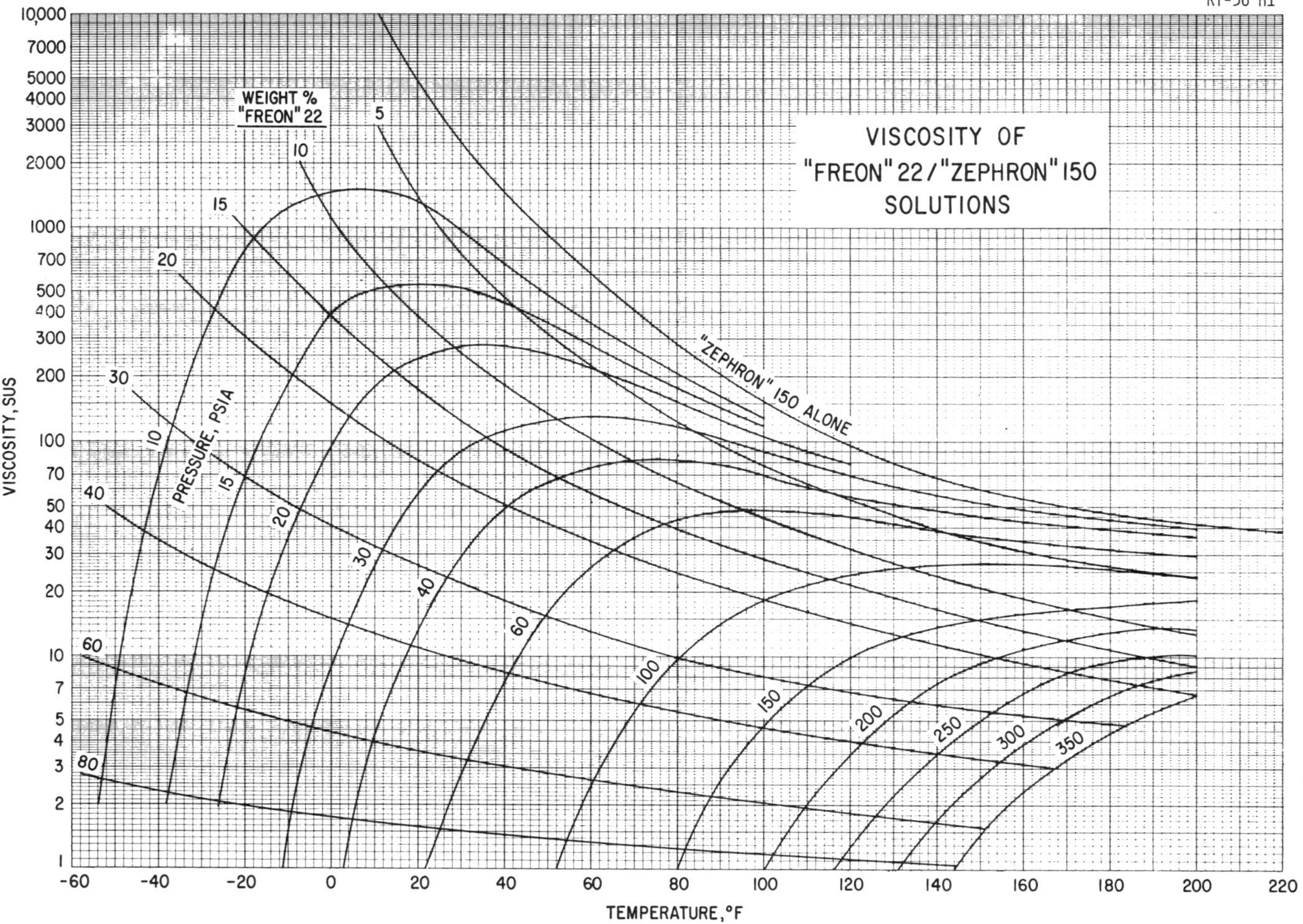

Figure 9–31 Viscosity and solubility relationships of R-502/alkyl benzene oil solutions (SUS). (Reprinted by permission of the DuPont Company.)

TABLE 9–34 VISCOSITY OF R-22/SUNISO 3GS OIL SOLUTIONS

Temp. (°F)	R-22 (wt %)	Density (g/cm³)	Viscosity cP	Viscosity cSt	Viscosity SUS	R-22 (wt %)	Density (g/cm³)	Viscosity cP	Viscosity cSt	Viscosity SUS
60	5	0.9326	60	64.3	290	10	0.9457	35	37.0	163
70		0.9290	43	46.3	224		0.9424	26	27.6	133
80		0.9250	32	34.6	162		0.9388	19.7	21.0	113
90		0.9221	22.2	24.1	116		0.9353	15.2	16.3	89
100		0.9184	18.4	20.0	99		0.9318	12.2	13.1	71
110		0.9154	14.6	15.9	82		0.9280	9.9	10.7	62
120		0.9123	11.8	12.9	70		0.9241	8.2	8.9	55
130		0.9090	9.6	10.6	61		0.9204	6.8	7.4	50
140		0.9057	8.0	8.8	55		0.9165	5.8	6.3	46
60	15	0.9581	18.8	19.6	98	20	0.9748	11.8	12.1	67
70		0.9552	14.6	15.3	80		0.9703	9.5	9.8	58
80		0.9520	11.5	12.1	67		0.9655	7.7	8.0	52
90		0.9478	9.3	9.8	59		0.9611	6.3	6.6	47
100		0.9436	7.6	8.1	52		0.9566	5.3	5.5	44
110		0.9397	6.4	6.8	48		0.9527	4.5	4.7	41
120		0.9357	5.4	5.7	44		0.9485	3.9	4.1	39
130		0.9317	4.6	4.9	42		0.9438	3.4	3.6	38
140		0.9275	4.0	4.3	40		0.9387	3.0	3.2	37
60	25	0.9889	7.4	7.5	50	30	1.0045	4.9	4.9	42
70		0.9846	6.1	6.2	46		0.9988	4.1	4.1	40
80		0.9798	5.1	5.2	43		0.9930	3.5	3.6	38
90		0.9744	4.3	4.4	40		0.9876	3.0	3.1	36
100		0.9689	3.6	3.8	38		0.9821	2.7	2.7	35
110		0.9647	3.2	3.3	37		0.9773	2.4	2.4	35
120		0.9601	2.8	2.9	36		0.9719	2.1	2.2	34
130		0.9553	2.5	2.6	35		0.9668	1.9	2.0	33
140		0.9500	2.2	2.3	34		0.9612	1.7	1.8	33

Source: Ref. 9–26.

TABLE 9–35 REVISED AND SMOOTHED VISCOSITY–SOLUBILITY DATA FOR R-22/SUNISO 3GS OIL SOLUTIONS

Temp. (°F)	R-22 (wt %)	Viscosity		
		SUS	cSt	cP
		At 16 psia		
−30	27	1960	423	439
−20	15	2960	641	637
−10	10.5	3480	754	735
0	8.1	3140	680	655
5	7.4	2740	593	569
10	6.8	2180	471	450
20	5.8	1460	315	299
30	5.2	1020	218	206
40	4.6	700	151	142
50	4.2	530	118	110
60	3.9	400	86	80
70	3.6	300	65	60
		At 25 psia		
−16	42	200	43.1	46.9
−10	31.5	460	101	105
−6	25	600	132	134
−2	21	720	158	159
0	19	760	167	167
2	17.5	800	173	172
5	15.5	850	183	180
10	13	920	202	197
14	11	890	196	189
18	9.6	850	183	176
23	8.3	800	173	165
30	7.1	710	156	148
40	5.9	570	125	118
50	5.1	450	96.8	90.7
60	4.5	350	75.3	70.1
70	4.1	300	64.5	59.8
		At 35 psia		
−2	43	40	4.2	4.5
0	39	100	20.4	21.7
5	30	200	42.8	44.1
10	24	270	58.7	59.2
		At 35 psia		
15	19.5	340	74.3	73.7
20	16	370	80.9	79.1
25	13.5	380	83.1	80.5
30	12	390	85.3	82.1
35	10.7	380	83.1	79.6
40	9.6	370	80.9	77
45	8.7	350	75.3	71.4
50	8.0	320	69.8	65.9
60	6.8	260	55.9	52.5
70	5.9	210	45.3	42.2
		At 45 psia		
7	57	40	4.2	4.8
10	48	65	11.7	12.8
15	37.5	110	22.8	24.1
20	30.5	140	29.6	30.4
25	25.5	170	36	36.3
30	21.5	200	42.8	42.5
35	18.5	215	46.5	45.7
40	16	221	47.8	46.5
45	14	220	47.2	45.5
50	12.5	210	45.3	43.4
55	11	200	42.8	40.8
60	10	195	42.0	39.8
70	8.4	160	34	31.9
		At 60 psia		
23	50	40	4.2	4.6
30	38	70	13	13.6
35	32	90	18.1	18.5
40	27.5	100	20.4	20.5
45	24	110	22.8	22.6
50	21	120	25	24.5
60	16.5	120	25	24.1
70	13	110	22.8	21.7

Source: Ref. 9–25.

REFERENCES

9–1. American Society of Heating, Refrigerating and Air-Conditioning Engineers, *ASHRAE Guide and Data Book*, "Fundamentals and Equipment." Atlanta, Ga.: ASHRAE (1965), p. 635, Fig. 1.

9–2. G. H. Green, "Influence of Oil on Boiling Heat Transfer and Pressure Drop in Refrigerants 12 and 22," *ASHRAE J.*, 7 (Dec. 1965), 57.

9–3. R. D. Bennett, L. J. V. Earthy, and H. Heckmatt, "Lubrication of Refrigerant 22 Refrigeration Plant," *J. Refrig.* (Sept. 1966), 211.

9–4. H. Heckmatt, "Some Lubrication Problems of Refrigerating Machines Using Refrigerant 12," *J. Refrig.* (Jan.–Feb. 1960).

9–5. R. C. Downing, "Oil in Refrigeration Systems," *Service Application Manual.* Des Plaines, Ill.: Refrigeration Service Engineers Society (1967), Sec. 4.

9–6. W. D. Cooper, "Influence of Oil-Refrigerant Relationships on Oil Return," ASHRAE Symposium (Jan. 1971).

9–7. American Society of Heating, Refrigerating and Air-Conditioning Engineers, *ASHRAE Systems Handbook.* Atlanta, Ga.: ASHRAE (1984), Chap. 29.

9–8. Exxon Corporation, booklet of oil properties published in 1973.

9–9. Sun Oil Company, data in a letter to H. M. Parmelee, DuPont Company, Freon Products Division (Jan. 9, 1961).

9–10. Sun Oil Company, Bulletin A-5860.

9–11. G. Bambach, Abhandlung Nr. 9 des Deutschen Kaltetechnischen Vereins, C. F. Muller, Karlsruhe (1955).

9–12. University of Colorado, unpublished data (1949).

9–13. H. O. Spauschus, "Thermodynamic Properties of Refrigerant/Oil Solutions," *ASHRAE J.*, 5 (1963), 63.

9–14. DuPont Company, Freon Products Division, unpublished data.

9–15. L. F. Albright and A. S. Mandelbaum, "Solubility and Viscosity Characteristics of Mixtures of Lubricating Oils and Freon 13 or 115," *Refrig. Eng.*, 64 (Oct. 1956), 37.

9–16. H. O. Spauschus, "Vapor Pressure, Volumes, and Miscibility Limits of Refrigerant 22-Oil Solutions," *ASHRAE Trans.*, 70 (1964), 306.

9–17. W. O. Walker, A. A. Sakhnovsky, and S. Rosen, "Behavior of Refrigerant Oils and Genetron-141," *Refrig. Eng.*, 65 (Mar. 1957), 38.

9–18. H. O. Spauschus, "Migration Rates for Refrigerant 22," *ASHRAE Trans.*, 71, pt II (1965), 34.

9–19. H. M. Parmelee, "Viscosity of Refrigerant–Oil Mixtures at Evaporator Conditions," *ASHRAE Trans.*, 70 (1964), 173.

9–20. W. D. Cooper, "Oil Return in Low Temperature Systems," discussion prepared for use in refrigeration meetings, DuPont Company, Freon Products Division.

9–21. W. D. Cooper, R. C. Downing, and J. B. Gray, "Alkyl Benzene as a Compressor Lubricant," *Proceedings of the 1974 Purdue Compressor Technology Conference* (July 1974), 88.

9–22. K. S. Sanvordenker, "A Review of Synthetic Oils for Refrigeration Use," ASHRAE Symposium, Nassau (June 1972).

9–23. R. C. Downing, "Viscosity of Refrigerant/Oil Solutions," DuPont Company, Freon Products Division, unpublished results.

9–24. R. C. Downing, "Viscosity of Oil Solutions," DuPont Company, Freon Products Division, Bulletin D-34L.

9–25. R. C. Downing, "Viscosity of Freon 22/Suniso 3GS Oil Solutions," DuPont Company, Freon Products Division, Bulletin X-223 (1975).

9–26. Sun Oil Company, private communication (1975).

10

EFFECT ON POLYMERS

The effect of refrigerants on elastomers and plastics is often judged by the amount of linear swelling that occurs. A strip of the elastomer or plastic and the refrigerant are sealed in a glass pressure tube and stored at a relatively constant temperature. The change in length of the test piece is occasionally measured until no further change occurs; usually a day or two but sometimes several days may be required to reach equilibrium. The following general comments apply to both plastics and elastomers and should be considered only as trends rather than definitive rules. When considering a polymer for use with a refrigerant, the combination should be thoroughly tested. Any data given here should be used only as guidelines.

Polymer swelling will be just as great if it is exposed to saturated vapor as to liquid. Gas at less than saturated pressure will cause swelling about in proportion to the pressure. Swelling may be considered an attempt by the polymer to dissolve the refrigerant. The refrigerant dissolved on one side of a polymer will tend to evaporate on the other side. The degree of swelling will give some idea of the relative amount of diffusion through the polymer that might be expected.

Products that are chemically similar tend to dissolve each other: for example, oil and water do not mix, but alcohol and water are soluble in each other in all proportions. So also, polymers will tend to swell more in the presence of refrigerants that are similar in nature than in contact with widely different refrigerants. Fluorinated polymers will be more likely to swell in the presence of fluorinated refrigerants than in the presence of hydrocarbons.

Polar polymers may be affected by polar refrigerants but not as much by nonpolar compounds. Polar compounds are those with a definite dipole moment. They may have different chemical groups such as alcohol, ester, amine, nitrile, and so on. Part of the molecule may be polar and part nonpolar, so it may be difficult to estimate its performance. It is always best to test. Nonpolar polymers generally have uniform, symmetrical molecules such as polyethylene or polypropylene. In the same way, nonpolar refrigerants tend to have balanced molecules such as hydrocarbons and completely halogenated hydrocarbons as R-11 or R-12. Refrigerants containing a hydrogen atom such as R-21 or R-22 tend to be polar. Hydrocarbon lubricating oils tend to be nonpolar.

As a general rule, compounds containing fluorine cause less swelling than compounds containing chlorine. Compounds containing bromine have more effect than those containing chlorine. Sometimes combinations of polymers and refrigerants may have characteristics pointing in conflicting directions. In this case it is more important than ever to test each combination that is considered.

Many measurements of swelling have been made and the results are very useful in studying refrigeration systems or other applications of refrigerants. However, some caution should be used when applying the data. Factors other than swell may also be important. Refrigerants may tend to extract plasticizers or low-molecular-weight parts of polymers. Test pieces should be weighed before and after exposure to measure the extent of extraction or perhaps retention of refrigerant. The test piece should be examined for hardness, flexibility, deterioration, and so on. The application should be considered. For example, O-ring gaskets may seal better when they swell a little. Elastomers in refrigeration service are often plasticized with lubricating oil and equilibrium develops between oil extracted by the refrigerant and oil present in the circulating refrigerant. The same gasket in pumping service might become hard and brittle as the plasticizer is removed. The quality of an elastomer is affected not only by the nature of the basic polymer but also by the formulation, the compounding, the plasticizer, the temperature, and the duration of curing. Plastic categories may not be clearly defined and properties altered by blending different monomers and by methods of formation and treating.

ELASTOMERS

The linear swelling of a number of elastomers by various refrigerants is given in Table 10–1. The tests were made at room temperature and show the maximum percentage increase in length of the test piece while it is in contact with the refrigerant. Some of the elastomers are identified in Table 10–2, page 274. The chemical composition of the elastomer is the most important factor in determining the resistance of an elastomer toward swelling, but the curing temperature, time and the plasticizers used in compounding are also of great importance. These details were not always recorded in swelling tests but should be considered.

In choosing an elastomer for use with a refrigerant, the swelling data in Table 10–1 are a very good guide but may not be the only consideration. The properties of the elastomer must be suitable for the application. For example, polysulfide-type elastomers are low in swelling but their physical properties may not be adequate.

TABLE 10–1 LINEAR SWELLING OF ELASTOMERS[a]

		Elastomer											
Refrigerant	Chemical formula	Neoprene type W	Neoprene type GN	Buna N	Natural rubber	Buna S	Poly-sulfide	Butyl	Silicone	Hypalon 40	PVA	Viton	Urethane
R-10	CCl_4		36	11	44	31	14	50					
R-11	CCl_3F	9	17	6	23	21	2	41	38	3	0	4	20
R-12	CCl_2F_2	−1	0	2	6	3	1	6		1	−8	10	5
R-13	$CClF_3$	3	0	1	1	1	0	0		1	−1	3	
R-13B1	$CBrF_3$	0	2	1	1	1	0	2		2	1	5	
R-20	$CHCl_3$		43	54	45	32	91	45					
R-21	$CHCl_2F$	11	28	48	34	49	28	24		24	9	26	
R-22	$CHClF_2$	0	2	26	6	4	4	1	20	3	6	37	28
R-30	CH_2CL_2		37	52	34	26	59	23					
R-31	CH_2ClF	4	9	38	12	10	8	3		9	2	40	
R-32	CH_2F_2	1	0	3	0	0	0	0		3	0	21	
R-40	CH_3Cl		22	35	26	20	11	16					
R-112	CCl_2FCCl_2F	9	10	7	32	23	2	43		9	0	2	
R-113	$CClF_2CCl_2F$	−1	3	1	17	9	1	21	34	1	−1	7	7
R-114	$CClF_2CClF_2$	0	0	0	2	2	0	2		1	−2	9	
R-115	$CClF_2CF_3$		0	0	0	0	0	0				6	
R-142b	CH_3CClF_2	0	3	3	5	4	1	3		4	−3	25	
R-152a	CH_3CHF_2	0	2	1	2	1	3	1		16	−4	34	
R-502	[b]		1	7	4	3						10	
R-218	$CF_3CF_2CF_3$		0	0	0	0	0	0				2	
R-C316	$C_4Cl_2F_6$		0	0	1	0	0	2					
R-C318	C_4F_8	0	0	0	0	0	0	0		2	−4	5	
R-113B2	$CBrClFCBrF_2$		28	17	36	25	7	43					
R-114B2	$CBrF_2CBrF_2$	3	7	7	26	15	1	22		4	0	9	
R-30B1	CH_2BrCl		39	53	34	25	99	14					
n-Butane	C_4H_{10}		3	1	16	8	0	20			−1		
n-Pentane	C_5H_{12}		4	1	20	9	0	29			0		

[a]Maximum percent increase in length at room temperature (about 70°F).

[b]Azeotropic mixture of R-22/R-115 (48.8/51.2 wt %).

TABLE 10–2 DESCRIPTION OF ELASTOMERS

Elastomer	Description
Buna N	Copolymer of butadiene and acrylonitrile
Buna S	Copolymer of butadiene and styrene
Butyl	Copolymer of isobutylene and butadiene (or isoprene)
Hypalon	Chlorosulphonated polyethylene
Neoprene	Polymers of 2-chlorobutadiene-1,3
PVA	Polyvinyl alcohol
Natural rubber	Polymer of isoprene
Polysulfide	Thiokol, dichlorodiethylformal and an alkali polysulfide
Viton	Copolymer of hexafluoropropylene and vinylidene fluoride

In general, elastomers most often used with the fluorinated refrigerants are neoprene, Hypalon 40, and Buna N (not with R-22). If exposure to higher temperatures is likely, Viton might be used despite higher swelling.

The measured swelling of an elastomer may be offset somewhat by extraction of the plasticizer and perhaps some other materials used in formulating the elastomer. The degree of extraction can be determined by changes in weight and length of the elastomer after it has been air-dried at room temperature. Several days or more may be necessary for the evaporation of residual refrigerant and arrival at a state of equilibrium. As mentioned above, lubricating oil is often used as a plasticizer for the elastomer in refrigeration service. Extraction of the oil is reduced by the presence of oil in the circulating refrigerant. For applications other than refrigeration, the plasticizer should have low solubility in liquids that may be in contact with the elastomer.

The effect on swelling of fluorine atoms in refrigerant molecules is illustrated by comparing related series of compounds. In the following examples, the percent swell is given below each product.

Neoprene:

CCl_4	CCl_3F	CCl_2F_2	$CClF_3$
36	17	0	0
CH_3Cl	CH_2ClF	$CHClF_2$	$CClF_3$
22	9	2	0

Buna N:

CCl_4	CCl_3F	CCl_2F_2	$CClF_3$
11	6	2	1
CH_3Cl	CH_2ClF	$CHClF_2$	$CClF_3$
35	38	26	1

The tests in Table 10–1 involved only the elastomer and refrigerant. When oil is added, the results may differ depending on the elastomer, the refrigerant, and the oil. If the elastomer is not appreciably affected by the oil alone, swelling in the oil–refrigerant mixture is usually not much different from that in refrigerant alone (except for R-22). If the oil does affect the elastomer, swelling in the mixture may be much greater than in refrigerant alone. These effects are illustrated in Table 10–3. The Suniso 3G oil used in these tests is naphthenic based with an aniline point of about 172°F.

TABLE 10–3 SWELLING OF ELASTOMERS BY SUNISO 3G OIL AT ROOM TEMPERATURE[a]

	Percent linear swell									
	Buna N		Neoprene[b]		Hypalon 40		Butyl		Natural	
R	No oil	With oil	No oil	With oil	No oil	With oil	No oil	With oil	No oil	With oil
None	—	−1	—	1	—	0	—	5	—	13
R-11	6	4	9	7	3	6	41	—	23	26
R-12	2	4	−1	3	1	3	6	45	6	23
R-22	26	19	0	7	3	10	1	34	6	25
R-113	1	4	1	1	1	4	21	50	17	23
R-114	0	3	0	0	1	2	2	45	2	22

[a]50% oil by liquid volume.
[b]Neoprene W, specialty oil-resistant product.
Source: Ref. 10–4.

The nature of the oil also has a bearing on the swelling of elastomers as illustrated in Table 10–4. The aniline point is the temperature at which equal volumes of aniline and oil form one solution. A high aniline point indicates a higher paraffinic hydrocarbon content in the oil and less solvent tendency. This direction is also found with the refrigerants, that is, oils with higher paraffinic content are less soluble in the refrigerants than those with smaller amounts of paraffinic components. With

TABLE 10–4 EFFECT OF OIL TYPE ON SWELLING AT ROOM TEMPERATURE[a]

	Buna N			Neoprene		
		With oil aniline point (°F)			With oil aniline point (°F)	
R	No oil	172	232	No oil	172	232
R-12	2	0	0	0	8	1
R-22	26	12	11	2	15	4
CCl_4[b]	11	5	—	36	23	—

[a]20 vol % refrigerant.
[b]50 vol %.
Source: Ref. 10–4.

R-12 there is no difference in the effect of the two oils on Buna N, but with neoprene the less paraffinic oil caused more swelling. With R-22, both oils reduced swelling of Buna N but caused an increase with neoprene.

At modest temperatures small changes have little effect on the swelling of elastomers. Tests at 75°F and 130°F gave essentially the same results. At high temperatures the suitability of an elastomer depends on its stability as well as the effect of the refrigerant. In one series of tests with R-114 at 350°F the following elastomers were judged suitable [10–1]. Linear swelling is shown but evaluation also included hardness and stability measurements.

Elastomer	Description	Linear change (%)
Epcar EPDM	Ethylene–propylene–diene terpolymer	5.3
Fluorel Cpd C	Fluoroelastomer	5.5
Vamac[a]	Copolymer of ethylene and methyl acrylonitrite	3.2 (4.2)
Krynac[a]	50% acrylonitrile	0.8
Rulon A[b]	Compounded material based on Teflon	−0.5

[a]Needs more testing.

[b]Not an elastomer; included as a reference.

When elastomers are used in diaphragms, hose, and so on, outward diffusion loss of refrigerant and inward infusion of air and water may be important. Some diffusion rates with water, R-12, and R-22 are given in Table 10–5.

TABLE 10–5 DIFFUSION OF WATER, R-12, AND R-22 THROUGH ELASTOMERS

	Diffusion rate		
Elastomer	Water[a]	R-12[b]	R-22[b]
Neoprene	350	6.5	40
Buna N	53	2	600
Hypalon 40	223	5	16
Adiprene[c]		15	450
Butyl	21	45	9
Viton	—	30	110
Polyethylene	60	—	—
Natural	697	—	—

[a]0.003-in. film, 100% relative humidity at 100°F (grams of water per 100 m^2/h).

[b]Film thickness, 1 mil; temperature, 77°F ($cm^3 \times 10^{-4}$ gas at 1 atm and 32°F/day-m^2.

[c]Isocyanate elastomer.

Source: Ref. 10–4.

Some additional data on the permeability of plastics and elastomers to R-12, R-22, R-134a, and water have recently been published [10–2] and are shown in Tables 10–6 to 10–8. The units are given as published and so cannot be compared directly with earlier data.

TABLE 10–6 PERMEABILITY CONSTANTS FOR R-12, R-22, AND R-134a

	1×10^{-10} (cm³)(cm)/(s)(cm²)(cm Hg)			
Polymer	R-22 (200°F)	R-12 (200°F)	R-12 (150°F)	R-134a (200° F)
Thermoplastic elastomers				
Hytrel polyester	29	12	2.2	
Cytor polyether urethane	37	9	1.9	
Pellethane polyester urethane	90	16	10	
PVC-NBR	124	39	26	
Cross-linked elastomers				
Epichlorohydrin	183	9	1.5	4
Polyester urethane	135	33	14	53
Ethylene-acrylic	276	70	34	64
Polynorbornene	211	229	140	76
Current polymers				
Polyamide	16	3	1.2	
Chlorosufonated polyethylene	34	12	4	3
Nitrile	81	19	14	
Plastics				
PET polyester	0 (72 h)	0 (72 h)	—	
Polyamide 11	3	—	—	
Plasticized Polyamide 11 (12%)	6	—	—	
Polypropylene	11	1.5	0 (7 d)	
Plasticized Polyamide 12 (5%)	11	1.7	—	
Plasticized Polyamide 12 (15%)	19	5	—	

Source: Ref. 10–2.

R-12 has become the standard refrigerant for auto air conditioning. One problem in using lower-boiling refrigerants that would give more capacity is finding an elastomeric flexible hose that would have low permeability for the refrigerant. In one study [10–3], R-12, R-22, R-500, and R-502 were tested with hose liners made of nylon, neoprene, and CSM (chlorosulfonated polyethylene). The hoses were of the standard construction used in automotive air conditioning. The thickness of the nylon liner was 0.025 in. The elastomer hoses were 0.140 in. thick. The test hoses were 0.50 in. in internal diameter and 18 in. long. Loss of refrigerant was measured by weighing the hose containing the refrigerant. Each test lasted 2 weeks. Duplicate tests were made with and without oil (500 SUS). Some results are given in Table 10–9, page 279, as pounds of refrigerant lost per year per linear foot of hose.

TABLE 10–7 WATER PERMEABILITY DATA

Polymer	Weight loss at 73°F ($g/h\text{-}m^2$)	Hours tested
Polynorbornene	0.0946	742
PET polyester	0.1577	754
Polypropylene	0.5363	742
Ethylene-acrylic	0.9148	1237
Hytrel polyester	1.672	726
Plasticized Polyamide 12 (5%)	2.934	507
Pellethane polyester urethane	3.975	1237
Polyester urethane elastomer	5.394	742
Plasticized Polyamide 12 (15%)	5.647	507
Cytor polyether urethane	6.183	1237
Plasticized Polyamide 11 (12%)	7.256	432
Polyamide 11	0.222	726

Source: Ref. 10–2.

Both nylon and CSM appear to be suitable for use with R-500 and R-502. The permeation rate for R-22 was least with nylon and CSM but still considerably greater than for the other refrigerants. The permeation of all refrigerants through neoprene was excessively high. In all tests the loss rate increased with temperature due, in part, to the increase in vapor pressure of the liquid refrigerant in the hose.

TABLE 10–8 IDENTIFICATION OF POLYMERS

PET polyester	DuPont's Mylar
Polyamide 11	Rislan's Nylon 11 BMNO
Plasticized Polyamide 11 (12%)	Rislan's Nylon BMNO P40 containing 12% plasticizer
Plasticized Polyamide 12 (5%)	Huels' L 2121 containing 5% plasticizer
Plasticized Polyamide 12 (15%)	Huels' L 2121 containing 15% plasticizer
Polypropylene	Hercules' polypropylene 6523
Hytrel polyester	DuPont's Hytrel 6346
Cytor polyether urethane	American Cyanamid's Cytor 7090
Pellethane polyester urethane	Upjon's Pellethane 2102 (caproester)
PVC-NBR	Uniroyal's TPR 5370

Source: Ref. 10–2.

TABLE 10–9 PERMEATION OF REFRIGERANTS THROUGH AUTO AIR-CONDITIONING HOSE[a]

Hose	°F	Oil present	Refrigerant			
			R-12	R-22	R-500	R-502
Nylon	75	+	0.0022	0.4	—	0.065
	75	−	0.0012	—	—	—
	100	+	0.0050	0.62	0.0038	0.13
	100	−	0.0032	—	0.0027	—
	140	+	0.017	1.2	0.027	0.32
	140	−	0.013	—	0.050	—
	180	+	0.048	2.0	0.15	0.75
	180	−	0.038	—	0.60	—
CSM	75	+		0.53		
	100	+		0.88	0.01	0.025
	140	+	0.043	1.8	0.09	0.16
	180	+	0.33	3.3	0.62	0.77
Nitrile	75	+	0.028	—	0.23	1.4
	75	−	0.09	—	0.38	—
	100	+	0.068	—	0.43	2.0
	100	−	0.19	—	0.75	—
	140	+	0.24	—	1.2	3.5
	140	−	0.48	—	1.9	—
	180	+	0.70	—	2.5	5.6
	180	−	1.1	—	4.2	—

[a] Loss rate (lb/ft-yr).
Source: Ref. 10–3.

The presence of oil had a variable effect on the permeation but was not a significiant factor.

The measured refrigerant loss was converted to a permeation constant by the following equation:

$$K = \frac{(\mathrm{Pr})(t)}{(A)(P)}$$

where Pr = observed loss rate, g/day
A = inside area of hose (28.25 in.^{2})
P = refrigerant vapor pressure, psia
t = thickness of hose liner, 0.025 in. for nylon, 0.14 in. for the others
K = permeation constant, (g)(in.)/(day)(psi)

For example, the permeation constants for the various elastomers and refrigerants at 140°F are as follows:

Elastomer	Refrigerant	$K \times 10^5$	Elastomer	Refrigerant	$K \times 10^5$
CSM	12	—	NBR	12	1.02
	22	5.03		22	—
	500	0.34		500	4.08
	502	0.40		502	8.50
Neoprene	12	3.59	Nylon	12	0.01
	22	7.83		22	0.55
	500	4.46		500	0.02
	502	2.57		502	0.13

During the tests the composition of the two mixed refrigerants, R-500 and R-502, was measured to see if there was preferential permeation of one of the components. Changes in composition were negligible. The results are compared in Table 10–10.

TABLE 10–10 COMPOSITION CHANGE FOR ONE YEAR

Hose	Refrigerant		Composition (%) Initial	Final
CSM	R-500			
		R-12	73.7	73.9
		R-152a	26.3	26.1
CSM	R-502			
		R-22	48.8	47.3
		R-115	51.2	52.7
Nylon	R-500			
		R-12	73.7	73.5
		R-152a	26.3	26.5

PLASTICS

The effect of refrigerants on plastics is described in the following article from the Dupont Company, Freon Products Division Bulletin B-41.

EFFECT OF REFRIGERANTS ON PLASTICS

Some generalizations about the compatibility of refrigerants and plastics at room temperature follow.

1. Most plastic materials are not appreciably affected by the refrigerants and can be used satisfactorily in a majority of applications.
2. An exception is polystyrene which is dissolved by R-11, R-21, and R-22, but is only slightly affected by R-12 and R-113.

3. Methyl methacrylate polymers are not affected by short exposure to most fluorocarbons, but tend to be crazed by some products on longer exposure. These polymers are dissolved by R-22 and R-21. Cast acrylics appear to be more resistant than extruded or molded material.
4. Fluorinated polymers tend to swell slightly in highly fluorinated compounds and in some cases may be unsuitable.
5. In general, the effect on plastics becomes less as the amount of fluorine in the molecule is increased. For example, R-12 has less effect than R-11, while R-13 is almost entirely inert. (Note: The reverse may be true for fluorinated polymers.)
6. Plastic materials should be tested for compatibility for each application. In many cases, the plastic is a mixture of different types or is copolymerized from different monomers. Sometimes a plasticizer is used. These factors may affect the compatibility.

A brief summary of the effect of the refrigerants on various plastic materials is given below, but compatibility should be tested for specific applications. Differences in polymer structure and molecular weight, plasticizer, temperature, etc., may alter the resistance of the plastic.

Teflon TFE Fluorocarbon Resin. No swelling observed when submerged in the liquid but some diffusion found with R-12 and R-22.

Polychlorotrifluoroethylene. Slight swelling but generally suitable for use with the refrigerants.

Polyvinyl Alcohol. Not affected by the refrigerants (except R-21), but very sensitive to water. Used especially in tubing with an outer protective covering.

Vinyl. Resistance to the refrigerants depends on vinyl type and plasticizer and considerable variation is found. Samples should be tested before use.

Orlon Acrylic Fiber. Generally suitable for use with the refrigerants.

Nylon. Generally suitable for use with the refrigerants but may tend to become brittle at high temperatures in the presence of air or water. Tests at 250°F with R-12 and R-22 showed the presence of water or alcohol to be undesirable. Adequate testing should be carried out.

Polyethylene. May be suitable for some applications at room temperature but should be thoroughly tested since greatly different results have been found with different samples.

"Lucite" Acrylic Resin (Methacrylate Polymers). Dissolved by R-22 but generally suitable for use with R-12 and R-114 for short exposure. On long exposure tends to crack and craze and become cloudy. Use with R-113 may be questionable and probably should not be used with R-11.

TABLE 10–11 EFFECT OF REFRIGERANTS ON PLASTICS AT ROOM TEMPERATURE[a]

Plastic	R-11		R-12		R-13B1		R-21		R-22		R-112		R-113		R-114	
	S	W	S	W	S	W	S	W	S	W	S	W	S	W	S	W
Delrin acetal resin	0	0	1	2	0	0	—	—	3	2	—	—	0	−1	0	0
Cellulose acetate	0	0	0	13	1	0	D	—	—	—	0	−16	0	0	—	0
Cellulose nitrate	1	−2	0	0	—	—	D	—	—	—	—	—	0	1	—	0
Chlorotrifluoroethylene polymer	0-3	—	2	—	—	—	—	—	1	—	0	0	0	0	—	0
Lucite acrylic resin	0	0	0	0	0	0	D	—	D	—	—	—	0	0	1	-1-1
Mylar polyester film	—	—	—	—	0	−1	—	—	—	—	—	—	—	—	—	—
Nylon (Zytel 101)	0	0	0	0	—	—	0	1	1	1	0	0	0	0-5	0	0
Phenol formaldehyde resin	0	0	0	0	—	—	0	0	—	—	0	0	0	0	—	—
Polyethylene	6	1	1	0	3	0	5	1	2	0	—	—	2	1	1	—
Polyethylene, linear	—	—	—	—	1	0	—	—	1	0	4	13	2	9	—	—
Polypropylene	—	—	—	—	1.9	1.4	—	—	—	—	—	—	—	—	—	—
Polystyrene	D	—	0	2	—	—	D	—	—	—	D	—	0	0	—	—
Polyvinyl alcohol	0-3	0	−1	0	1	−5	13	5	—	—	—	—	0	0	—	—
Polyvinyl chloride	0	10	0	0	0	−3	15	10	—	—	—	—	0	0	—	—
Polyvinylidene chloride	0-3	0	0	0	2	0	1	1	4	—	—	—	0	0	—	—
Teflon TFE fluorocarbon resin	0	0	0	3	2	1	0	0	1	0	0	—	0	0	—	—

[a]S, maximum percent linear swell when submerged in liquid phase; D, disintegrated; W, percent increase in weight after drying in air for about 2 weeks; —, not tested.

Cast "Lucite" acrylic resin is much more resistant to the effect of solvents than extruded resin and can probably be used with most of the refrigerants.

Polystyrene. Considerable variation found in individual samples but generally not suited for use. Some applications might be all right with R-114.

Phenolic Resins. Usually not affected by the refrigerants. However, composition of resins of this type may be quite different and samples should be tested before use.

Epoxy Resins. Resistant to most solvents and entirely suitable for use with the refrigerants unless highly plasticized.

Cellulose Acetate or Nitrate. Suitable for use.

Delrin Acetal Resin. Suitable for use with the refrigerants under most conditions.

Examples of the effect of refrigerants on plastic materials are shown in Table 10–11. The tests were conducted by immersing strips of the plastic in the liquid product at room temperature. The linear swelling of the plastic was observed and is reported as a percent of the original length. The results should be used as a guide only. Other samples of the same general types of plastics may give different results depending on the degree of polymerization, presence of plasticizers, or other polymers.

In some applications, exposure to refrigerants may be for very short periods. In this case nearly all plastics are unaffected and present no compatibility problems. The effects of 5-min exposures to three compounds are summarized in Table 10–12.

TABLE 10–12 EFFECT OF REFRIGERANTS ON PLASTICS[a,b]

Plastic	R-11	R-112	R-113
Alathon 7050 linear polyethylene resin	0	—	0
Alathon 9140 polypropylene resin	0	—	0
Delrin acetal resin	0	—	0
Epoxy resin	0	0	0
Ethyl cellulose	—	—	4
Kralastic ABS polymer	0	—	1
Lexan polycarbonate resin	0	4	0
Lucite methylmethacrylate resin (cast)	0	—	0
Polyvinyl alcohol	—	—	0
Polyvinyl chloride (unplasticized)	0	—	0
Styron 475 polystyrene	—	4	2
Surlyn A ionomeric resin	—	—	0
Teflon TFE resin	0	0	0
Zytel 101 nylon resin	0	0	0

[a]Exposed for 5 min at the boiling point.

[b]0, no visible effect; 1, very slight effect; 2, compatibility should be tested; 3, probably not suitable; 4, disintegrated or dissolved.

TABLE 10–13 PERMEABILITY OF PLASTIC FILMS[a]

Plastic film	Temp. (°F)	Permeability (ml @ STP) (mil)/(day) (100 in.^{2}) (atm)								Reference
		N_2	O_2	R-11	R-12	R-13	R-21	R-22	R-114	
Chlorinated polyether	130	0.65	—	—	< 0.07	—	—	—	—	10–6
Mylar polyester	77	1.0	10.3	—	<0.004	—	—	—	—	10–6
	95	1.7	—	—	<0.004	—	—	—	—	10–6
	130	3.8	37	—	<0.004	—	—	—	—	10–6
Hostaphen polyester	68	0.008	—	0.03	<0.03	0.09	0.04	0.1	0.004	10–7
Mylar polyester, aluminized	130	0.29	1.3	—	<0.007	—	—	0.24	—	10–6
Nylon	130	5	—	—	0.04	—	—	—	—	10–8
Polyethylene	78	130	460	—	1300	—	—	—	—	10–6
	76	80	—	—	—	—	10[b]	—	—	10–8
	76	120	—	—	—	—	0.6[b]	—	—	10–8
Polyethylene, Suprathen N (HP)	—	—	—	339	42	4.7	834	45	18	10–7
Polyethylene, V56 (LP)	—	—	—	670	8.3	1.5	217	14.7	2.4	10–7
Polyethylene, V57 (LP)	—	—	—	97	3.7	0.8	75	7.2	5.9	10–7
Polypropylene, PP11	—	—	—	3050	0.08	0.2	108	2.9	0.04	10–7
Polypropylene	77[c]	—	—	—	—	—	—	—	—	10–8
	130[d]	—	—	—	—	—	—	—	—	10–8
Polyvinyl alcohol	130	0.65	—	—	< 0.07	—	—	—	—	10–6
Polyvinyl chloride Genotherm U.G.	—	—	—	0.15	<0.03	0.11	15.5	0.13	<0.03	10–7
Polyvinyl fluoride	130	1.8	—	—	1.4	—	—	20	—	10–8
Polyvinylidene chloride Saran	77	0.3	0.9	—	< 0.004	—	—	—	—	10–6
	95	1.0	2.5	—	0.34	—	—	—	—	10–6
	130	2.3	5.4	—	1.9	—	—	—	—	10–6
Teflon 100-X FEP	130	>47	—	—	—	>47	—	—	—	10–8
Urethane, polyester	76	0.9	—	0.04	—	—	—	—	—	10–8
	130	14	—	1	0.1	—	—	—	—	10–8
Urethane, polyester	130	24	—	0.7	0.13	—	—	—	—	10–8

[a] —, not tested.
[b] R-C318.
[c] No diffusion with R-116 or R-C-318.
[d] Quite permeable to R-116 and R-C-318.

Permeability

Some typical measurements of the permeability of various plastic films to the refrigerants and nitrogen and oxygen are listed in Table 10–13. The values are representative, but the diffusion rates may vary quite widely for a particular type of plastic depending on its molecular weight and the presence of plasticizers or other additives.

REFERENCES

10–1. B. Toekes, "Dichlorotetrafluoroethane Sensitivity of Various Elastomers in Geothermal Energy Conversion Applications," 179th National Meeting of the American Chemical Society, Houston, Texas (Mar. 24, 1980).

10–2. H. E. Trexler, "Resistance of Polymers to Permeation by Air-Conditioning Refrigerants and Water," *Rubber Chem. Technol.*, 56 (1983), 105.

10–3. R. F. Goldman and D. D. Rudy, "Permeation of Refrigerants through Elastomeric and Plastic Hoses," Society of Automotive Engineers, SAE Congress and Exposition, Detroit (1971).

10–4. B. J. Eiseman, Jr., "The Use of Elastomers in Refrigeration Systems," *Bull. Int. Inst. Refrig.*, 41, no. 3, (1966).

10–5. DuPont Company, Freon Products Division, Bulletin B-41.

10–6. H. M. Parmelee, "Permeability of Plastic Films to Refrigerant 12 and Nitrogen," *Refrig. Eng.*, 66 (Feb. 1958), 35.

10–7. H. Braunisch, "Untersuchungen uber die Frigen-Durchlassigkeit von Kunstoff-Folien," *Kaltetechnik*, 13 (Feb. 1961), 59.

10–8. Dupont Company, Freon Products Division, unpublished data (1982).

11
LEAK DETECTION

Leak detection is a very important part of the manufacturing and servicing of refrigeration and air-conditioning equipment and, indeed, any type of equipment where leakage can be a factor. Fluorocarbon refrigerants may be involved in three different areas of application. The analytical method used often depends on the application, the sensitivity required, ease of use, cost, and so on.

Fluorocarbon detection is important in:

1. Finding leaks in refrigeration and air-conditioning equipment during manufacture and after installation and operation in the field. In both cases the test method must be sensitive, accurate, reliable, safe, easily operated, and inexpensive. For the serviceman it must also be portable.
2. Finding leaks in tanks or cylinders or other containers that must be gastight. A fluorocarbon is often used as a tracer gas mixed with air or nitrogen to reduce cost and permit higher pressurization. The choice of detection instrument is usually about the same as in the refrigeration industry.
3. Monitoring the atmosphere to check on fluorocarbon levels to be sure that they do not exceed any codes or generally acceptable concentrations for human exposure. Fluorocarbons are also sometimes used to study wind patterns and dispersion of gases in the atmosphere.

REFRIGERANT LEAKAGE

The need to find and fix leaks in refrigeration systems is evident. If enough refrigerant leaks out, the equipment will not operate properly. Many small systems have a fairly critical charge of refrigerant, so the loss of even a small amount is serious. The capacity of the system is reduced and in hermetic units where refrigerant gas is used to help cool the motor, overheating and failure may occur.

METHODS OF LEAK DETECTION

Water Immersion

One of the first methods of detecting leaks was to pressurize the unit and put it in a tank of water. Any gas will do for pressurizing but usually air or nitrogen is used. Fluorocarbons can also be used and, of course, are already in place if charged refrigeration units are tested. A detergent is usually added to the water to reduce surface tension and encourage bubbles to rise to the surface where they can be seen. Low-surface-tension liquids such as kerosene and perchloroethylene have also been used.

If submersion of the unit is impractical, painting joints and other suspect areas with a soap or detergent solution can be effective. It is said [11–1] that a soap bubble 1 mm in diameter observed to form in 5 s is equivalent to 0.0001 cm/s of escaping gas. A viscous solution used as a toy for making bubbles is especially good since it stays in place much longer than ordinary soap solutions.

Halide Torch

The halide torch has been the basic tool for detecting fluorocarbon refrigerant leaks for many years. Early torches used methyl alcohol as a fuel, but propane or sometimes acetylene are used now. The flame is used to heat a copper block or screen. Escaping refrigerant is drawn into the flame through a flexible sniffer tube and is decomposed. Copper salts—fluorides and chlorides—are formed and produce a greenish-blue color in the flame. When R-12 is tested the color is faint and on the green side at low concentrations [11–2]. At higher concentrations the color becomes deeper and bluer. Chloride salts tend to be greener and fluorides more blue. An approximate correlation between color and concentration of halogen compound is shown in Table 11–1. In this case the halogen compound is methyl bromide, CH_3Br, but similar color changes are said to be observed with R-12, R-22, and similar compounds.

Since leak detection depends on the observation of a color change, the sensitivity may depend on the operator. A limitation has been reported for the method due to operator color fatigue after prolonged use. This problem has been reduced by supplying reference colors which can be optically compared.

The sensitivity of the halide torch also depends on its construction and the temperature of the flame. Higher temperatures are better since the color depends on the destruction of the refrigerant. Under ideal conditions, the torch is reported

TABLE 11–1 COLOR CHANGE WITH CONCENTRATION

Methyl bromide (ppm by volume)	lb/1000 ft^3	Flame color in daylight
25	0.00625	Faint fringe of green
50	0.0125	Moderate green
125	0.031	Green
250	0.0625	Strong green
500	0.125	Strong green-blue fringe
800	0.20	Strong blue-green
1000	0.25	Blue

Source: Ref. 11–1.

able to detect an R-12 leak of 0.002 g/h (18 g or 0.6 oz/yr) [11–2]. Another reference [11–3] mentions a limit of 5 oz/yr with most torches able to detect down to 9 oz/yr. Color blindness, bright sunlight, inexperience, and the hazard of an open flame are other conditions suggested as difficulties in using the torch. Nevertheless, it has been a useful tool and is still going strong.

Electronic Detector

Electronic has been used to describe two different types of units; one using ion conductivity and the other thermal conductivity.

The first type is represented by the H line from General Electric. A platinum wire is heated to about 900°C (1650°F) and serves as an anode. It is surrounded by a platinum tube acting as a cathode. A fan draws air through the sniffer tube and through the detection unit. When the refrigerant contacts the wire it is ionized, increasing the ion current between the electrodes. This change can be signaled by light, tone, and/or leak-rate meter [11–3]. With the H-10 model, the flashing rate of a neon lamp is in direct proportion to the size of the leak.

Sensitivity control has two settings: high and low. The high sensivity corresponds to a leakage rate of 0.1 oz/yr of R-12 and is recommended for checking air-conditioning units and all devices where small leaks will affect performance in a short time. The low range is recommended for all other applications such as auto air conditioning and refrigeration [11–4]. The sensitivity varies with the refrigerant concentration and the following correlation has been suggested by General Electric:

R-12 concentration (%)	Sensitivity (oz/yr)
100	½
50	1
10	5
1	50

Some advantages for these leak detectors are that they are positive, quiet, clean, and very sensitive. Some disadvantages might be the need for an electrical outlet in some models—or a life limit of about 100 hr for batteries in portable units. Maintenance needs may be a problem. The sensor is in the body of the unit and may need replacement after upward of 100 hr or so of operation. The electric motor used to drive the blower needed to draw air through the sniffer tube is one more possible complication. Because of the high temperature of the anode, it should be used only in well-ventilated areas. Overall, however, it has a very good record and is widely used.

The second type of electronic device is represented by the TIF detector. The sensing element is at the tip of the flexible probing tube, so it is not necessary to draw a gas sample into the hand-held case. This instrument is said to be rugged, easily maintained, and very sensitive [11–5].

Dye

A small amount of an oil-soluble red dye can be added to a refrigeration system and pinpoint leaks by the appearance of a bright red color on the outside of tubing, valves, connections, or wherever the leak occurs. The dye should also be soluble in the refrigerant, so it will not separate from the liquid solution at any point or at any temperature encountered in the refrigeration cycle. In the parts of a system where the refrigerant is a gas the dye is dissolved in the oil that is always present to some extent. In fact, it is even more effective in these areas than in other parts of a system where the refrigerant is a liquid. Part of the oil travels as a film around the wall of the piping, where it is pushed along by the flowing refrigerant gas. The concentration of dye dissolved in the oil is much greater than in the liquid refrigerant. When the oil film reaches a point of leakage, the distinctive red color on the outside appears much sooner and stronger than at points where the liquid refrigerant leaks out.

A red dye has been used for many years to find leaks in refrigeration systems and is still in general use. It is an azo-type dye. Its solubility in refrigerant and thermal stability are not as good as would be desirable. A new dye has been developed and promoted by DuPont in recent years with the trade name Dytel. It is an anthraquinone-type dye and has better solubility and thermal stability properties than the older dye [11–6]. The older dye has been known to clog screens, filter, and capillary tubing. Dytel does not have these problems.

The indicator dye has been especially useful in situations where other leak detectors are difficult to use, such as with rooftop units and other equipment located outdoors. On windy days leaks that are impossible to find by other methods readily show up with a dye that sticks to the outside of piping, and so on. Leaks have been found in gages and piping walls with the red dye. It also shows up easily through the ice in skating rinks. Other specialty applications have been developed in addition to good response in more usual refrigeration and air-conditioning installations.

Refrigerant Tracer

Leak testing is required in all kinds of pressure vessels at welds, seams, valves, joints, and so on. Often the test is made with a fluorocarbon refrigerant (usually R-12) mixed with air or nitrogen and leaks are found using the same detectors described above for refrigeration and air-conditioning use. Mixtures are used for two reasons: (1) to reduce the cost, and (2) to reach higher pressures than with the refrigerant alone.

With refrigerant alone, the highest pressure obtainable is the vapor pressure. Attempts to add more refrigerant will cause liquid to form without any increase in the pressure. Higher pressures can be reached by mixing with air or nitrogen since each gas tends to act as if the other were not there and pressures are additive. One exception to this rule of independent action is found when trying to prepare a mixture in the vessel to be leak tested.

The first thought in using a mixture of gases is to put one in the container to be tested at a certain pressure and then add the other gas to the final pressure. However, in practice, one gas tends to push the other away to the other end of the tank. Unless some form of agitation is provided, the gases will mix very slowly. In a round vessel several hours may be needed to get uniform mixing. If piping, valves, or other small passageways are involved, mixing may not occur for many days. It is best to prepare the mixture outside the tank to be leak checked. One method of mixing is illustrated in Figure 11–1.

A supply of R-12, for example, is put in a cylinder with a standpipe. A standard 145-lb shipping cylinder with a standpipe is available. Larger or smaller containers would need to be constructed. The regular valve opening is connected to a source of nitrogen. Another valve is installed in the body of the regular valve as an outlet for the gas mixture. A pressure regulator is installed at the outlet of the nitrogen cylinder and is set at the total pressure desired for the gas mixture. A pressure relief valve is used since the pressure of the nitrogen may be higher than could safely be admitted to the R-12 cylinder. The available cylinder fitted with a standpipe and extra valve has a working pressure of 300 psig and this total pressure should not be exceeded. The nitrogen is bubbled through the liquid R-12 and is mixed with the R-12 vapor at its vapor pressure. The concentration by volume of the R-12 in the mixture will be approximately the same as the pressure percent. Since the vapor pressure of the R-12 is a function of temperature, the temperature must be controlled—often by adding heat. As the flow of nitrogen removes R-12 vapor from the cylinder, liquid is evaporated to take its place and the remaining liquid tends to be cooler. In Table 11–2, page 292, the vapor pressure of R-12 is shown and the total pressure necessary to produce a given mixture. For example, if the liquid R-12 is maintained at 60°F, and the nitrogen pressure is kept at 652 psia, the gas mixture will contain about 10% R-12. It has been found that mixtures with about 10 to 20% R-12 are sufficient if care is used in testing. Of course, using a higher concentration of R-12 does no harm, but costs more. When the testing has been completed it is often possible to reclaim the gaseous mixture using a gas-phase compressor. The reclaimed gas can be stored at a higher pressure than the test pressure. However, the storage pressure should not be so high that the R-12 will liquefy or the composition of the mixture will change. Suppose that the R-12 liquid is maintained at 0°F and a nitrogen mixture containing 20% R-12 is prepared with a total pressure of 119 psia. To conserve storage space, it will be reclaimed at a higher pressure. What is the maximum storage pressure? The storage container

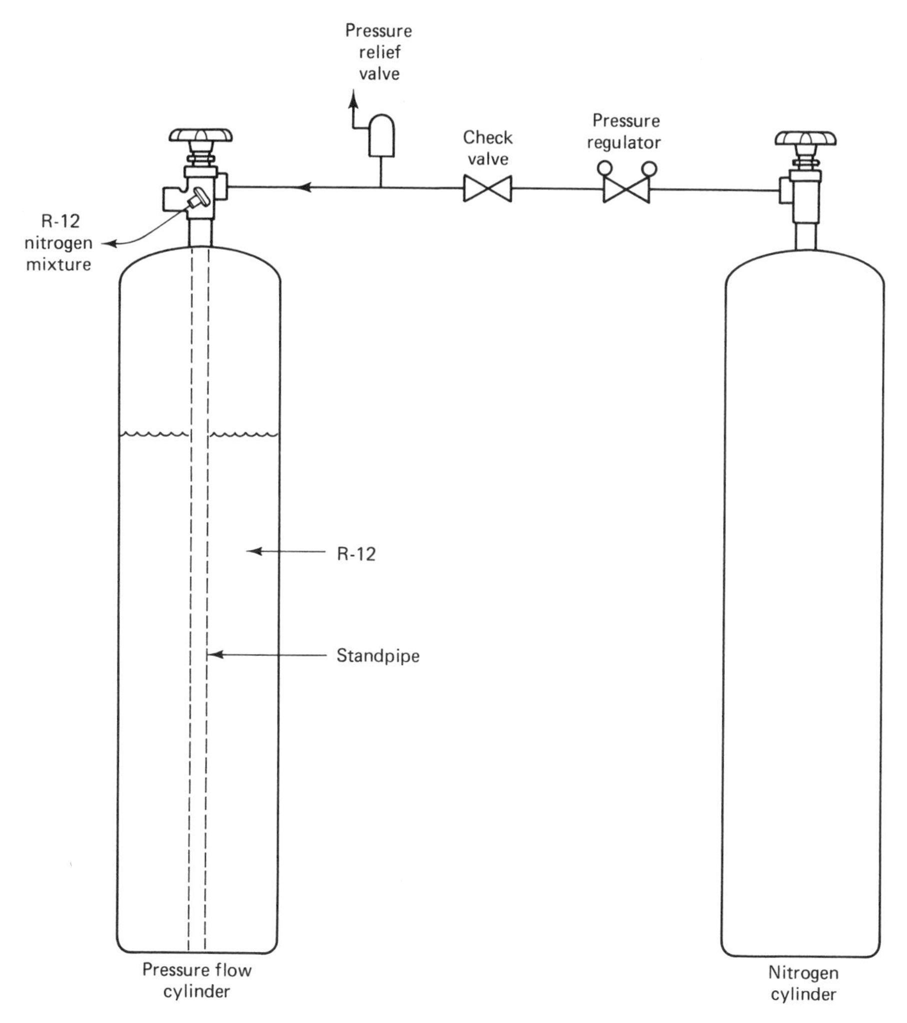

Figure 11–1 Sketch showing connection of nitrogen cylinder to pressure-flow cylinder. (Reprinted by permission of the DuPont Company.)

will be at a temperature of 80°F. The vapor pressure of R-12 at 80°F is 98.9 psia, a fourfold increase. So if the total pressure of 119 psia is increased four times to 476 psia, the partial pressure of R-12 at 80°F will not be exceeded.

In many cases it may be desirable to check the composition of fluorocarbons and air or nitrogen, especially if the mixture is reused. DuPont has developed an inexpensive analyzer for such mixtures with an accuracy of 2 to 5%—adequate for this service [11–8]. It is based on Poiseuille's law for gases flowing in small-diameter tubing with a fixed length under laminar flow conditions.

$$P = \frac{(32)(L)(u)(V)}{(D)^2 g}$$

where P = pressure drop
L = length
u = viscosity
V = velocity
D = diameter
g = gravitational constant

If the gas flow is maintained at a constant or limiting rate, for example by reaching acoustic velocity in an orifice or nozzle downstream from the test section, the resulting pressure drop in the tubing will be a function only of the physical properties of the gas. If two gases in a mixture are sufficiently different in the properties affecting pressure drop, the composition of the gas mixture can be related to the pressure drop. This is the case with air or nitrogen and fluorocarbons.

TABLE 11–2 PRESSURE AND COMPOSITION OF MIXTURES

		Total pressure (psia)				
	Vapor presure (psia)	R-12 in mixture (vol %)				
°F		5	10	15	20	25
−40	9.3	186	93	62	47	37
−20	15.3	306	153	102	76	61
0	23.8	476	238	159	119	95
20	35.7	714	357	238	178	143
40	51.7	1034	517	345	258	207
60	72.4	1448	724	483	362	290
80	98.9	1978	989	659	495	396

Source: Ref. 11–7.

MONITORING FLUOROCARBONS IN THE ATMOSPHERE [11–10]

It is becoming more important for users and handlers of fluorocarbon refrigerants as well as other organic vapors to monitor the concentration of these impurities in the air. For many gases a threshold limit value (TLV) has been established by the American Conference of Governmental Industrial Hygienists [11–9]. Some comments are offered here on the subject of monitoring, but developments in this field move rapidly and for those especially interested, current information should be obtained.

Several different analytical methods can be used to determine the amount of foreign gases present in air. Some methods can be used to identify specific products. Others will measure the total amount of organic vapors present. Some methods are better suited for occasional spot checks of air quality, while others can be adapted for continuous monitoring of several locations. The minimum concentration level that it is desired to detect may also affect the choice of method or the refinement (and cost) of a particular method.

Some analytical methods and suggestions for their use are listed in Table 11–3. For example, infrared absorption can be used to determine the presence and amount of a particular compound if it has a characteristic absorption wavelength. On the other hand, the hydrogen flame method will measure the total amount of organic vapors present. In addition to monitoring total air quality, this method may also be useful for a specific application, such as responding to a refrigerant leak in an operating system.

TABLE 11–3 MEASURING FLUOROCARBONS IN THE AIR

Objective	Method
Quantitative analysis of a specific compound or class of compounds	Infrared
Quantitative identification of all gases present	Gas chromatograph a. Direct analysis of air b. First adsorbed on charcoal
Measurement of total organic vapors present	a. Hydrogen flame ionization b. Ultraviolet Absorption

Charcoal Adsorption

Gaseous contaminants are adsorbed on charcoal in a small cartridge in a portable unit for individual use. The inlet of the sampling tube can be attached to the collar, shirt front, or wherever convenient. Air is drawn through the charcoal by a small pump with a constant-flow controller. The pump can be carried in a pocket or attached to a belt. At the end of the sampling period—for example, a normal workday—adsorbed contaminants are washed from the charcoal. The nature and amount of collected material are determined by gas chromatography.

Infrared

Infrared analysis is probably the simplest and easiest to use of the methods available. It can be calibrated for use with any compound having a distinctive absorption band. Portable units are available (but weigh in the neighborhood of 40 lb). Infrared units can be used for spot checks or for continuous analysis. Levels of detection range from a few parts per billion to a few percent. In general, the greater the sensitivity, the greater the cost.

Hydrogen Flame Ionization

This instrument is known as the organic vapor analyzer and is portable, easy to use, and fairly simple to maintain. It is also available for fixed installation and remote sampling of one or more spots. It will detect all organic vapors—except carbon monoxide and carbon dioxide—so must be corrected for background if gases other than the one being monitored are present. A gas chromatographic attachment is available for quantitative analysis. The operation of the instrument depends on a small hydrogen flame. The supply of hydrogen in the portable unit is enough for about 13 hours of continuous operation.

Gas Chromatograph

In the past, air was often analyzed by bringing samples to the laboratory for gas chromatographic analysis. Now more-or-less portable machines are available for direct analysis in the field. For fluorocarbons the three most common methods of detection used with the gas chromatograph are thermal conductivity, flame ionization, and electron capture. Of these three, thermal conductivity is the least sensitive and electron capture the most sensitive. However, for use with fluorocarbons, the thermal conductivity detector is generally satisfactory. The lower limit of detection (depending on the instrument, sample size, operating conditions, etc.) is about 10 to 20 ppm.

Other Methods

Other instruments use thermal conductivity directly, ultraviolet absorption, and perhaps other methods. These instruments are generally less sensitive than those mentioned above but may be adequate for specific applications.

SENSITIVITY

The fluorocarbon refrigerants respond at different rates to leak detection devices where chemical composition and stability are involved, such as the halide torch and electronic devices. Some differences are given in Table 11–4 in terms of flow rate needed to give the same response as R-12. For example, R-11 must leak at a rate 1.25 times R-12 for the same response.

TABLE 11–4 RESPONSE TIME FOR DIFFERENT REFRIGERANTS

	Response time relative to R-12	
R	By volume	By weight
12	1	1
11	1.25	0.75
13	35 (47)	–
13B1	3.3	–
14	207	145
22	1 (1.33)	0.75 (0.95)
114	1 (1.66)	1.25 (2.43)

Notes

Standard conditions are a temperature of 77°F and a pressure of 14.7 psia or 25°C and a pressure of 760,000 μm. In Table 11–5, all numbers on the same line (reading across) are approximately the same at the same pressure through the same physical leak. Experimental data indicate that no visible water will leak when dry air at the same pressure will leak at the rate of 1.8×10^{-3} cm/s—probably because of surface tension.

TABLE 11–5 COMPARISON OF THE LUSEC WITH OTHER UNITS

	lusec	liter-atm/h	μm-ft³/hr	cm³-atm/min	mm³-atm/s
liter-atm/h	210	1	2.7×10^4	16.1	277
cm³-atm/min	12.67	0.06	1580	1	0.178
mm³-atm/s	0.76	0.0036	96	0.06	1
μm-ft³/hr	0.0078	0.000037	1	6.3×10^{-4}	0.0104
lusec	1	0.0047	128	0.079	1.32

The volumetric leakage rate of helium is about the same as that for air, R-12, R-22, R-114, and so on, for leaks of approximately 1×10^{-6} cm/s or larger. For smaller leaks, the volumetric rate of leakage for helium will be about the same as for air.

Units and Leakage Rates [11–11—11–13]

The *lusec* is defined as the size of leak in a volume of 1 liter causing a pressure change of 1 micrometer of mercury in 1 second. Some relationships illustrating the size of a lusec and comparison with other units are given here and in Tables 11–5 and 11–6. Leaks that are just visible—for example, water leaking from a pail—are

TABLE 11–6 OTHER COMPARISONS

Air				Refrigerant-12 leakage[a]	
Standard atmospheric (cm³/s)	Standard atmospheric (in.³/day)	μm-ft³/hr	Lusec (μm-liter/s)	oz/yr	Time for 1 lb to leak (yr)
1.8×10^{-2}	94.6	1720	13.7	100	0.16
1.8×10^{-3}	9.46	172	1.37	10	1.6
1.8×10^{-4}	0.946	17.2	0.14	1	16
9×10^{-5}	0.473	8.5	0.07	0.5	32
1.8×10^{-5}	0.0946	1.72	0.014	0.1	160
1.8×10^{-6}	0.00946×10^{-3}	0.17	0.0014	0.01	1600
1×10^{-8}	5.6×10^{-5}	1×10^{-3}	7.6×10^{-6}	6×10^{-5}	270,000
1×10^{-10}	5.6×10^{-7}	1×10^{-5}	7.6×10^{-8}	6×10^{-7}	27 million

[a]R-12, R-22, and R-114, under the same conditions of pressure and temperature, will pass through a given leak at approximately the same volumetric rate.

Source: Ref. 11–13.

about 10 lusec or larger. The halide torch has a sensitivity of about 10^{-2} lusec if the apparatus is kept in good condition, not too much air is drawn in, and it is used with great care and attention.

The General Electric Tester (probably type H-2) has a sensitivity of about 2×10^{-4} lusec or 0.28 g of R-12 per year. The General Electric detector, type H-6, has a sensitivity of about 1×10^{-2} lusec or 15 g of R-12 per year.

COMPARATIVE LEAK RATE

The leakage rate is essentially the same process as the flow rate through an orifice and can be calculated from the following relationships:

$$G = \sqrt{(2)(g)(P)(d)}$$

$$W = (G)(\text{area of orifice})(\text{constant, } C)$$

where G = mass velocity, lb/sec-ft^2
g = gravitational constant = 32.174 ft/sec^2
P = change in pressure when gas passes through the orifice, lb/ft^2
d = density, lb/ft^3
W = flow rate, lb/sec
C = about 0.65 for sharp edge orifices and about 0.68 for something like a hole drilled in a plate; C would be somewhat different for the orifice in a pressure regulator where the orifice is changed by something entering the orifice area

The equations are more useful for engineering and laboratory use than for field use since the size of the orifice is not usually known. However, if a reasonable estimate could be made, an approximate leakage rate could be calculated.

When the size of the orifice is the same and the change in pressure is the same, the flow rates of two different gases are related by the square root of the molecular weights. For example, the value of G for R-116 and nitrogen can be compared.

$$\frac{G_{\text{R-116}}}{G_{\text{N2}}} = \frac{\sqrt{\text{molecular weight of R-116}}}{\sqrt{\text{molecular weight of N}_2}}$$

$$\frac{G_{\text{R-116}}}{G_{\text{N2}}} = \frac{\sqrt{138.01}}{\sqrt{28}} = \frac{11.748}{5.292} = 2.22$$

The mass velocity G and the flow rate W in lb/sec are 2.22 times the values for nitrogen if conditions are otherwise the same. The volumetric flow rates can be calculated using the specific volumes. For example, at a temperature of 80°F and a pressure of 1 atm, the volumetric flow rates for the two gases are:

$$1 \text{ lb of nitrogen} = 14.073 \text{ ft}^3/\text{lb} \times 1 \text{ lb/sec} = 14.073 \text{ ft}^3/\text{sec}$$

$$\text{R-116} = 2.835 \text{ ft}^3/\text{lb} \times 2.22 \text{ lb/sec} = 6.294 \text{ ft}^3/\text{sec}$$

REFERENCES

11–1. W. G. Sylvester, ''Use of Compressed Gas for Leak Detection,'' Supplement to the Forty-second Annual Report of the Compressed Gas Association, Arlington, Va.

11–2. DuPont Company, Freon Products Division, Bulletin B-35 (1957).

11–3. R. A. Dalferro, ''Leak Detection—No Holds Barred,'' *Refrig. Serv. Contract.* (Feb. 1975), 12.

11–4. General Electric, Service Bulletin HBK-8232 (Jan. 25, 1965).

11–5. T. H. Gerard, ''Leak Detection—One More Time,'' *Refrig. Serv. Contract.* (Nov. 1975), 8.

11–6. C. S. Booz, Jr., ''Updating with Dyed Refrigerant,'' *Refrig. Serv. Contract.* (Feb. 1975), 13.

11–7. DuPont Company, Freon Products Division, ''Preparation of 'Freon' 12/Nitrogen Mixtures in a Pressureflow Cylinder,'' Bulletin B-35C (1983).

11–8. DuPont Company, Freon Products Division, ''DuPont Leak Test Gas Analyzer,'' Bulletin B-35D (1979).

11–9. American Conference of Governmental Industrial Hygienists, *Threshold Limit Values for Chemical Substances in the Work Environment*. Cincinnati, Ohio: ACGIH (1983–84).

11–10. DuPont Company, Freon Products Division, ''Measuring Fluorocarbons in the Air,'' Bulletin X-253 (1977).

11–11. J. A. Knobfout, ''Finding Leaks in Refrigerating Installations,'' The Hague, *J. Refrig.* (Nov.–Dec. 1960), 136.

11–12. DuPont Company, Freon Products Division, Bulletin B-35A (circa 1965).

11–13. General Electric Company, ''Halogen Leak Detector Control Unit,'' Bulletin HBK-8231 (Jan. 11, 1965).

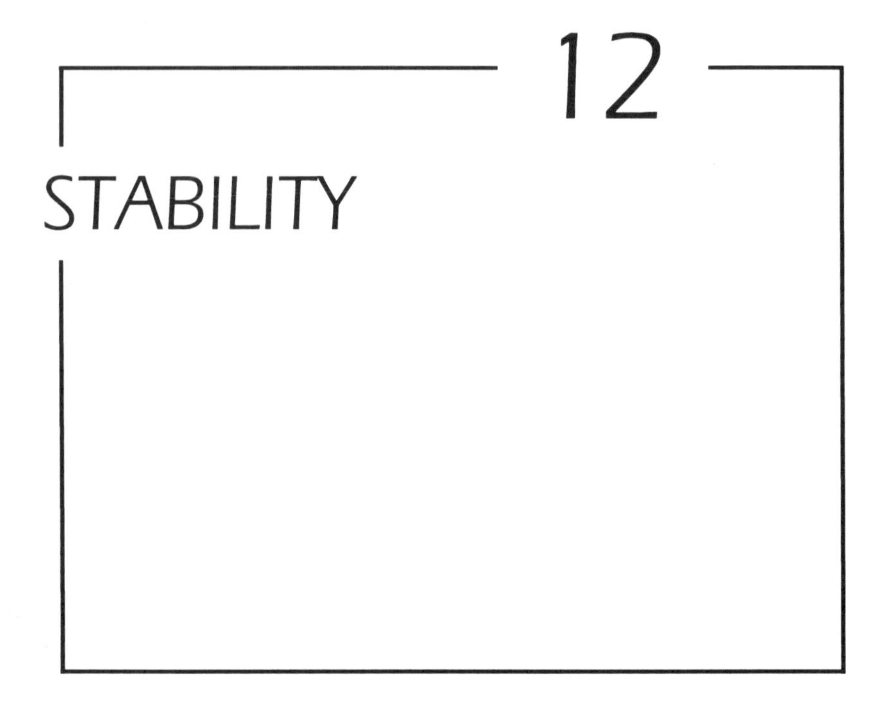

12 STABILITY

One of the attributes of a good refrigerant is *stability*—being able to perform without deterioration over a wide range of temperatures and in the presence of other materials necessary for the operation of the system. The fluorocarbon refrigerants are unusually stable but do have some limitations. Some are more stable than others. As a general rule, compounds containing more fluorine atoms are more stable than those with fewer such atoms. Compounds containing chlorine are more stable than those containing bromine. One-carbon compounds are usually more stable than those with two or more carbon atoms when the relative numbers of other atoms is about the same.

Chemical compounds have an inherent level of thermal stability that is adversely affected by nearly all other materials that might be present. The ''sealed-tube'' test has been developed to determine and measure the effect that conditions and materials may have on the stability of refrigerants. An ASHRAE committee [12–1] is now in the midst of standardizing sealed-tube procedures. Glass (or sometimes metal) tubes are charged with refrigerant and, usually, oil and metal strips and other materials, such as elastomers, plastics, solders, wire coatings, and so on—anything that may be of concern in a refrigeration system. The tubes are stored at various temperatures and for periods of time that may be a week or two or for much longer. Test results are evaluated visually by color of liquids, deposits on metals, appearance of solids, and by analysis for acids, chloride ion, and the presence of decomposition products.

Thermal decomposition of the refrigerant alone may be studied by passing the vapor through heated metal tubes at various flow rates and various temperatures.

The exit gases are analyzed to determine the nature of the decomposition and reaction rates. Stability of the fluorocarbons is of great importance not only in refrigeration but in many other applications of these versatile compounds.

INHERENT STABILITY

The ultimate temperature limitation for a material is when the chemical bonds in a molecule are no longer strong enough to hold the molecule together. Borchardt studied this question and arrived at the summary presented in Table 12–1. Of course, in use, other materials must be present and affect the stability of the refrigerant. The comparisons in Table 12–1 are interesting but may not carry over to practical situations. For example, R-22 has the lowest inherent stability, but in combination with metals, oils, water, and so on, is often more stable than some of the others.

TABLE 12–1 INHERENT STABILITY OF FLUOROCARBONS

Refrigerant	Temperature where decomposition is readily observed in laboratory[a] (°F)	Maximum use temperature[b] (°F)	Major gaseous decomposition products
R-22	800	480	CF_2CF_2[c], HCl
R-11	1100	>570[d]	R-12, Cl_2
R-114	1100	710	R-12
R-115	1160	740	R-13
R-12	1400	930	R-13, Cl_2
R-13	1550	>1000[e]	R-14, Cl_2, R-116

[a]Decomposition rate is about 1% per minute.

[b]Decomposition is about 1% per year.

[c]A variety of side products are also produced here and with other refrigerants, some of which may be quite toxic.

[d]Conditions were not found at which this reaction proceeds homogenously.

[e]Rate behavior is too complex to permit extrapolation to 1% per year.

Source: Ref. 12–2.

Borchardt found that the stability of R-22 can be represented by the following equation [12–2]:

$$k = 10^{12.36}e^{-51,400/RT}$$

and R-114 [12–3] by

$$k = 10^{13.3}e^{-244,460/RT}$$

where k is in reciprocal seconds, R is 1.987 cal/mol-deg, and T = kelvin.

Some of the products of decomposition found by Borchardt for some of the fluorocarbons are summarized below.

$$2CCl_3F \leftrightarrows [CCl_2F_2 + CCl_4] \longrightarrow CCl_2F_2 + Cl_2 + \text{solids}$$
R-11 R-12 R-12

$$2CCl_2F_2 \longrightarrow [CClF_2CClF_2 + Cl_2]$$
R-12 R-114

$$2CCl_2F_2 \longrightarrow [CClF_3 + CCl_3F] \longrightarrow CClF_3 + Cl_2 + \text{solids}$$
R-12 R-13 R-11 R-13

$$2CClF_3 \leftrightarrows Cl_2 + CF_3CF_3$$
R-13 R-116

$$2CHClF_2 \longrightarrow CF_2{=}CF_2 + 2HCl$$
R-22

SEALED-TUBE TESTS

The importance of sealed-tube studies for evaluating the stability of refrigeration systems has been covered in an excellent discussion in the *ASHRAE Systems Handbook* [12–1]. Interpretation of the results of the tests is, of course, very important. As described in the *Handbook*, oil color was suggested by Elsey, Flowers, and Kelley [12–4] as a means of judging the degree of reaction in the test. Later, Spauschus and Doderer [12–5] found that a major chemical reaction involved a direct interchange of chlorine and hydrogen atoms between the refrigerant and the oil.

$$CCl_2F_2 + \text{oil} \longrightarrow CHClF_2 + \text{Cl-oil}$$
R-12 R-22

or

$$CHClF_2 + \text{oil} \longrightarrow CH_2F_2 + \text{Cl-oil}$$
R-22 R-32

Measurement of the product formed in the oil reaction has become the standard method of judging the stability of the system.

Still later, Borchardt [12–6] correlated oil color with the formation of R-22 (when R-12 is the refrigerant in the test). The absorption of light at 450 nm was arbitrarily chosen as a measure of color. The amount of light absorbed by a sample varies with the concentration of the color absorbing material and the length of the light path through the sample. Another quantity, called the *absorptivity*, is defined as follows:

$$\text{absorptivity} = \log\left[\frac{\text{fraction of light transmitted}}{(\text{concentration})(\text{path length})}\right]$$

The oil sample is diluted with an inert, transparent solvent such as ethylene chloride. Concentration is expressed as volume of oil per volume of solution and is dimensionless. The path length is the inside diameter of the cell used for the measurement.

The results of a number of tests using oil and R-12 in the presence of steel, copper, and aluminum at 400°F are shown in Figure 12–1. The mineral oil used was a typical 150 SUS oil such as Suniso 3GS. The synthetic oil was 150 SUS alkylated benzene. It can be seen that the mineral oil develops considerably more color than the synthetic oil for the same conversion of R-12 to R-22. This comparison does not reflect the much longer time needed for the synthetic oil to reach the same stage of degradation as the mineral oil.

EFFECT OF AIR

There is no doubt that the presence of air in a refrigeration system is harmful. Walker and Rosen [12–7] conducted an extensive study of the factors influencing stability. They found that the rate and degree of color development and the rate and percent occurrence of wall deposits, corrosion, and copper plating markedly increased in the presence of air. The test temperatures were 250°F and 375°F and the tests continued for 360 days. About 1 wt % (4 mol %) air was used. Parmelee [12–8] also found that air increased the decomposition of the refrigerant.

Norton [12–9] found that iron and copper oxides greatly increased the rate of decomposition of R-22 and R-12. Air will certainly tend to promote the formation of these oxides and thus indirectly increase decomposition. It has been shown [12–

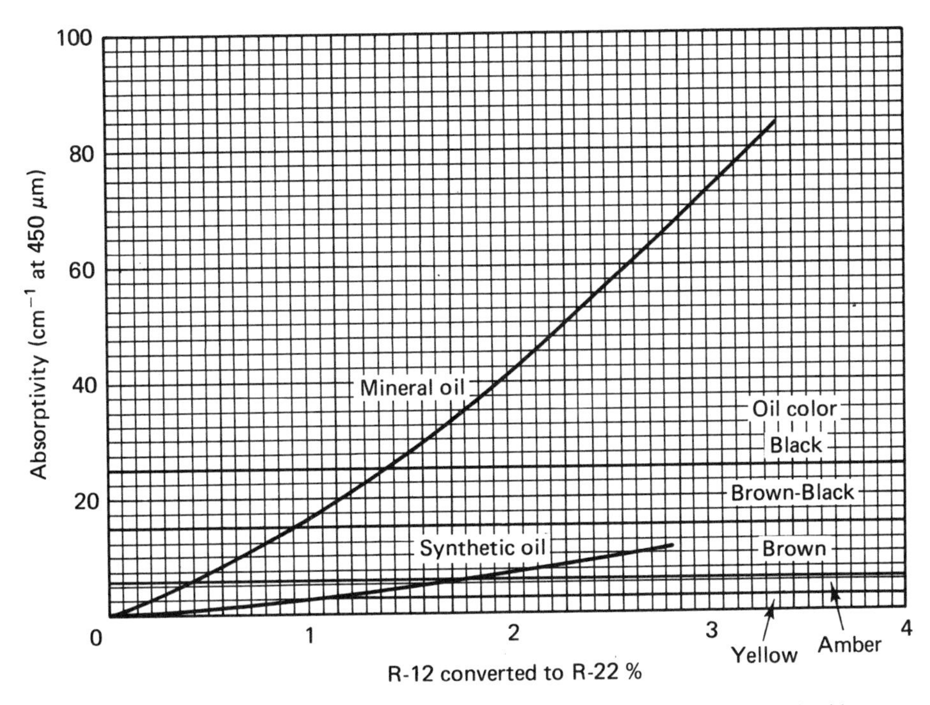

Figure 12–1 Absorptivity of mineral and synthetic oils that had been heated with equal volumes of R-12 in the presence of steel, copper, and aluminum at 400F as a function of R-22 formed from the refrigerant-oil reaction.

3] that concentrations of 2.5 to 3 mol % air based on the refrigerant will cause erratic operation and smaller amounts cause increases in pressure and temperature—conditions that favor decomposition.

EFFECT OF METALS

Most of the commonly-used construction metals, such as steel, cast iron, brass, copper, tin, lead, and aluminum, can be used satisfactorily with the fluorocarbon refrigerants under normal conditions of use. At high temperatures some of the metals may act as catalysts for the breakdown of the refrigerant. The tendency of metals to promote thermal decomposition of these refrigerants is in the following general order [12–10]:

least decomposition inconel $<$ 18–8 stainless steel $<$ nickel $<$ 1040 steel $<$ aluminum $<$ copper $<$ bronze $<$ silver most decomposition

This order is only approximate and exceptions may be found for individual compounds or for special conditions of use.

Magnesium alloys and aluminum containing more than 2% magnesium are not recommended for use in systems where water may be present.

Zinc is not recommended for use with R-113. Experience with zinc and other fluorinated compounds has been limited but no unusual reactivity has been observed. However, it is somewhat more chemically reactive than other common metals and it would seem wise to avoid its use unless adequate testing is carried out.

Metals that may be questionable for use in applications requiring contact with fluorocarbon refrigerants for long periods of time or unusual conditions of exposure, however, can be cleaned safely with fluorinated solvents. Cleaning applications are usually for short exposures at moderate temperatures.

Most halocarbons may react violently with highly reactive materials, such as the alkali and alkaline earth metals, sodium, potassium, barium, and so on, in their free metallic form. Materials become more reactive when finely ground or powdered, and in this state, magnesium and aluminum may react with fluorocarbons, especially at higher temperatures. Highly reactive materials should not be brought into contact with fluorocarbons until a careful study is made and appropriate safety precautions are taken.

Borchardt [12–11] studied the effect of metals on the reaction of R-12 with oil. He found that steel accelerates the reaction, whereas aluminum has only a minor effect and copper none. Combining copper with either steel or aluminum reduces the effect of the steel. Degradation is more pronounced in the presence of steel and aluminum then with either metal alone.

The role of aluminum in refrigeration applications has been somewhat ambiguous. The reaction of aluminum with methyl chloride to form products spontaneously ignitable in air has long been known, but similar reactivity with the fluorocarbons has not been found. Aluminum has been used widely and successfully in evaporators and compressors. The discovery of a destructive reaction between R-12 and an aluminum impeller in a centrifugal compressor was very surprising. Several such incidents were reported. In one case the refrigerant was R-13B1. The reaction was self-sustaining and continued until either the aluminum or the refrigerant was used up. A satisfactory explanation of this phenomenon was not found until Eiseman [12–12] studied the problem in the laboratory. He found that reaction between the fluorocarbons and aluminum was very slow—even at high temperatures—probably because of a protective coating of aluminum oxide or fluoride. If this film were ruptured, however, a vigorous reaction occurred similar to that reported.

In the laboratory test, the film was broken by three different methods: (1) sudden melting of the metal above a pool of liquid fluorocarbon, (2) heating the metal in a stream of fluorocarbon gas, and (3) production of a fresh metal surface by abrasion. Positive reactions were obtained with R-12 and R-22 but not with R-11—evidently due to the much lower pressures developed with R-11. It was shown that reactions in field equipment were probably caused by rubbing of the aluminum impeller against a steel casing, thus rupturing the protective coating on the aluminum and the development of local high temperatures.

Borchardt [12–13] later showed that the acidic products formed in the reaction of R-12 with oil could also destroy the aluminum coating and initiate the direct reaction between R-12 and aluminum.

From available data, Downing summarized the stability of R-11 and R-114 as shown in Tables 12–2, 12–3, and 12–4.

EFFECT OF OIL

Spauschus and Doderer [12–5] discovered that R-12 (CCl_2F_2) reacts with oil to produce R-22 ($CHClF_2$) and in fact that the exchange of a chlorine atom for a hydrogen atom is a rather general reaction. Parmelee [12–8] confirmed the formation of R-22 from R-12 and also showed that R-22 can be converted to R-32 (CH_2F_2), R-13B1 ($CBrF_3$) to R-23 (CHF_3), and R-115 ($CClF_2CF_3$) to R-125 (CHF_2CF_3). It is most likely that the chlorine from the refrigerant enters an oil molecule, eventually forming HCl and unsaturated oil molecules. The latter then polymerize to form deposits of varnish, sludge, and coke. Borchardt [12–2] found that the tendency of the refrigerants to react with oil in the presence of steel decreases in the order R-11 > R-12 > R-114 > R-115 > R-22.

The nature of the oil affects the rate of reaction with refrigerants. In general, paraffinic-type oils are least reactive and aromatic oils most reactive, with naphthenic oils in between (see Chapter 9). For example, Parmelee [12–8] found that reaction with R-22 at 400°F was 0.06% for refined white oil (53 days) compared with 0.69% with naphthenic oil (160 days). Even allowing for the difference in time, the white oil was less reactive.

ACIDITY IN REFRIGERATION SYSTEMS

In checking on the performance of refrigeration systems or in attempting to diagnose the cause of difficulties, a test for acidity is often made on the refrigerant or oil. The test is performed by titrating an alcoholic solution of the refrigerant or oil

TABLE 12–2 STABILITY OF R-114 ALONE AND WITH STEEL

°F	Decomposition (%/year) Alone	Decomposition (%/year) Steel	Test days	Test	Reference
800	9.8		6.3	Quartz tube, flowing gas	12–14
800	8.7		6.7	Quartz tube, flowing gas; rate was 7.3% per year for last 5 days	12–14
800	45			Equation 1[a]	12–3
800		20.3	3.6	Quartz tube, flowing gas	12–15
800		23.1	1.9	Quartz tube, flowing gas	12–14
800		19.2	3.8	Quartz tube, flowing gas	12–14
752		38.4	23	Steel tube, flowing gas	12–15
752		21.3	40	Steel tube, flowing gas	12–15
752		32.1	35	Steel tube, flowing gas	12–15
752		30.8	20	Steel tube, flowing gas	12–15
752		89.7	2.4	Steel tube, flowing gas	12–15
734		153	8.6	Steel bomb	12–15
710	1.07			Equation 1[a]	12–3
700	0.68			Equation 1[a]	12–3
700	1.04		42	Quartz tube, flowing gas	12–14
700		0.73	30	Quartz tube, flowing gas	12–14
662		7.8	6	Steel bomb	12–3
650	0.063			Equation 1[a]	12–3
600	0.0046			Equation 1[a]	12–3
600		0.45	41	Quartz tube, flowing gas	12–15
600		0.46	40	Quartz tube, flowing gas	12–14
572		120	6	Steel bomb	12–3
550	0.00026			Equation 1[a]	12–3
500	0.000011			Equation 1[a]	12–3
482		26	6	Steel bomb	12–3
450	3×10^{-7}			Equation 1[a]	12–3
400		12	6	Steel bomb	12–3
400		0.5	730	Sealed glass tube	12–8
300		0.006	730	Sealed glass tube	12–8

[a] $k = 10^{13.3}e^{-244,460/RT}$, where k = reciprocal seconds, R = 1.987 cal/mol-deg, and T = kelvin.

with standardized alcoholic potassium hydroxide. The result is reported as milligrams of KOH per gram of sample (acid number). In nearly all cases, a measurable amount of acidity is found in operating systems. The significance of the acidity value is often not clear and recommendations are difficult to make.

Acidity tends to develop in all refrigeration systems using fluorinated refrigerants. Acidity may be caused by thermal decomposition or hydrolysis of the refrigerant, by degradation of the oil to form various organic acids, and by degradation of insulation materials. Very little work has been done toward identifying the source of the acidity determined by titration with alcoholic potassium hydroxide. However, some information is available.

During normal operation, there is some evidence that the formation of acidity is not due to breakdown of the refrigerant. In several cases, measurable acidity

TABLE 12–3 STABILITY OF R-11 ALONE AND WITH STEEL

	Decomposition (%/year)				
°F	Alone	Steel	Test days	Test	Reference
800	73		3	Quartz tube, flowing gas	12–14
800		1000	1.8	Quartz tube, flowing gas	12–14
800		544	1.8	Quartz tube, flowing gas	12–14
700	9.1		32	Quartz tube, flowing gas	12–14
700		20.5	16	Quartz tube, flowing gas	12–14
662		4700	6	Steel bomb	12–3
572		600	6	Steel bomb	12–3
570	1.0			Platinum tube	12–2
482		120	6	Steel bomb	12–3
350	0.21		35	Sealed glass tube	12–3
350	0.0011		324	Sealed glass tube	12–3
392		27	6	Steel bomb	12–3
300		0.54	730	Sealed glass tube	12–8
300		0.17	730	Sealed glass tube	12–8
250		0.021	730	Sealed glass tube	12–3
250		0.015	730	Sealed glass tube	12–3

was found by titration, but no indication of chloride ion was found using the silver nitrate test. In nearly all cases where data are available, the oil from a system shows very much more acidity than the refrigerant. Evaporation and recovery of the refrigerant from a solution of refrigerant and oil is sufficient in most cases to reduce the acidity of the refrigerant to a very low level and concentrate acidity in the oil.

TABLE 12–4 STABILITY OF R-11 AND R-114 WITH COPPER AND ALUMINUM

	°F	Metal	Test days	Decomposition (%/year)
R-11	250	Copper	208	15
	250	Copper	189	46
	250	Aluminum	722	<0.001
	250	Aluminum	733	0.004
	300	Copper	208	170
	300	Copper	17	472
	300	Aluminum	722	0.021
	300	Aluminum	733	0.002
R-114	300	Copper	727	0.075
	300	Copper	727	0.105
	300	Aluminum	727	<0.001
	400	Copper	727	0.27
	400	Aluminum	727	0.014
	400	Aluminum	727	0.016

Source: Ref. 12–8.

The significance of acidity in a refrigeration system must be evaluated in connection with other factors, such as length of operation, recent history, odor and color of oil and refrigerant, and so on. The absolute value of the acidity may not be as significant as the trend over a period of time and comparison with the original condition of the materials.

The development of unusual amounts of acidity is a symptom of abnormal operation. Merely replacing the refrigerant or oil would be effective in curing the symptom but not the disease. When acidity is found in centrifugal units, the causes may be different than in reciprocating compressors. The purge unit may not be operating correctly and the water and air concentrations may be higher than they should be. If excessive oil is found in the refrigerant, the shaft seals may be defective.

In reciprocating units, acidity may be caused by abnormally high operating temperatures, excessive amounts of water or alcohol, electrical shorts, or failure to clean up a system properly after a motor burnout. Whenever high acidity is found, an attempt should be made to determine the cause as well as removing the acidity.

With the present state of knowledge about the nature and causes of acidity, attempts to quantitatively measure its significance are for the most part subjective and arbitrary. However, the following classification might be considered.

1. Acidity of refrigerant less than 0.01 mg of KOH per gram is in the normal range of acidity. However, color of oil and water content should be noted to see if future difficulties might be expected.
2. Acidity of refrigerant in the range 0.01 to 0.1 mg of KOH per gram is higher than normal and an activated alumina drier should be used to reduce the acidity. The cause of the acidity should be determined if possible and steps taken to correct it. This condition would not be considered alarming but should be watched.
3. If the acidity of the refrigerant is above 0.1 mg of KOH per gram, something is probably wrong. An activated alumina drier should be used and the source or cause of the acidity determined. When the acidity of the refrigerant is high, the acidity of the oil is probably much higher. If the oil is dark or if it has a strongly acrid odor, the oil should be replaced. In extreme cases, it may be necessary to replace the refrigerant and, as a general precaution, would be desirable. In many cases, however, replacing the oil and using a good drier should be sufficient to reduce the acidity to a normal level.

Jones et al. [12–16] have evaluated the acidity of a large number of oil samples from field systems and have selected a value of 0.063 mg of KOH per gram of oil as the dividing line between satisfactory and unsatisfactory acidity.

STABILITY OF R-12 AND R-22

In studying stability using sealed tubes, results are usually interpreted by comparison with a standard or control system treated under the same conditions as the test. The standard consists of refrigerant, a naphthenic-based oil (150 or 300 SUS viscos-

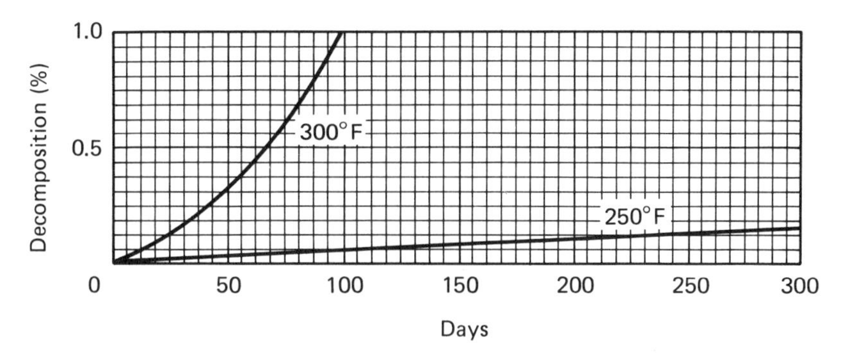

Figure 12–2 Decomposition of R-12 in the presence of oil, copper, and steel.

ity), and test pieces of copper and mild steel. The curves in Figures 12–2 and 12–3 illustrate the decomposition of R-12 and R-22 for such control systems [12–17]. It is evident that the decomposition rate for R-12 is much higher than for R-22 under these conditions. With R-12 at 300°F and R-22 at both 250°F and 300°F, the rate is linear. The results of sealed-tube tests are helpful but must be interpreted in relation to realistic systems. For example, the decomposition of R-12 at 300°F is very high—but in most systems temperatures never get that high, or if they do, only for very short periods.

These sealed-tube tests are either neutral or become acidic if decomposition of the refrigerant or oil occurs, and under these conditions R-22 is more stable than R-12. However, under alkaline conditions, the reverse is true. In fact, R-22 reacts so rapidly with alkaline material at room temperature compared with R-12 that this difference can be used to distinguish between the two. R. B. Ramsey of DuPont [12–3] developed several analytical methods for rapid testing in the field or laboratory to determine if a given refrigerant is R-22 or R-12.

1. Bubbling R-22 through a solution containing 8 wt % sodium methylate in methyl alcohol caused immediate breakdown and the formation of a white

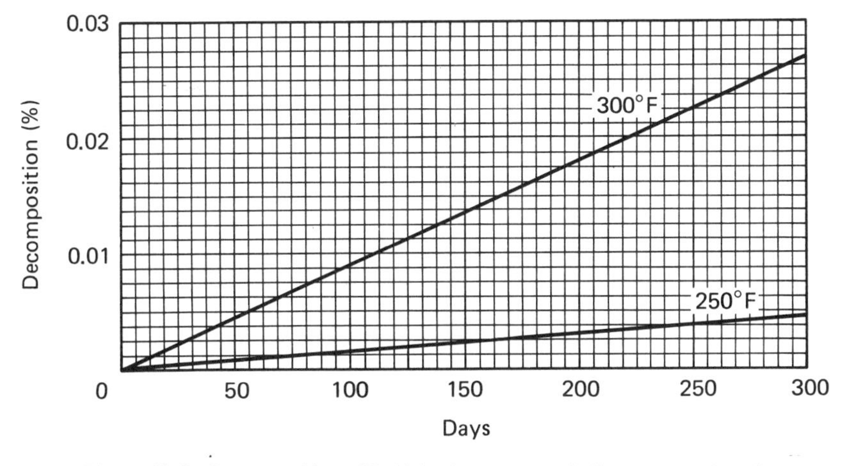

Figure 12–3 Decomposition of R-22 in the presence of oil, copper, and steel.

deposit of sodium chloride. R-12 is not affected. Mixtures of R-22 and R-12 give a positive result at R-22 concentrations down to about 5 wt %.

2. When R-22 is bubbled through an aqueous sodium hydroxide solution it is decomposed and the addition of a silver nitrate solution causes the precipitation of silver chloride. R-12 does not react under these conditions. The presence of hydrochloric acid will give a false result and oil may yield a cloudy or milky solution that will obscure the analysis.
3. Both R-22 and R-12 react with pyridine in the presence of alcoholic potassium hydroxide at room temperature. With R-22 a yellow color is produced, while with R-12 a pink or red solution is formed.

REFERENCES

12–1. American Society of Heating, Refrigerating and Air-Conditioning, Engineers, *ASHRAE Systems Handbook*. Atlanta, Ga.: ASHRAE (1984), Chap. 27.

12–2. H. J. Borchardt, "New Findings Shed Light on Reactions of Fluorocarbon Refrigerants," *DuPont Innovation*, 6 (Winter 1975).

12–3. DuPont Company, Freon Products Division, unpublished information.

12–4. H. M. Elsey, L. C. Flowers, and J. B. Kelley, "A Method of Evaluating Refrigerator Oils," *Refrig. Eng.*, 60 (July 1952), 737.

12–5. H. O. Spauschus and G. C. Doderer, "Reaction of Refrigerant 12 with Petroleum Oils," *ASHRAE J.*, 3 (Feb. 1961), 65.

12–6. H. J. Borchardt, "Darkening of Synthetic and Mineral Oils due to Reaction with Refrigerant 12," DuPont Company, Freon Products Division, Bulletin X-226.

12–7. W. O. Walker and S. Rosen, "Stability of Mixtures of Refrigerants and Refrigerating Oils," *ASHRAE J.*, 4 (1962), 59.

12–8. H. M. Parmelee, "Sealed-Tube Stability Tests on Refrigeration Materials," *ASHRAE Trans.*, 71, pt 1 (1965), 154.

12–9. F. J. Norton, "Rates of Thermal Decomposition of $CHClF_2$ and $CClF_3$," *Refrig. Eng.*, 65 (Sept. 1957), 33.

12–10. DuPont Company, Freon Products Division, Bulletin B-2 (1971).

12–11. H. J. Borchardt, "Effect of Metals on the Reaction of Refrigerant 12 with Oil," in *Some Interactions of Refrigerants with Oil and Water*. Atlanta, Ga.: American Society of Heating, Refrigerating and Air-Conditioning Engineers, Inc. (1973).

12–12. B. J. Eiseman, Jr., "Reactions of Chlorofluorohydrocarbons with Metals," *ASHRAE J.*, 5 (May 1963), 63.

12–13. H. J. Borchardt, "The Reaction of Aluminum with Refrigerant 12 and Oil," in *Lubricants, Refrigerants, and Systems—Some Interactions*. Atlanta, Ga.: American Society of Heating, Refrigerating and Air-Conditioning Engineers, Inc. (1972).

12–14. K. L. Snider, "Thermal Stability of Several Fluorocarbons," *ASHRAE J.*, 9 (Nov. 1967) 54.

12–15. J. A. Callighan, "The Thermal Stability of Fluorocarbons 114 and 216," *ASHRAE J.*, 11 (Sept. 1969) 65.

12–16. E. Jones, A. A. Krawetz, T. Tovrog, and A. E. Thompson, ''Field Evaluation of Compressor Lubricant Acidity,'' *ASHRAE J.*, 8 (Dec. 1966), 54.

12–17. D. E. Kvalnes and H. M. Parmelee, ''Behavior of Freon-12 and Freon-22 in Sealed-Tube Tests,'' *Refrig. Eng.*, 65 (Nov. 1957), 40.

12–18. R. C. Downing, ''Thermal Stability of R-114 and R-11 in Solar Applications,'' in *Solar Engineering*, Book No. H00212, W. D. Turner, ed. New York: American Society of Mechanical Engineers (1982) (presented at the ASME meeting, Albuquerque, N.M., Apr. 27, 1982).

13

SOLAR APPLICATIONS

Fluorocarbons are used in solar applications for the same reasons that they are used in refrigeration and air conditioning.

Nonflammable and low in toxicity
Stable and noncorrosive
Nonfreezing
Heat transfer using phase change and latent heat

Properties of special interest in solar application include:

Faster response to heat from the sun
Can be used without a liquid pump for some applications
Simplified controls due to phase change
No reverse heat flow
Small pipe size

In operation, liquid fluorocarbon is evaporated in the solar collector by heat from the sun. The hot gas is condensed in a heat exchanger, giving up latent heat to water or air. The condensed liquid is returned to the collector, either by gravity or with a liquid pump, and reevaporated.

Most of the fluorocarbon refrigerants can be used for transferring heat from the sun, although the lower-boiling ones have somewhat better properties, as illustrated in Figure 13–1. R-12 would be a good choice if higher vapor pressures are not objectionable, since it is used in most refrigerators and is readily available. It does have a fairly low critical temperature (234°F) and would not operate at temperatures near or above the critical. R-114 or R-11 is usually selected because pressures are moderate and other properties are satisfactory. The atmospheric boiling point of R-113 (118°F) is a little high and if leakage should occur, air and water vapor could leak in at lower temperatures.

All of the fluorocarbons shown in Figure 13–1 can be used with steel, aluminum, and copper, with the single exception of R-11 with cooper. At temperatures that may be reached during stagnation, some decomposition of R-11 has been observed. Other metals, such as steel and aluminum, have not been affected. The reaction has not been observed in copper lines carrying R-11 to and from the collector.

One important difference between the fluorocarbons and water or other liquids is the pressure. For those familiar with refrigeration this difference will not be a problem but should be remembered and understood. In general, the pressure is determined by the temperature at which liquid and vapor are in equilibrium in the coolest part of the system. Under some conditions, if the fluorocarbon charge in the system is too high, hydrostatic pressure could develop—producing pressures much higher than the vapor pressure.

Ordinarily, the pressure throughout the system will be the same—governed by the temperature of the liquid—but temperatures may vary. When no heat is being transferred, temperatures in the collector may, on rare occasions, reach 300°F or more.

The liquid heat capacity of the fluorocarbons is about one-fourth as great as that of water. An amount of heat that will raise the temperature of water 1°F will raise the temperature of the same weight of refrigerant liquid 4°F.

The use of solar energy depends on the transfer of heat from the collector piping to the fluid, so heat-transfer coefficients are important. A chart of heat flux versus coefficient is shown in Figure 13–2. This relationship is for the film of liquid on the inside of the pipe in the collector. It is assumed that heat transfer

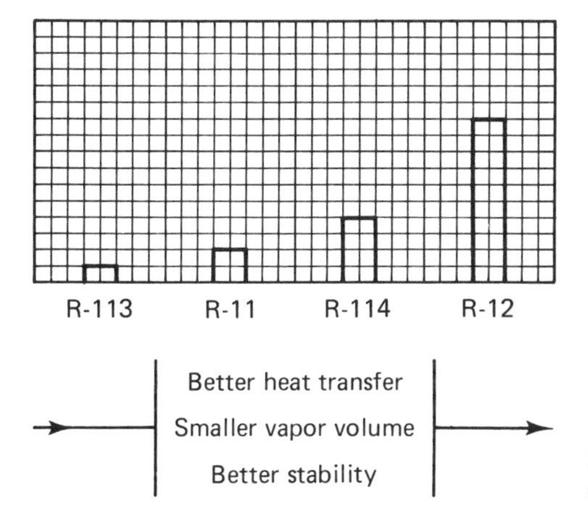

Figure 13–1 Comparative properties of solar fluids.

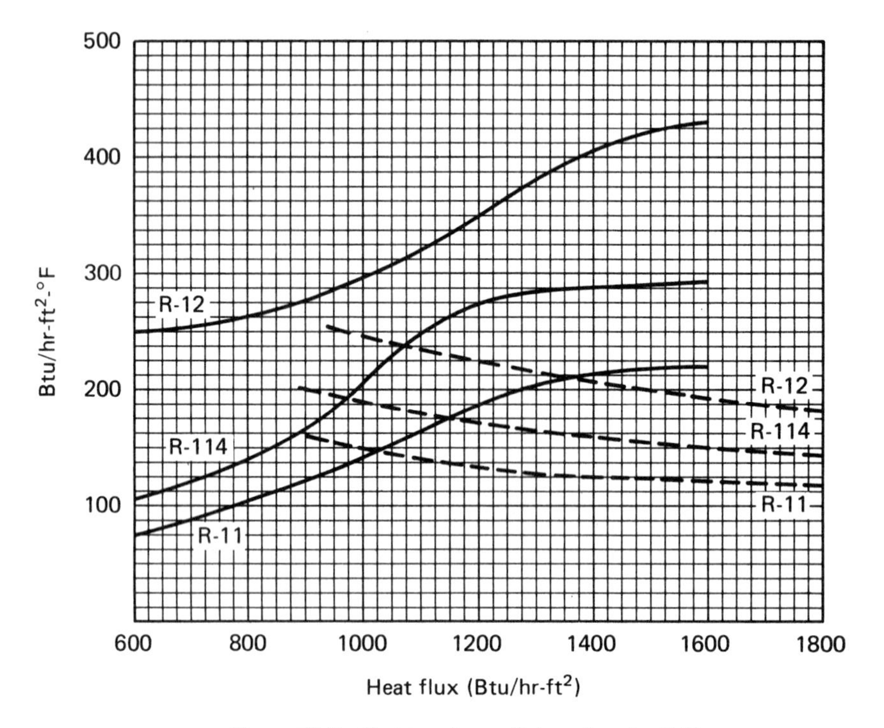

Figure 13–2 Heat transfer coefficients for solar fluids.

through the walls of the pipe will be faster than transfer through the film. The values in Figure 13–2 were measured and specifically apply only to equipment similar to that used in the measurement. It is believed, however, that the experimental conditions were sufficiently general that the results can be used, without serious error, for most solar applications. The following example illustrates the use of the boiling heat-transfer coefficient for R-114.

Assume that each square foot of collector panel contains 2-ft sections of copper pipe with internal diameter of 0.5 in. Assume that the effective heat flow into the liquid in the pipe is 250 Btu/h-ft^2 of collector surface. The inside area of the pipe is

$$(0.04167 \text{ ft})(\pi)(2 \text{ ft}) = 0.2618 \text{ ft}^2$$

$$\text{Heat flux} = \frac{250}{0.2618} = 955 \text{ Btu/hr-ft}^2 \text{ of pipe surface}$$

From Figure 13–2 the heat-transfer coefficient for R-114 is 130 Btu/hr-ft^2-°F. The temperature difference across the pipe wall is

$$\frac{955 \text{ Btu/hr-ft}^2}{130 \text{ Btu/hr-ft}^2\text{-°F}} = 7.3\text{°F}$$

To decrease the temperature difference, a larger-diameter pipe could be used or another length of pipe added to the 1-ft^2 area of the collector.

To illustrate the use of condensing coefficients, assume that the heat flow is 250 Btu/hr for each square foot of collector surface and that water pipe with an outside diameter of 0.625 in. (0.05208 ft) is used. The condensing surface area per foot of pipe will be 0.1636 ft^2.

$$\text{Heat flux} = \frac{250 \text{ Btu/hr}}{0.1636 \text{ ft}^2} = 1528 \text{ Btu/hr-ft}^2$$

From Figure 13–2, the condensing coefficient for R-114 is 150 Btu/hr-ft^2-°F, so the temperature difference in the condenser will be

$$\frac{1528 \text{ Btu/hr-ft}^2\text{-°F}}{150 \text{ Btu/hr-ft}^2 \text{ -°F}} = 10.2\text{°F}$$

So a 1-ft length of copper pipe (0.625 in. OD) will be sufficient to transfer heat at a rate of 250 Btu/hr from the saturated vapor of R-114 condensing on the outside of the pipe to water flowing with turbulent flow inside of the pipe with a temperature drop of about 10°F. For a collector area of 60 ft^2, about 60 linear feet of pipe would be needed. The condensing area would be $60 \times 0.1636 = 9.8$ ft^2. Since this calculation is for bare pipe, if finned tubing were used, the linear footage would be much less.

The following relationship is useful:

$$L = \frac{I}{(A)(n)(dT)}$$

where L = length of condenser pipe in feet for 1 ft^2 of collector surface
I = effective heat input from solar radiation actually entering liquid of heat-transfer fluid, Btu/hr-ft^2 of collector area
n = condensing coefficient, Btu/hr-ft^2-°F from Figure 13–2
dT = temperature difference between saturated vapor of refrigerant and water in the pipe

A typical solar collection system using fluorocarbons for heat transfer is illustrated in Figure 13–3. Other arrangements may work as well or better. In Figure 13–3 the collector is shown with both a top and a bottom header. Liquid enters the tubes from the bottom header. As solar energy generates vapor it moves upward to the top header and at some level of heat flux entrains liquid, which must be separated from the vapor in the header. The liquid should be returned to the bottom header by a separate passageway.

The condensor or heat exchanger shown in Figure 13–3 is simply an example; other configurations can be used. The condensing fluorocarbon should be on the shell side. Heat transfer is just as important in the condenser as in the collector. For 60 ft^2 of collector surface, about 10 ft^2 of condensing surface is adequate. About 3 ft^2 of surface is sufficient on the inside of the finned tubing. Most condensers are built with finned tubing with an effective surface area much greater than bare pipe, so the linear feet needed would be much less. For much of the time, the effective incoming solar radiation will be less than the maximum, so a relatively small temperature drop will be found.

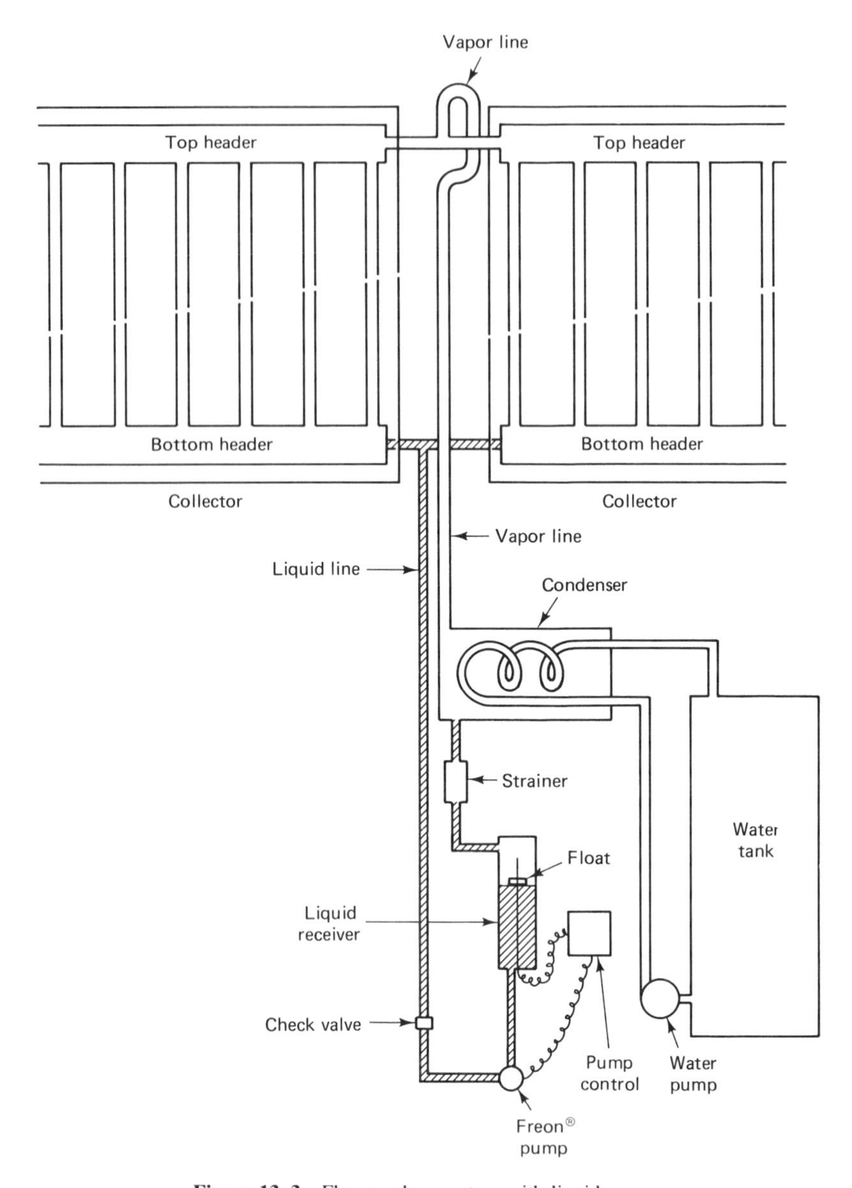

Figure 13–3 Fluorocarbon system with liquid pump.

Condensate should leave the condensing surface as quickly as possible. Most refrigerants are poor thermal conductors and any buildup of liquid film on the condensing surface will seriously impede the heat transfer. A strainer in the liquid line from the condenser to the receiver will help protect the liquid pump by removing pieces of metal that may be left in the system during manufacture and assembly.

The liquid receiver should be located at least 2 ft above the pump to provide a good liquid head. With a float control, it serves to operate the pump when needed and a reservoir for the liquid when the system is not operating. The size of the receiver is not critical. A pump cycle should return 2 to 2.5 lb of liquid to about 60 ft^2 of collector surface.

Cavitation occurs when the pump cavities do not get filled with liquid on the intake side. As the pump rotates and an unfilled cavity sees the high-pressure discharge side, the vapor in the unfilled cavity collapses and the impact is noisy and damaging. The slower a pump rotates and the higher the suction head, the better the chance that the cavities will be filled with liquid. Another physical condition that promotes cavitation is the temperature difference between the liquid to be pumped and the ambient temperature. The condensate leaving the heat exchanger may be 10 to 15°F colder than the ambient temperature if the hot water storage tank happens to be essentially at the temperature of the water supply. In this case, heat will flow into the cold liquid, causing it to boil and vapor to enter the pump. To offset this condition, the condenser should be as high as possible above the pump. Seven feet of liquid R-114 head will raise the boiling point about 7°F and reduce the tendency toward pump cavitation.

One system design parameter is the volume of vapor that must be transferred for a given amount of latent heat. Some comparisons of this factor are given in Table 13–1. A system designer can choose between higher pressures or larger vapor line sizes. Vapor lines from the collector should be sloped properly to avoid liquid traps that increase the pressure drop in the lines or even block the flow of vapor to the condenser.

It is essential that a refrigerant-charged heat-transfer system be as leak-free as possible. Flushing of all piping systems to remove fluxes, metal particles, and so on, is recommended before charging with a refrigerant. It is suggested that $\frac{3}{4}$-in. soft-copper tubing be used for the vapor lines for runs under 40 ft and 60 ft^2 of collector surface (Tables 13–2 and 13–3). The maximum pressure drop would be about 2 psi—and most of the time, much less.

The dimensions of soft-annealed copper tubing and hard-drawn copper tubing (pipe), type L, are given in Table 13–2. Both types of tubing should be referred to by the actual outside diameter. In the past there has been some confusion by using a so-called "nominal size." For soft tubing this size is the same as the outside diameter, but for hard tubing the nominal size may be $\frac{1}{8}$ inch smaller than the outside diameter.

TABLE 13–1 FLUOROCARBON VOLMETRIC EFFICIENCY

Fluorocarbon	Liquid temperature (°F)	System pressure (psig)	Volumetric efficiency (Btu/ft^3)
113	78	6.6 (psia)	14
	212	49	100
11	78	0.9	30
	212	104	170
114	78	17	55
	212	190	255
12	78	81	139
	212	470	425

TABLE 13–2 COPPER PIPE AND TUBING

Nominal size (in.)	Soft copper tubing[a] Outside diameter (in.)	Inside diameter (in.)	Inside cross section (in.^{2})	Type L copper tubing[b] Outside diameter (in.)	Inside diameter (in.)	Inside cross section (in.^{2})
$\frac{1}{4}$	0.250	0.190	0.02835	0.375	0.315	0.07793
$\frac{3}{8}$	0.375	0.311	0.07596	0.500	0.430	0.1452
$\frac{1}{2}$	0.500	0.436	0.1493	0.625	0.545	0.2333
$\frac{5}{8}$	0.625	0.555	0.2419	0.750	0.666	0.3484
$\frac{3}{4}$	0.750	0.666	0.3632	0.875	0.785	0.4840
$\frac{7}{8}$	0.875	0.785	0.4840			
1				1.125	1.025	0.8252

[a]American Society of Heating, Refrigerating and Air-Conditioning Engineers, ASHRAE Handbook, Equipment volume, Atlanta, Ga., 1983.

[b]ARI, "Refrigeration and Air-Conditioning," Prentice-Hall, Inc. (1979) p. 34; R. Miller, "Refrigeration and Air Conditioning Technology," Bennett Publishing Company, Peoria, Il., 1983, p. 61.

TABLE 13–3 WEIGHT OF REFRIGERANT IN TYPE L COPPER PIPE

Nominal pipe size (in.)	OD (in.)	Weight in pounds in 100 ft of pipe: R-11 80°F	R-11 140°F	R-114 80°F	R-114 140°F	R-12 80°F	R-12 140°F
			Liquid				
$\frac{1}{4}$	0.375	4.97	4.69	4.90	4.53	4.41	3.94
$\frac{3}{8}$	0.500	9.26	8.74	9.13	8.43	8.21	7.33
$\frac{1}{2}$	0.625	14.9	14.0	14.7	13.6	13.2	11.8
$\frac{5}{8}$	0.750	22.2	21.0	21.9	20.2	19.7	17.6
$\frac{3}{4}$	0.875	30.9	29.1	30.4	28.1	27.4	24.5
1	1.125	52.7	49.7	51.9	47.9	46.7	41.7
			Saturated Vapor				
$\frac{1}{4}$	0.375	0.022	0.057	0.056	0.14	0.13	0.30
$\frac{3}{8}$	0.500	0.040	0.011	0.10	0.26	0.25	0.57
$\frac{1}{2}$	0.625	0.065	0.17	0.17	0.42	0.39	0.91
$\frac{5}{8}$	0.750	0.10	0.25	0.25	0.63	0.59	1.36
$\frac{3}{4}$	0.875	0.14	0.35	0.35	0.87	0.82	1.89
1	1.125	0.23	0.60	0.60	1.49	1.39	3.22

14

CONVERSION FACTORS AND SHIPPING

UNITS

English units are used by the refrigeration industry in the United States, although most of the rest of the world uses the new International System (joule) or the older metric system (calorie). Some conversion factors are given in Table 14–1 and more complete tables and discussions of units are readily available [14–2—14–4].

Other Conversions

Temperature

$$°K = °C + 273.15 = \frac{°F + 459.67}{1.8}$$

$$°R = °F + 459.67 = (°C + 273.15)(1.8)$$

$$°C = °F - 32 \times \frac{5}{9}$$

$$°F = °C \times \frac{9}{5} + 32$$

Vacuum. It is regretable that the artificial division of pressure into absolute and gage in English units was adopted by the refrigeration industry. Using only

TABLE 14–1 CONVERSION FACTORS

To convert from:	To:	Multiply by:
atm	bar	1.01325
	in. Hg (32°F)	29.921
	psia	14.696
	kPa, absolute	101.325
	kg/cm^2	1.03323
	mm Hg (0°C)	760
	torr	760
Btu (thermochemical)	cal	251.996
	ft^3-atm	0.36747
	J	1,054.35
	hp-hr	0.00039275
	liter-atm	10.4053
	W-hr	0.292875
Btu (IT)	Btu (thermochemical)	1.00065
	J	1055.056
Btu/lb	cal/g	0.55556
	ft^3-atm/lb	0.367471
	cm^3-atm/g	22.9405
	J/g	2.32444
	kW-h/g	6.4568×10^{-7}
	W-h/g	0.00064568
Btu/min	cal/min	251.996
	kcal/min	0.251996
	J/s	17.5725
	W	17.5725
	Hp	0.023565
	ton of refrigeration	0.0049967
Btu/ft^3	cal/cm^3	0.00889915
	cal/liter	8.89940
	Btu/m^3	35.31467
	J/ft^3	1,054.35
	J/m^3	37,234.02
Btu/lb-°F	cal/g-°C	1
	J/g-°C	4.184
Btu/hr	kcal/hr	0.251996
	kW	0.000292875
	Hp	0.000392752
	ton of refrigeration	8.32789×10^{-5}
	J/s	0.292875
Btu-ft/hr-ft^2-°F or Btu/hr-ft-°F	cal/hr-cm-°C	14.8816
	W/cm-°C	0.017296
	J/hr-cm^2-°C	62.2636
	cal/s-cm-°C	0.0041338
	J/s-cm-°C	0.017296
Calorie	Btu	0.003967
	W-h	0.00116222
	J	4.184

TABLE 14–1 (Cont.)

To convert from:	To:	Multiply by:
Cal/g	Btu/lb	1.8
	J/g	4.184
	W-h/g	0.00116222
	hp-hr/lb	0.000706953
cal/g-°C	Btu/lb-°F	1
cal/m^3	Btu/ft^3	0.000112370
	cal/liter	0.001000028
	J/ft^3	0.118478
	J/liter	0.00418412
cal/min	Btu/min	0.0039683
	J/s	0.069733
	W	0.069733
	hp	9.35139×10^{-5}
cm^3	ft^3	3.53147×10^{-5}
	m^3	1×10^{-6}
	gal (U.S. liquid)	0.00026417
	liter	0.000999972
cm^3/g	ft^3/lb	0.01601846
	gal(U.S.liquid)/lb	0.119826
cm^3/min	ft^3/min	0.00003531
	gal(U.S. liquid)/min	0.00026417
	liter/hr	0.0599983
ft^3	m^3	0.0283168
	gal (U.S.liquid)	7.48052
	liter	28.31605
	in.^{3}	1728
ft^3/hr	liter/hr	28.31605
	gal(U.S.liquid)/min	0.124675
	cm^3/s	7.86579
ft^3/lb	cm^3/g	62.42796
	liter/kg	62.426204
	m^3/kg	0.062428
	gal(U.S.liquid)/lb	7.48052
m^3	cm^3	1×10^6
	ft^3	35.3147
	gal (U.S.liquid)	264.172
	liter	999.972
m^3/s	ft^3/min	2,118.88
	liter/min	59,998.32
	gal(U.S.liquid)/min	15,850.32
m^3/kg	ft^3/lb	16.01846
	liter/kg	999.972
	cm^3 /g	1,000
	gal(U.S.liquid)/lb	119.8264
°K	°R	1.8
gal (U.S./liquid)	ft^3	0.13368
	m^3	0.0037854
	liter	3.7853

TABLE 14–1 (Cont.)

To convert from:	To:	Multiply by:
ft	cm	30.48
	m	0.3048
	μm	304,800
	mil	12,000
g	lb	0.00220462
g/cm^3	lb/ft^3	62.4280
	lb/gal(U.S.liquid)	8.3454
	kg/m^3	1,000
hp	Btu/min	42.4356
	cal/s	178.227
	J/s	745.70
	kW	0.74570
	ton of refrigeration	0.21204
in.	cm	2.54
	mm	25.4
	μm	25,400
in. Hg (32°F)	atm	0.033421
	kPa	3.3864
	bar	0.033864
	kg/cm^2	0.034532
J	Btu	0.00094845
	cal	0.239006
	W-h	0.00027778
	W-s	1
J/s	Btu/min	0.056907
	cal/min	14.3403
	hp	0.00134102
	W	1
J/g	Btu/lb	0.43021
	cal/g	0.23901
	W-h/g	0.00027778
J/g·°C	Btu/lb-°F	0.23901
	cal/g-°C	0.23901
	W-h/g-°C	0.00027778
kcal/m^3	Btu/ft^3	0.11237
kg/m^3	lb/ft^3	0.062428
kPa	atm	0.0098692
	bar	0.01
	psia	0.145038
	kg/cm^2	0.010197
kg/cm^2	atm	0.96784
	bar	0.980665
	lb/ft^2	2,048.16
	psi	14.2233
	in. Hg (32°F)	28.959
	kPa	98.06806

TABLE 14–1 (Cont.)

To convert from:	To:	Multiply by:
kW	Btu/hr	3,414.43
	kcal/min	14.3403
	hp	1.3410
	J/s	1,000
liter	ft^3	0.035316
	cm^3	1000.028
	gal (U.S.liquid)	0.026418
lb	g	453.59
lb/ft^3	kg/m^3	16.01846
	g/liter	16.01891
	lb/gal(U.S.liquid)	0.13368
psia	atm	0.068046
	bar	0.068948
	in. Hg (32°F)	2.03602
	kg/cm^2	0.070307
	kPa	6.89473
ton of refrigeration	Btu/min	200.1309
	kcal/min	50.43215
	hp	4.7161
	J/s	3,516.80
mil	cm	0.02454
	in.	0.001
	μm	25.4

absolute pressures would have been much less confusing. Conversion of English units and the old metric units to the newer SI system would seem like a good opportunity to simplify the use of pressure—but does not seem likely to happen. To avoid misunderstanding, it is best (though cumbersome) to be explicit, as:

$$\text{kilopascals, absolute} - 101.325 = \text{kilopascals, gage}$$

A similar distinction is used with English units:

$$\text{lb/in.}^2\text{, absolute} - 14.696 = \text{lb/in.}^2\text{, gage}$$

or

$$\text{psia} - 14.7 = \text{psig}$$

Units of vacuum are usually expressed as "inches of mercury, vacuum" or "inches of mercury below 1 atmosphere." These vacuum units are related to pressure below. The chemical symbol for mercury is Hg and is sometimes used.

$$\text{in. Hg, vac} = 29.921 - \text{in. Hg, pressure (abs)}$$

$$\text{in. Hg, vac} = 29.921 - (29.921)(\text{atm})$$

$$\text{in. Hg, vac} = 29.921 - (2.936)(\text{psia})$$

$$\text{in. Hg, vac} = 29.921 - (28.96)(\text{kg/cm}^2)$$

$$\text{atmospheres} = \frac{29.921 - \text{in. Hg, vac}}{29.921}$$

$$\text{psia} = (0.491)(29.921) - \text{in. Hg, vac})$$

$$\text{kg/cm}^2\text{, abs} = (0.0345)(29.921 - \text{in. Hg, vac})$$

$$\text{inches of mercury} = \frac{\text{centimeters of mercury}}{2.54}$$

$$\text{kPa, absolute} = (3.3864)(29.921 - \text{in. Hg, vac})$$

in. of Hg, vac	kPa, absolute
2	94.55
4	87.78
6	81.01
8	74.23
10	67.46
12	60.69
14	53.92
16	47.14
18	40.37
20	33.60

Heat Units

In the measurement and calculation of refrigerant properties, values for the calorie and the British thermal unit were usually based on the International Steam Tables (abbreviated IT). The calorie (IT) is defined as 1/860 W-h (Int). The calorie has been recently redefined as equal to 4.184 absolute joules and is called the thermochemical calorie. The thermochemical Btu is defined by the calorie.

$$\text{Btu(thermochemical)/lb-°F} = \text{cal(thermochemical)/g-° C}$$

The difference between the IT and the thermochemical definitions is slight, but the use of the thermochemical unit is recommended. Originally, a ton of refrigeration was defined as equal to the heat required to melt a ton of ice in 24 hours or 288,000 Btu/day or 200 Btu/min. With the new definition of calorie and Btu, a ton of refrigeration is equivalent to the melting of 2009.1 lb of ice per day or 200.1309 Btu/min.

Abbreviations

abs = absolute

atm = atmosphere

Btu = British thermal unit

°K = kelvin

kg = kilogram

kPa = kilopascal

°C = Celsius
cal = calorie
cm = centimeter
g = gram
h, hr = hr
Hg = mercury
hp = horsepower
in. = inch
IT = based on international steam tables
kW = kilowatt
lb = pound
liq = liquid
m = meter
min = min
mm = millimeter
°R = Rankine
s, sec = second
U.S. = United States

COLOR

Many fluorocarbon refrigerant cylinders are painted in distinctive colors to help in identifying the contents. The use of color in this way is not mandatory but is entirely voluntary by the refrigerant manufacturers. It is not required by any code. The label on the cylinder is the only primary and legal method of identifying the product on the inside. Depending only on color for product identification is not recommended, for a number of reasons.

More than 100 industrial gases are stored and transported in cylinders, so working out a color scheme for all of them would be complicated. This concern might be of little interest in areas where only a few refrigerants are used, but in plants handling dozens of different compressed gases, depending on color could be a real safety problem. In addition, some people are color blind, which has obvious complications in identification. Atmospheric conditions over a period of time may also affect the paint on a cylinder, causing it to fade or be altered so that color recognition is difficult.

On the other hand, after checking the lable, cylinder color is a handy way of identifying refrigerants. Many refrigerant manufacturers use the following color code for the fluorinated refrigerants:

R-11	Orange	R-115	Olive green
R-12	White	R-500	Yellow
R-22	Light green	R-502	Orchid
R-113	Purple	R-503	Aquamarine
R-114	Dark blue		

A combination of colors is sometimes used for less-common refrigerants; for example,

R-13	Powder blue base with a royal blue band
R-13B1	Red base with a white band
R-14	Orange base with a yellow band
R-116	Gray base with a green band

To add to the confusion, some users have special requirements. For example, all cylinders, regardless of refrigerant, used in military service in the United States must be painted orange [14–1]. White cannot be used in India, so a different nonconflicting color must be used. DuPont, for example, uses a gray base with an orchid top.

Color may also be used in other ways. The United States Military Standard [14–1] requires pipelines carrying "physically dangerous materials" to be colored gray. Included in this category are R-11, R-12, R-13, R-14, R-21, R-22, R-112, R-113, R-114, R-115, R-124a, R-C316, R-C317, and R-C318.

Refrigerant hardware such as expansion valves is sometimes color coded, using the same colors as those used for the refrigerants. An exception is R-12, where yellow is used for expansion valves.

SHIPPING

The manufacture, testing, filling, and shipment of compressed gas containers is regulated by the U.S. Department of Transportation. Many years ago, concern about safety in shipping hazardous materials prompted the Association of American Railroads to establish a Bureau of Explosives to conduct tests and issue regulations regarding shipment. Although the DOT has assumed most of the responsibilities initiated by the Bureau, publication of the regulations is still a function of the Bureau of Explosives. The title page of the current regulations effective May 31, 1984 reads (in part) as follows:

Bureau of Explosives
TARIFF No. BOE-6000-D
publishing
Hazardous Materials Regulations of the
Department of Transportation
by
AIR, RAIL, HIGHWAY, WATER
and
MILITARY EXPLOSIVES BY WATER
including
SPECIFICATIONS FOR SHIPPING CONTAINERS

Issued by: Thomas A. Phemister, Agent
Laura L. Smith, Alternate Agent
Association of American Railroads
Bureau of Explosives
1920 L St. N.W.
Washington, D.C. 20036
Telephone 202-835-9500

Changes or additions to the Regulations are occasionally issued by the Bureau. A current copy should be consulted for up-to-date information.

The following comments and excerpts from the Regulations should not be considered as official statements.

Page 261, 173.300 Definitions

(a) Compressed gas. The term "compressed gas" shall designate any material or mixture having in the container an absolute pressure exceeding 40 psi at 70°F or, regardless of the pressure at 70°F, having an absolute pressure exceeding 104 psi at 130°F; or any liquid flammable material having a vapor pressure exceeding 40 psi absolute at 100°F as determined by ASTM Test F-323.

(b) Flammable compressed gas. Any compressed gas as defined in paragraph (a) of this section shall be classed as "flammable gas" if any one of the following occurs:

(1) Either a mixture of 13 percent or less (by volume) with air forms a flammable mixture or the flammable range with air is wider than 12 percent regardless of the lower limit. These limits shall be determined at atmospheric temperature and pressure. The method of sampling and test procedure shall be acceptable to the Bureau of Explosives and approved by the Associate Director for HMR.

(2) Using the Bureau of Explosives Flame Projection Apparatus (see Note 1), the flame projects more than 18 inches beyond the ignition source with valve opened fully, or, the flame flashes back and burns at the valve with any degree of valve opening.

(3) Using the Bureau of Explosives Open Drum Apparatus, there is any significant propagation of flame away from the ignition source.

(4) Using the Bureau of Explosives Closed Drum Apparatus, there is any explosion of the vapor-air mixture in the drum.

(c) Non-liquefied compressed gas. A "non-liquefied compressed gas" is a gas, other than gas in solution, which under the charged pressure is entirely gaseous at a temperature of 70°F.

(d) Liquefied compressed gas. A "liquefied compressed gas" is a gas which, under the charged pressure, is partially liquid at a temperature of 70°F.

(h) Service pressure. The term "service pressure" shall designate the authorized pressure marking on the container. For example, for cylinders marked DOT-3A1800, the service pressure is 1800 psig (pounds per square inch gauge).

(i) Refrigerant gas or Dispersant gas. The term "Refrigerant gas" or "Dispersant gas" applies to all flammable or nonflammable, nonpoisonous refrigernat gases, dispersant gases (fluorocarbons) listed in ¶ 172.101, ¶ 173.304(a)(2), ¶ 173.314(c), ¶ 173.315(a)(1) and ¶ 173.315(h), and mixtures thereof, or any other compressed gas meeting one of the following:

(1) A nonflammable mixture containing not less than 50% fluorocarbon content, having a vapor pressure not exceeding 260 psig at 130°F.

(2) A flammable mixture containing not less than 50% fluorocarbon content, not over 40% by weight of a flammable component, having a vapor pressure not exceeding 260 psi at 130°F.

Note: R-11, R-113, and R-114 are not regulated since they are nonflammable and have vapor pressures lower than 40 psia at 70°F. In these Regulations gage pressures are used unless absolute is specifically indicated.

Only cylinders listed in ¶ 173.304, page 265 of the Regulations, can be used for shipping the fluorinated refrigerants and other compressed gases. If a refrigerant is not specifically listed, it can be shipped as Refrigerant gas, n.o.s. (not otherwise specified). The liquid content cannot completely fill a container at 130°F. The maximum permitted filling density for each listed product is also given in ¶ 173.304 in terms of percent by weight relative to water. The volume of water is reported as 27.737 in.^{3}/lb at 60°F. Cylinder manufacturers are required to measure and report the weight of water that will completely fill each type of cylinder. From this information the volume of the cylinder can be calculated and from refrigerant tables, the liquid level at various temperatures and loadings. The filling densities and identifcation numbers for some fluorinated refrigerants are shown in Table 14–2. All are classified as nonflammable gases.

Smaller quantities of refrigerant (less than about 60 lb) are usually shipped in DOT-39 nonreturnable cylinders. Some specifications for this type of cylinder are listed below.

Page 437, ¶ 178.65 Specification 39; nonreusable, nonrefillable cylinder.

178.65-2 Type, size, service pressure, and tests pressure.

(a) Type: Each cylinder must be seamless, welded, or brazed construction. Spherical pressure vessels are authorized and covered by references to cylinders in this specification.

(b) Size limitation: Maximum water capacity may not exceed:

(1) 55 pounds (1,526 cubic inches) for a service pressure of 500 p.s.i.g. or less, and

(2) 10 pounds (277 cubic inches) for a service pressure in excess of 500 p.s.i.g.

(c) Service pressure: The marked service pressure may not exceed 80 percent of the test pressure.

(d) Test pressure: The minimum test pressure is the maximum pressure of contents at 130°F or 180 p.s.i.g. whichever is greater.

(e) The term "pressure of contents" as used in this specification means the total pressure of all the materials to be shipped in the cylinder.

178.65-14 Marking. (a) The markings required by this section must be durable and waterproof. The requirements of 173.24(c)-(1)(ii) and (iv) of this chapter do not apply to this section.

TABLE 14–2 SHIPPING INFORMATION

Product	Shipping name	Maximum permitted filling density (% of water weight)	DOT identification number
R-11	Not regulated	—	UN 1956
R-12	Dichlorodifluoromethane	119	UN 1028
R-13	Chlorotrifluoromethane	100	UN 1022
R-14	Compressed gas, n.o.s.	[a]	UN 1956
R-22	Chlorodifluoromethane	105	UN 1018
R-23	Compressed gas, n.o.s.	[b]	UN 1956
R-114	Not regulated	[c]	UN 1956
R-115	Chloropentafluoroethane	110	UN 1020
R-116	Compressed ga, n.o.s.	[b]	UN 1956
R-500	Refrigerant gas, n.o.s.	[c]	UN 1078
R-502	Refrigerant gas, n.o.s.	[c]	UN 1078
R-503	Refrigerant gas, n.o.s.	[b]	UN 1078
R-13B1	Bromotrifluoromethane	124	UN 1009

[a]Filling density is based on rated service pressure of the cylinder.

[b]For products with critical temperatures in the vicinity of ambient temperature the filling density is based on the rated service pressure of the cylinder even though it may become liquid-full.

[c]Cylinder must not become liquid-full below 130°F.

(b) Required markings are as follows:
(1) DOT-39
(2) NRC
(3) The service pressure.
(4) The test pressure.
(5) The registration number of the manufacturer.
(6) The lot number.
(7) The data of manufacture if the lot number does not establish the data of manufacture.
(8) The following statement: Federal law forbids transporation if refilled—penalty up to $25,000 fine and 5 years imprisonment (49 U.S.C. 1809),
(c) The markings required by paragraph (b)(1) through (5) of this section must be in number and letters at least $\frac{1}{8}$ inch high and displayed sequentially. For example:
DOT-39 NRC 250/500 M1001.
For complete specifications, see the Regulations.

As an example of a cylinder that may be returned and refilled (only by the refrigerant manufacturer—or other owner), some specifications for DOT-4BA are shown.

Page 411, 178.51 Specification 4BA; welded or brazed steel cylinders made of definitely prescribed steels.
178.51-2 Type, size and service pressure. (a) Type. Cylinders may be spherical or cylindrical in shape. Closures made by the spinning process are not authorized.
(1) Spherical type cylinders must be made from two seamless hemispheres joined by the welding of one circumferential seam.
(2) Cylindrical type cylinders must be of circumferentially welded or brazed construction.
(b) Size. The capacity of the cylinder must be 1,000 pounds water capacity or less.
(c) Service pressure. The service pressure must be at least 225 and not over 500 pounds per squre inch gauge.
178.51-19 Marking. (a) Marking on each cylinder stamped as follows:
DOT-4BA240).
(2) A serial number and an identifying symbol of the maker. The symbol must be registered with the Associate Director for HMR. Duplications unauthorized. Lot numbers in place of serial numbers authorized for cylinders not over 2 inches outside diameter or for cylinders with volumetric capacity not exceeding 60 cubic inches.
(3) Inspector's official mark.
(4) Date of test (such as 4-50 for April 1950).
(5) Additional markings are permitted.

Suggestions and rules for the safe handling of compressed gas containers are discussed in Pamphlet P-1, available from the Compressed Gas Association, Inc., 1235 Jefferson Davis Highway, Arlington, VA 22202.

REFERENCES

14–1. ''Color Code for Pipelines and for Compressed Gas Cylinders,'' Military Standard 101A (Mar. 16, 1954).

14–2. American Society for Testing and Materials, "Metric Practice Guide," ASTM E 380, Philadelphia, Pa.: ASTM (1976).

14–3. American Society of Heating, Refrigerating, and Air-Conditioning Engineers, *ASHRAE SI Metric Guide*. Atlanta, Ga.: ASHRAE (1978).

14–4. O. T. Zimmerman and I. Lavine, *Conversion Factors and Tables*. Dover, N.H.: Industrial Research Service, Inc. (1961).

15 SAFETY

Perhaps the outstanding feature of the fluorocarbon refrigerants is their safety in use. Earlier refrigerants such as ammonia, sulfur dioxide, methyl chloride, and the hydrocarbons have good thermodynamic and physical properties but suffer from undesirable chemical properties—some are toxic, some are flammable, and some are both. The fluorocarbons are for the most part nonflammable and low in toxicity and can be used with fewer restrictions than other refrigerants. On the other hand, they can be misused. Table 15–1 outlines some of the potential hazards and includes methods of avoiding them and treatment if overexposure occurs.

When the fluorocarbon refrigerants were introduced they were labeled "nontoxic"—and indeed they were in comparison with other products. However, hardly anything is "nontoxic" given the right circumstances and dosage, and the fluorocarbons are not exceptions. A more fitting description is "low in toxicity." When these products were first used there was a tendency to include them all in the same category as "safe" or "nontoxic." However, it gradually became apparent that some are safer than others. Every organic compound is a separate entity with different properties. The fluorocarbon refrigerants as a group can properly be called low in toxicity, but the level of physiological effect may be different for individual refrigernats.

In 1933, the Underwriters' Laboratories [15–1] reported studies on the hazardous properties of a number of refrigerants, including R-11, R-12, and R-114. Since then, other fluorocarbon refrigerants have been evaluated in the same way. These studies were directed toward the effect of acute inhalation, that is, the inhalation

TABLE 15–1 POTENTIAL HAZARDS OF FLUOROCARBONS

Condition	Potential hazard	Safeguard
Vapors may decompose in flames or in contact with hot surfaces.	Inhalation of toxic decomposition products.	Good ventilation. Toxic decomposition products serve as warning agents.
Vapors are 4 to 5 times heavier than air. High concentrations may tend to accumulate in low places.	Inhalation of concentated vapors can be fatal.	Avoid misuse. Forced-air ventilation at the level of vapor concentration. Individual breathing devices with air supply. Lifelines when entering tanks or other confined areas. Do not administer epinephrine or other similar drugs.
Deliberate inhalation to produce intoxication.	Can be fatal.	(same safeguards as above)
Some fluorocarbon liquids tend to remove natural oils from the skin.	Irritation of dry, sensitive skin.	Gloves and protective clothing.
Lower-boiling liquids may be splashed on skin.	Freezing of skin.	Gloves and protective clothing.
Liquids may be splashed into eyes.	Lower-boiling liquids may cause freezing. Higher-boiling liquids may cause temporary irritation and if other chemicals are dissolved, may cause serious damage.	Wear eye protection. Get medical attention. Flush eyes for several minutes with running water.
Contact with highly reactive metals.	Violent explosion may occur.	Test the proposed system and take appropriate safety precautions.

of relatively large amounts of refrigerant in periods of time ranging up to 2 hours. Weight changes, external examination, condition of various organs, and other observations were the basis for judging the effect of exposure to the refrigerant. The results of the UL tests are briefly discussed since they form the basis for some of the recommendations and restrictions in the Safety Code for Mechanical Refrigeration [15–2] and in city, state, and national codes throughout the world.

As the impact of the fluorocarbon refrigerants increased, many other investigations studied the results of both acute and chronic exposure on human beings and various types of animals.* In general, the results agreed—at least qualitatively—with those of the Underwriters' Laboratories. However, there were exceptions. In some cases different types of tests revealed physiological activity not found in acute exposures. For example, R-21, $CHCl_2F$, passed the UL-type acute exposure tests very well, but in long-term chronic exposure was found to seriously affect the organs of test animals. It is no longer manufactured.

Other tests that help to define the properties of refrigerants include cardiac sensitization, carcinogenicity, and mutagenic and teratogenic potential. All of these

*Sanders' *Handbook of Aerosal Technology* [15–3] contains an excellent review of fluorocarbon toxicity.

possible reactions should be considered in defining the toxicity of a product. At the same time, the application and possible exposure should also be factors in determining the degree of safety associated with use of the refrigerant.

It is beyond the scope of this book to include all of the pertinent tests, but a few results are summarized.

UNDERWRITERS' LABORATORIES

The first major study of the toxicity of the fluorocarbon refrigerants was conducted by the Underwriters' Laboratories [15–1] and sponsored by Kinetic Chemicals, Inc. (a forerunner of the Freon Products Division of the DuPont Company). The tests were designed to compare the hazards of different refrigerants by exposing guinea pigs for relatively brief periods of time to high concentrations. These conditions were intended to represent what might happen if a leak should develop in a household refrigerator or air-conditioning unit. The products tested were classified by the UL in six groups. A description of the six groups and examples of refrigerants found to belong in each group are shown in Table 15–2. Very toxic products such as sulfur dioxide are in Group 1, while the least toxic materials such as R-12 and R-114 were placed in Group 6. No attempt was made to reproduce or imitate all the various conditions that may exist when a leak or break in a refrigeration system occurs. However, the fundamental conditions affecting the hazards, as well as the inherent properties of the refrigerants, were carefully considered. The chronic or cumulative effects of prolonged or recurrent exposures to the vapors or gases were not included in the investigation.

The conditions maintained in the tests were severe but not identical to those found in a given exposure. The size of the enclosure, the ventilation, and other variables are important variables. The data obtained in the studies are comparable and do serve as practical measures of the comparative hazards of refrigerants under working conditions.

Comparative Life Hazard in the Absence of Heat

It is generally accepted that small warm-blooded animals are quite reliable indicators of the presence of dangerous concentrations of toxic gases. The general metabolism of small animals is more rapid in relation to their weight than in large animals or people and the rate at which they absorb volatile substances through inhalation is probably also more rapid. For this reason guinea pigs were selected for the tests at the Underwriters' Laboratories. Tests on these animals do not serve as an exact measure of the toxic action on man, but a sufficient body of information on the physiological response of both men and animals to industrial gases has been developed to permit group correlation.

The guinea pigs were exposed to concentrations of the gases in air ranging from 0.5 vol % to 20 vol % or more. If no definite effect was found at low concentrations, greater amounts were used in succeeding tests. None of the animals were used in more than one test. Observations of unnatural behavior, changes in weight,

TABLE 15–2 CLASSIFICATION OF COMPARATIVE LIFE HAZARD OF GASES AND VAPORS

Group	Definition	Examples
1	Gases or vapors which in concentrations of the order of $\frac{1}{2}$ to 1% for durations of exposure of the order of 5 min are lethal or produce serious injury	Sulfur dioxide
2	Gases or vapors which in concentrations of the order of $\frac{1}{2}$ to 1% for durations of exposure of the order of $\frac{1}{2}$ hr are lethal or product serious injury	Ammonia, Methyl bromide
3	Gases or vapors which in concentrations of the order of 2 to $2\frac{1}{2}$% for durations of exposure of the order of 1 hr are lethal or produce serious injury	Bromochloromethane, Carbon tetrachloride, Chloroform, Methyl formate
4	Gases or vapors which in concentrations of the order of 2 to $2\frac{1}{2}$% for durations of exposure of the order of 2 hr are lethal or produce serious injury	Dichloroethylate, Methyl chloride, Ethyl bromide
Between 4 and 5	Appear to classify as somewhat less toxic than Group 4	Methylene chloride, Ethyl chloride
	Much less toxic than Group 4 but somewhat more toxic than Group 5	Freon 113
5a	Gases or vapors much less toxic than Group 4 but more toxic than Group 6	Freon 11, Freon 22, Freon 114B2, Freon 500, Freon 502, Carbon dioxide
5b	Gases or vapors which available data indicate would classify as either Group 5a or Group 6	Ethane, Propane, Butane
6	Gases or vapors which in concentrations up to at least about 20 vol % for durations of exposure of the order of 2 hr do not appear to produce injury	Freon 13B1, Freon 12, Freon 114, Freon 115, Freon 13[a], Freon 14[a], Freon 23[a], Freon 116[a], Freon C-318[a], Freon 503[a]

[a]Not tested by UL but estimated to belong in group indicated.

outstanding physiological symptoms, and so on, were recorded after 5, 30, 60, and 120 min of exposure. Two guinea pigs were removed at each time interval for further observation and examination. Autopsies and pathological studies were made by a toxicologist.

Comparative Life Hazard in the Presence of Flames or Hot Objects

Most halogenated organic compounds will undergo thermal decomposition at high temperatures. As a general rule, this reaction occurs most readily with bromine atoms attached to carbon, less readily with chlorine atoms, and least readily with fluorine atoms. The thermal stability of compounds containing fluorine atoms and chlorine (or bromine) atoms is directly related to the amount of fluorine present.

If a source of hydrogen is present (water, oil, gas, etc., or present in the original molecule), the products of decomposition will include the halogen acids corresponding to the halogen atoms in the compound. Thus if R-12, CCl_2F_2, is decomposed under the foregoing conditions, both hydrochloric and hydrofluoric acids will be formed. If water or some other source of oxygen is present, a smaller amount of carbonyl halide may also be formed.

In a second series of tests with guinea pigs, the UL determined the comparative toxic hazards if various refrigerant gases should be present in areas where they could by exposed to gas flames or electric heaters. The same facilities and conditions were used as for the tests with the pure gases, except for the presence of a gas or electric stove. In addition to the effects of guinea pigs, analyses of the atmosphere in the chamber at various intervals were obtained.

The comparative hazards resulting from exposure of the refrigerants to oil and wood fires were judged from analyses of the decomposition products. In these cases the presence of smoke and products of combustion would have greatly complicated direct tests with animals.

Small gas flames. The comparative life hazard of the following products appears to be of a lethal order for continuous exposures of 5 to 15 min at initial concentrations of 0.5 to 2.5 vol %. The test chamber was vapor-tight and had a volume of 500 ft^3. A blue flame in a gas stove was burning throughout the tests.

Carbon tetrachloride	Ethyl chloride
Chloroform	R-11
Dichloroethylene	R-22
R-12	R-113
R-114	Methylene chloride
Ethyl bromide	Methyl bromide

A method was developed by Midgley and Henne [15–4] which makes it possible to calculate at any given time the acid concentration generated in a room from the decomposition of a known quantity of R-12 by a flame. They also demonstrated in a real kitchen that leakage of R-12 in a habitable room containing a flame does not create a hazard.

Large gas flames. Contact of the halogenated gases with larger gas flames than those found in kitchen stoves may be expected to result in somewhat higher concentrations of decomposition products.

Oil and wood fires. When in initial concentrations of 5 vol % in air surrounding oil and wood fires, no ventilation being provided, halogenated gases yield volatile toxic products of decomposition in lethal concentrations.

Nature of warning. It is to be noted that the volatile toxic decomposition products or fumes produced by the refrigerants listed above, including also chloroform and carbon tetrachloride, in the presence of flames are exceedingly irritating, and not only give definite warning of their presence even when in very small concentrations but cause people to make efforts to escape.

Humidity. The analytical results show that increasing humidity of the air in the test room caused a marked decrease in the resulting concentration of decomposition products in the case of dichlorodifluoromethane.

Practical significance of these data. Assuming a concentration of vapor or gas on the order of $\frac{1}{2}$ to $2\frac{1}{2}$ vol % in the presence of the flames from a gas range in an average-size kitchen of 1000-ft^3 capacity having absolutely no ventilation, the concentration of toxic products of decomposition would be about half (or less) of those obtained in the test room of 500-ft^3 capacity. Allowing for only moderate ventilation essential to normal activities, the concentrations of toxic fumes would be reduced still further. If 10 oz of R-12 was released all at once from a leaking refrigerator, the resulting concentration would be only 0.2 vol %. Considering all of these factors and the distinctive warning odor of the decomposition products, the life or toxic hazard of these compounds in the presence of flames under ordinary conditions is small.

In commercial and industrial establishments where refrigerating systems containing large charges, the life hazard when a break occurs will depend on the resulting concentration of fumes. The hazard, however, will ordinarily be of a low order due to larger room size and the use of forced ventilation. The air-conditioning machinery in large buildings is often located in the basement. If refrigerant leakage occurs, the vapors may tend to concentrate in low areas since they are four to five times heavier than air. The importance of maintaining adequate ventilation to prevent the possibility of accumulation of fumes in dangerous concentrations in commercial and industrial establishments where refrigerants are used is evident.

The oil fires employed in the tests were of relatively high intensity, and the wood fires were of moderate intensity. In both cases, the test conditions were severe and are not to be anticipated under normal working conditions. The toxic hazard presented will depend on the concentration of the refrigerant and the degree of confinement of the fumes. Under actual fire conditions convection currents of air rapidly dilute and disperse fumes, except under conditions of confinement. In closed places, particularly where the prompt exit of persons is not possible, there is danger.

Small hot objects. When in initial concentrations of the order of 5 vol % in the air surrounding hot surfaces (1382 to 1427°F, cherry to bright red) in the form of open coil resistance wire units of electric ranges such as are used in kitchens for cooking, for durations of exposure of the order of $\frac{1}{2}$ hr (no ventilation), no appreciable amounts of toxic decomposition products were formed with R-12 or R-114.

Under the conditions above with methyl chloride, an appreciable amount of decomposition products was formed but not in concentrations dangerous to life for durations of exposure of the order of $\frac{1}{2}$ hr.

Carbon tetrachloride, methylene chloride, R-11, R-22, and R-113 under the foregoing conditions (no ventilation) formed toxic decomposition products in lethal concentrations for durations of exposure of the order of $\frac{1}{2}$ hr.

The foregoing classification of the volatile decomposition products of R-12, R-114, R-22, R-113, methyl chloride, and R-11 is based on the results of both animal and analytical tests. In the case of carbon tetrachloride and methylene chloride, animal tests were not included, but analytical tests of the volatile decomposition products were conducted.

The difference between the results of the tests with carbon tetrachloride and dichlorodifluoromethane in the presence of hot wire is probably accounted for on the basis that the thermal decomposition temperature of the former compound is considerably lower than that of the latter. The temperature of decomposition of both carbon tetrachloride and R-12 is below the temperature of the wire, but owing to convection currents, neither vapor was heated to the temperature of the wire.

Large hot objects. When in initial concentrations of the order of 5 vol % and in contact with the inner surface of an iron cylinder (diameter 6 in., length 26 in.) heated to a temperature of 1022°F, the following refrigerants and other substances included in the investigation gave toxic products of decomposition in lethal concentrations:

Carbon tetrachloride	Ethyl chloride
Chloroform	Methyl bromide
Dichloroethylene	Methylene chloride
R-12	R-11
R-114	R-22
Ethyl bromide	R-113

The test conditions above are severe and not to be anticipated under normal habitable conditions. Under fire conditions, the toxic hazard presented will depend on the amount or concentration of the refrigerant and the confinement, if any, of the liberated fumes. Fumes are usually rapidly dissipated under fire conditions by the convection currents of air. In small closed places, particularly where the prompt exit of a person is not possible, there will be danger.

ACUTE TOXICITY [15–5]

The fluorocarbon refrigerants are less toxic than most other refrigerants, propellants, or solvents and therefore can be used with fewer restrictions. However, in common with all chemical products, inhalation of high concentrations is dangerous since breathing the concentrated vapors can be fatal. If exposed to refrigerant vapors, move immediately to fresh air. Avoid the use of epinephrine or similar drugs because they may produce cardiac arrhythmias, including ventricular fibrillation.

The early studies by the Underwriters Laboratories have been described above. Another example of acute inhalation tests is the work of Lester and Greenberg [15–6]. White rats were exposed to high concentrations of several fluorinated products. The exposure period was 30 min. The oxygen content was maintained at 20%. The concentration of gas under study could be varied from 0 to 80 vol %. The values in Table 15–3 were suggested as the maximum allowable concentrations for single short exposures of human beings.

TABLE 15–3 SUGGESTED MAC VALUES

Product	Chemical formula	Conc. in air (vol %)
Vinylidene fluoride	$CH{=}CF_2$	30
R-12	CCl_2F_2	20
Vinyl fluoride	$CH_2{=}CHF$	20
Difluoroethane	CH_3CHF_2	15
Chloridifluoroethane	CH_3CClF_2	10
R-11	CCl_3F	5

CHRONIC TOXICITY

Chronic inhalation studies have not been as extensive as acute exposures but are an indispensable part of a toxicological examination. Some examples of early chronic studies include the work of Sayers and coworkers [15–7], who studied the effect of R-12 on dogs, monkeys, and guinea pigs for 12 weeks at a concentration of 20 vol % and an average daily exposure of about 7 hr. The animals were not unconscious during exposure but exhibited generalized tremor and had difficulty walking. After studying the test results, including changes in weight, changes in red blood cells, hemoglobin, and white blood cells, the fatality record, and pathological examination, their conclusion was: "Insofar as the results of animal experimentation serve as a measure of hazards to persons, the investigation described in this report has shown that the possibility of public health and accident hazards resulting from exposure to dichlorodifluoromethane when used as a refrigerant is remote."

A similar study by Yant and coworkers [15–8] with dichlorotetrafluoroethane, R-114, led to the conclusion: "The results of the investigators show that the toxicity of dichlorotetrafluoroethane on a vapor volume basis is somewhat greater than dichlorodifluoromethane, but on a weight basis it is of the same order as the latter, and is an organic vapor of remarkably low toxicity."

Some chronic toxicity tests with R-12 were reported by Siegel et al. of the U.S. Navy [15–9]. Various animals, including rats, guinea pigs, rabbits, monkeys, and dogs, were exposed to low concentrations. In one test, exposure was at a level of 810 ppm by volume for 92 days, 24 hr/day. In another test, exposure was 8 hr/day for 30 days at about the same level. There were no visible signs of toxicity related to R-12, although some liver cell vacuolation and focal necrosis was detected.

Consult the medical and chemical literature for more recent studies of the toxicology of these and other fluorocarbon refrigerants.

ORAL TOXICITY [15–10]

Fluorocarbons have low oral toxicity as judged by single-dose administration or repeated dosing over long periods of time. When R-11 and R-114 were fed to rats and dogs for 90 days, there were no adverse effects relative to nutritional, biochemical, hematological, urine analytical, or histopathological indices, even at the highest dose levels tested. A 2-year feeding study with R-12 resulted in similar findings. In addition, there were no adverse effects of R-12 on mutagenic, teratogenic, or three-generation reproductive indices. For a further discussion of oral toxicity, see the section on water solutions.

EFFECT ON THE SKIN [15–5]

Most of the fluorocarbon liquids will dissolve and extract the natural oils present in the skin. If contact is prolonged, the skin may become dry and perhaps cracked. Neoprene gloves are recommended if exposure to the liquid phase is possible. There is very little evidence of absorption through the skin or effect on internal organs similar to that found with some chlorinated solvents such as carbon tetrachloride.

The vapors have little or no effect on the eyes. Exposure to the liquid phase of the higher-boiling liquids or to solutions of fluorocarbons dissolved in oils or other liquids may produce temporary redness [15–11]. No permanent damage was found in tests with rabbits. Eye protection should be used at all times.

If the liquid phase of the lower-boiling refrigerants comes in contact with the skin, the rapid evaporation may cause freezing. The presence of fluorocarbon vapors in the air surrounding the point of contact does not affect the skin or frozen area and treatment is the same as for frostbite from any other source. The following treatment is suggested:

1. If treatment is begun within 20 to 30 min after exposure, soak the exposed area in lukewarm water. Do not soak in ice-cold water or in hot water.
2. If treatment is begun more than 30 min after exposure, eliminate the preliminary soaking.
3. Apply a very light coating of any bland ointment, such as petroleum jelly.
4. Apply a light bandage if the exposure is at a location where the presence of ointment would be awkward.

5. If the frostbite is severe and the frozen area large, administration of an anticoagulant or vasodilator should be considered by a physician to avoid the development of gangrene.

HUMAN EXPOSURE

Human exposure to dichlorodifluoromethane for periods ranging from 11 to 80 min at concentrations from 4 to 11 vol % led to the following conclusions [15–12]:

1. The highest concentration of dichlorodifluoromethane vapor in air that can be inhaled for an hour or more by a previously unexposed healthy human subject, without inducing some degree of cortical dysfunction, is somewhat less than 4 vol %.

2. The immediate effects of the exposure of two subjects to concentrations approximating 4% were not negligible. General tingling sensations, most pronounced in the extremities, a slight humming in the ears, and an increased awareness of visceral sensations were experienced, associated with some degree of apprehension. This phase of the effects tended to pass in 15 to 20 min, especially if the attention of the subject was directed elsewhere. Objectively, there were slight evidences of impaired cortical function, including electroencephalographic changes, a slight thickness of speech, and a slight change in muscle tonus, with some indication of borderline impairment of both mental activity and muscular coordination. The objective abnormalities tended to diminish during exposure after some lapse of time.

3. Exposure to concentrations above 4% intensified the symptoms and the observed effects without producing striking qualitative differences in response. At 7%, consciousness was maintained but cerebral function was considerably impaired and some incoordination developed; at 11% a marked decrease in the level of consciousness occurred after about 10 min of exposure. The risk to life involved in maintaining a state of anesthesia in man has not been determined.

4. Recovery after brief exposure was prompt, in fact almost immediate. (Human exposure was not carried beyond a period of 80 min in these experiments.)

5. The after-effects of brief exposures were negligible. There were no objective signs of delayed intoxication, tissue injury, or fatigue, while the only subjective sequlae consisted of a slight sensation of muscular tension especially in the calves of the legs, together with a persistent taste in the mouth not unlike that of acetone. The latter persisted from 12 to 24 hr and during this period smoking was highly distasteful.

6. The compound disappears with great rapidity from the body, except for a small residium that is lost more gradually. Apparently, there is little or no decomposition of the chemical in the tissues, the elimination occurring chiefly by way of the expired air. The urinary excretion of fluorine shows but little increase during the period of 12 to 24 hr following exposure, while only a doubtful trace is demonstrable in the saliva during the period in which the taste of the compound persists.

ASPHYXIATION

When the supply of oxygen in the air is reduced either by changes in density at higher altitudes or by dilution with another gas, there is a marked physiological response. Some of the immediate symptoms are outlined in Table 15–4. Other more serious and longer-lasting changes in physiology may take place at the same time.

TABLE 15–4 RESPONSE TO OXYGEN DEFICIENCY

Oxygen (vol %)	At rest: signs and symptoms of oxygen deficiency
12–14	Respirations deeper, pulse up, coordination poor
10–12	Cheyne–Stokes respiration, giddiness, poor judgment, lips blue
8–10	Nausea, vomiting, unconsciousness, ashen face
6–8	8 min, 100% of animals die; 6 min, 50% die and 50% recover with treatment; 4–5 min, all recover with treatment
4	Coma in 40 sec, convulsions, respiration ceases, death

Source: Ref. 15–13.

The composition of dry air in percent by volume is as follows [15–14]:

Nitrogen	78.08	Helium	0.00052
Oxygen	20.95	Methane	0.0002
Argon	0.93	Sulfur dioxide	0–0.0001
Carbon dioxide	0.03	Hydrogen	0.00005
Neon	0.0018		

$$\text{molecular weight} = 28.9645$$

Properties of air at 77°F and 1 atm pressure:

$$\begin{aligned}\text{density} &= 24{,}458\ \text{cm}^3/\text{mol} = 0.863726\ \text{ft}^3/\text{mol (gram)}\\ &= 0.02980\ \text{ft}^3/\text{g} = 13.5261\ \text{ft}^3/\text{lb}\\ &= 0.073931\ \text{lb/ft}^3\end{aligned}$$

The weight of refrigerant vapors in air corresponding to several reduced oxygen levels is shown in Table 15–5.

The fluorocarbon vapors are heavier than air and tend to accumulate near the floor. Eventually, all gases will reach a uniform composition, but if left undisturbed, the mixing may take a long time. A fan or other mechanical agitation will greatly hasten the mixing process. The density of some refrigerants relative to air is illustrated in Table 15–6.

TABLE 15–5 WEIGHT OF DILUENT GAS IN POUNDS PER CUBIC FOOT AT SEVERAL REDUCED OXYGEN CONCENTRATIONS

	Oxygen in mixture (vol %)				
	12	10	8	6	4
Nitrogen	0.0305	0.0374	0.0442	0.0510	0.0579
R-11	0.1557	0.1905	0.2252	0.2600	0.2948
R-12	0.1346	0.1647	0.1947	0.2248	0.2549
R-13	0.1148	0.1404	0.1661	0.1917	0.2174
R-13B1	0.1646	0.2013	0.2381	0.2749	0.3117
R-22	0.0957	0.1171	0.1385	0.1598	0.1812
R-113	0.0970	0.1187	0.1407	0.1625	0.1842
R-114	0.1922	0.2352	0.2781	0.3211	0.3641
R-502	0.1236	0.1513	0.1789	0.2065	0.2342

TABLE 15–6 DENSITY OF FLUOROCARBON GASES RELATIVE TO AIR

	Density relative at 77°F and 1 atm
R-11	4.93
R-12	4.26
R-13	3.63
R-13B1	5.21
R-22	3.03
R-113[a]	3.08
R-114	6.09
R-502	3.91

[a]At 6.9 psia.

The concentration of foreign gases in air is often given in terms of parts per million by volume or volume percent. This relationship is usually based on Avagadro's law of combining volumes, which says that the total volume is equal to the sum of the volumes of each component gas all at the same total pressure. For example, if R-12 is present in air at a concentration of 10 vol % at 1 atm and 77°F, each cubic foot of gas would contain 0.1 ft^3 of R-12 and 0.9 ft^3 of air, both at 1 atm. If it is necessary to know the amount of foreign gas present in terms of weight, the relationships in Table 15–7 can be used.

WATER SOLUTIONS

When refrigerants are used to transfer heat from any source to water, the effect of leakage must be considered. If water pressure is higher than refrigerant pressure and water enters the refrigeration system, the resulting damage is well known and

TABLE 15–7 DENSITIES AND VOLUMES OF FLUOROCARBON GASES AT 77°F AND 1 ATM

	ft^3/lb	lb/ft^3	g/liter	cm^3/g	mg/m^3 equal to 1000 ppm by volume
R-11	2.744	0.3644	5.837	171.3	5600
R-12	3.175	0.3150	5.046	198.2	4950
R-13	3.722	0.2687	4.304	232.4	4300
R-13B1	2.596	0.3852	6.170	162.1	6100
R-22	4.464	0.2240	3.588	278.7	3500
R-113[a]	4.672	0.2277	3.647	291.7	7600
R-114	2.222	0.4500	7.208	138.7	7000
R-502	3.456	0.2894	4.636	215.8	4600

[a]At 6.9 psia.

understood. The reverse condition when refrigerant pressure is higher than water pressure and refrigerant enters the water system is not as well understood. Two factors relevant to assessing the potential hazard if the water is taken internally or perhaps used in cooking are the solubility of refrigerant in water and the effect of oral consumption of refrigerant.

The solubility of R-11, R-12, and R-114 in water as a function of temperature is shown in Table 15–8. As the temperature rises, the solubility goes down. The solubility also depends on the pressure of the gas. The amount of refrigerant dissolved depends only on the pressure of the refrigerant and is not affected by the presence of other gases. For example, in an open glass or metal container under static condi-

TABLE 15–8 SOLUBILITY IN WATER

		At 1 atm		At 1 psia	
	Temperature (°F)	wt %	mg per quart of water	wt %	mg per quart of water
R-11	80	0.122	1152	0.0083	78
	100	0.083	786	0.0057	54
	120	0.059	560	0.0040	38
	140	0.044	415	0.0030	28
R-12	70	0.040	377	0.0027	25.7
	80	0.027	255	0.0018	17.4
	100	0.018	171	0.0012	11.6
	120	0.015	140	0.0010	9.5
	140	0.013	124	0.0009	8.4
R-114	70	0.016	152	0.0011	10.3
	80	0.013	121	0.00087	8.3
	100	0.0088	83	0.00060	5.6
	120	0.0065	61	0.00044	4.2
	140	0.0051	49	0.00035	3.3

tions the vapor immediately above liquid water would be nearly pure refrigerant with a small amount of water vapor. The amount of refrigerant dissolved in the water would be almost that shown in the table for 1 atm of pressure. In a closed container where the refrigerant pressure might be 2 or 3 atm, the amount dissolved might be two or three times greater. On the other hand, water coming out of a faucet or being poured from one container to another would be in contact with refrigerant vapor at a low pressure and the amount in solution would be corresponding low. Under nonequilibrium conditions, the dissolved gas quickly escapes from solution in an effort to reestablish equilibrium at the low refrigerant pressure.

Oral Toxicity

Animal tests have been used to study the effect of swallowing fluorinated refrigerants. In most cases, the refrigerant is administered as a solution in oil. Solubility in oil is much greater than in water—again depending on pressure. This procedure limits the amount of refrigerant per dose, but the quantities tested were far in excess of any conceivable consumption of refrigerant in water. The dosage is reported as milligrams of refrigerant per kilogram of body weight. The tests are described below and the results summarized in Table 15–9.

R-11 [15–15,15–16]

1. Rats were fed 3725 mg/kg R-11 as a paraffin mixture. None of the rats died and no significant changes in the liver were seen.
2. R-11 was administered to rats and dogs for 90 days at daily dose levels of 250 to 450 mg/kg (rats) and 170 to 350 mg/kg (dogs). Except for a slight transient equivocal effect on urinary fluoride excretion in rats fed the highest level, there was no nutritional, clinical, hematological, biochemical, or histopathological evidence of toxicity.

R-12 [15–15,15–17]

1. R-12 was orally administered to male and female rats, starting with 29-day-old animals, at a dose level ranging from 160 to 379 mg/kg, 5 days per week for 18 weeks. No clinical or nutritional signs of toxicity were seen. There were no meaningful differences between the clinical laboratory data of the control and test groups. No histological changes were observed in any of the more than 25 organs and tissues examined.
2. R-12 was orally administered to male and female dogs for 18 weeks at a dose level of 84 to 95 mg/kg without any clinical nutritional or biochemical manifestations of toxicity. No histological changes were seen.
3. R-12 was orally administered to male and female rats at dose levels of 15 and 150 mg/kg per day for 2 years, starting with the offspring of rats that had been fed the same compound at the same dose levels for 90 days. Except for a slightly decreased rate of weight gain in those rats fed the higher dose,

TABLE 15–9 ORAL TOXICITY TESTS

	Animal	Period	Feed rate: mg/kg per day	Feed rate: mg/day for 150-lb man	Water temp. (°F)	Equivalent quarts of water per day	Result
R-11	Rats	Single dose	3725	253,445	80	220	No effect
					140	611	
	Rats	90 da	450	30,618	80	27	No effect
					140	74	
	Dogs	90 da	350	23,824	80	21	No effect
					140	57	
R-12	Rats	18 wk	160	10,886	70	424	[a]
					140	1294	
	Rats	18 wk	379	25,787	70	1005	[a]
					140	3066	
	Dogs	18 wk	84	5,715	70	223	No effect
					140	680	
	Dogs	18 wk	95	6,464	70	252	No effect
					140	769	
	Rats	2 yr	15	1,021	70	40	No effect
					140	121	
	Rats	2 yr	150	10,206	70	398	No effect
					140	1214	
	Dogs	2 yr	7.5	510	70	20	No effect
					140	61	
R-114	Rats	Single dose	2250	153,087	70	1007	[b]
					140	3156	
	Rats	3 mo	240	16,329	70	107	No effect
					140	337	
	Rats	3 mo	450	30,617	70	201	No effect
					140	631	
	Dogs	3 mo	150	10,206	70	67	No effect
					140	210	
	Dogs	3 mo	380	25,855	70	170	No effect
					140	533	
	Rats	2 wk	1289	87,202	70	574	[c]
					140	1798	
	Rats	33 da	2000	136,078	70	895	No effect
					140	2806	

[a]Slight elevation of phosphatase activity; otherwise, no effect.
[b]Transient diarrhea.
[c]No effect, except diarrhea for 3 days.

there was no clinical, biochemical, urine analytical, hematological, or histopathological evidence of toxicity.

4. Male and female dogs were fed up to 300 ppm of R-12 (7.5 mg/kg) in their diet for 2 years without any nutritional, hematological, urine analytical, biochemical, or histopathological evidence of toxicity.

R-114 [15–15,15–18]

1. Approximate lethal dose (rat) is greater than 2250 mg/kg (the largest possible dose) when administered in peanut oil. The only effect was transient diarrhea.
2. R-114 was administered orally (in corn oil) to male and female rats and dogs for 3 months at daily dose levels of 240 to 450 mg/kg for rats and 150 to 380 mg/kg for dogs without any nutritional, clinical, hematological, urinary, biochemical, or histopathological evidence of toxicity.
3. R-114 was administered orally to rats at 1289 mg/kg (in peanut oil) per dose for 10 doses over a 2-week period. There were no observable effects except for diarrhea lasting for about 3 days.
4. No ill effects were noted when rats were orally administered doses of 2000 mg/kg daily for 23 to 30 days.

Hazard

The water solubility and animal studies data were combined to show how much water would need to be consumed each day by an average man before reaching the refrigerant level used in the tests. Based on this information there would seem to be little if any hazard to humankind from the ingestion of water that has been exposed to R-11, R-12, or R-114. Fear of contamination in leaking condensers, chillers, water heaters, or even in direct contact applications seems unwarranted. The solubility in water for other refrigerants is known, but oral ingestion studies are not available at present.

CARDIAC ARRHYTHMIA

Arrhythmia is defined as any disturbance in the rhythm of the heartbeat and can be produced by the inhalation of relatively high concentrations of most halocarbons and hydrocarbons. In extreme cases cardiac arrest may occur. In animal studies, arrhythmia is produced more easily in dogs and monkeys than in rats or mice [15–19]. In a series of tests, beagle dogs were exposed to various levels of fluorocarbons for 5 min and then injected intravenously with adrenalin. The results are shown in Table 15–10. A ''marked response'' represents serious arrhythmia. The injection of adrenalin is intended to simulate (but with greater severity) the normal bodily increase in adrenalin due to emotional stress, fright, excitement, physical activity, and so on. This test is apparently valid for comparing one fluorocarbon product with another but produces arrhythmia at lower concentration levels than would be expected as a result of natural stimuli.

MUTAGENICITY

A simple test for bacterial mutation was developed by Ames and found to have a 90% correlation for identifying substances suspected of being carcinogenic toward man [15–19]. Although useful as a screening test, it is concluded that ''these tests

TABLE 15–10 SCREENING TESTS WITH DOGS

	Concentration (vol %)	Duration (min)	Number of dogs	Number of marked responses	Percent of marked responses
R-11	0.1	5	12	0	0
	0.5	5	12	1	8
	1.0	5	12	5	42
R-12	2.5	5	12	0	0
	5.0	5	12	5	42
R-22	2.5	5	12	0	0
	5.0	5	12	2	17
R-113	0.25	5	12	0	0
	0.50	5	29	10	34
	1.0	5	4	3	75
R-114	2.5	5	12	1	8
	5.0	5	12	7	58
R-115	15.0	5	13	1	8
	25.0	5	12	4	33
Isobutane	2.5	5	12	0	0
	5.0	5	12	4	33
	10–20	5	6	6	100
Propane	5	5	6	0	0
	10	5	12	2	17
	20	5	12	7	58

Source: Refs. 15–20 to 15–22.

[mutagenic] do not provide an adequate basis for quantitative evaluation of the carcinogenic risk for humans'' and they might have added ''and in many cases, not even a qualitative basis.'' The additional comment is reported by H. E. Stokinger in *Patty's Industrial Hygiene and Toxicology* [15–19].

In general, the fluorocarbon refrigerants do not give a positive result in the Ames test. An exception is R-22, with negative results in the standard test but positive in some variations [15–15]. Later animal studies of carcinogenicity were negative.

TERATOGENICITY

Teratogenicity is the possible effect of fluorocarbon (or other gases) inhalation on offspring. The question has not been studied extensively but is not believed to be a hazard—for example, to pregnant women working on a refrigerator assembly line [15–15].

In one test, a mixture of R-12 and R-11 (90/10) was administered at a concentration of 20% in air for 2 hr daily to rats from the fourth to the sixteenth day of gestation, and to rabbits from the fifth to the twentieth day. There were no adverse effects on the offspring of the exposed pregnant animals [15–19].

RETENTION IN BODY [15–3]

R-12 and R-114 are rapidly eliminated from the body without appreciable decomposition. In animal tests, about 10 to 12% retention was found 30 minutes after exposure. In comparable tests, retention was 23% for R-11 and 20% for R-113.

COMPARISONS OF TOXICITY

Aviado and Micozzi, reporting in *Patty's Industrial Hygiene and Toxicology* [15–19], offer some fluorocarbon comparisons. In experimental animals, R-12 is less toxic than R-11. In animal studies, R-113 is more toxic than R-12 but less toxic than R-11. Although all three fluorocarbons are cardiotoxic to dogs and monkeys, there is a difference in their effects on the respiratory system of monkeys. R-11 causes early respiratory depression and R-12 causes bronchoconstriction, whereas R-113 does not influence either respiratory activity or airway resistance.

Animal studies indicate that the lethality and cardiotoxicity of R-114 is less than R-11 and about equal to that of R-12. R-115 has one of the lowest levels of cardiotoxicity, sensitizing the dog heart to epinephrine-induced arrhythmias in concentrations 25 to 50 times greater than that of R-11. On the basis of early inhalation experiments in guinea pigs and rats, R-22 was shown to be two to three times less toxic than R-11. More recent studies indicate an eight- to tenfold difference in cardiotoxicity in mice and dogs. Both R-22 and R-12 cause early respiratory depression, bronchoconstriction, tachycardia, myocardial depression, and hypotension in approximately equivalent concentrations (5 to 10%) in dogs and monkeys. The difference between the two is that R-22 does not induce cardiac arrhythmia in the monkey, although it sensitizes the heart to epinephrine in the mouse, and that R-22 does not decrease pulmonary compliance in the monkey.

R-152a is regarded as the least toxic of 20 fluorocarbons examined. R-142b has a level of toxicity lower than R-11 and R-114 but higher than R-C318.

General References

In addition to specific references in the medical and chemical literature, general discussions can be found in *Patty's Industrial Hygiene and Toxicology* [15–19], Sanders' *Handbook of Aerosol Technology* [15–3], and review articles by J. W. Clayton, Jr. [15–23 through 15–25].

SAFETY CODE FOR MECHANICAL REFRIGERATION

The Safety Code For Mechanical Refrigeration was prepared by and is monitored by the American Society of Heating, Refrigerating and Air Conditioning Engineers and is approved by the American National Standards Institute [15–2]. A brief description of the Code is given here, but the original reference should be studied if appropriate. Questions about the Code can be addressed to the ASHRAE Standards Committee. Although the Standard itself is voluntary, it has been incorporated in codes and regulations of most cities and states and so is of great importance in all refrigeration and air-conditioning applications.

1. SCOPE AND PURPOSE

1.1 Scope. "The application of this Code is intended to assure the safe design, construction, installation, operation, and inspection of every refrigerating system employing a fluid which normally is vaporized and liquefied in its refrigerating cycle, when employed under the occupancy classifications listed in Section 3. The provisions of this Code are not intended to apply to the use of water or air as a refrigerant, nor to gas bulk storage tanks that are not permanently connected to a refrigerating system, nor to refrigerating systems installed on railroad cars, motor vehicles, motor drawn vehicles or on shipboard. (For shipboard installations see ANSI/ASHRAE 26–1978.)

1.2 Purpose. This Code is intended to establish reasonable safeguards to life, limb, health, and property; to define certain practices which are inconsistent with safety; and to prescribe standards of safety which will properly influence future progress and developments in refrigerating systems.

1.3 Application. This Code shall apply to refrigerating systems installed subsequent to its adoption and to parts replaced or added to systems installed prior to or subsequent to its adoption. In cases of practical, difficult or unnecessary hardship, the authority having jurisdiction may grant exceptions from the literal requirements of this Code or permit the use of other devices, materials or methods, but only when it is clearly evident that equivalent protection is thereby secured.

Equipment listed by an approved nationally recognized testing laboratory is deemed to meet the design, manufacture, and factory test requirements section of this Code or equivalent, for the refrigerant or refrigerants for which such equipment is designed. Listed refrigerating systems are not required to be field tested to comply with Section 12 of this Code."

In the Standard, refrigerants are divided into three groups. Group 1, including the fluorocarbons, are nonflammable refrigerants with low toxicity. In Group 2 are the more toxic refrigerants, such as ammonia and sulfur dioxide, and in Group 3 are the flammable hydrocarbon refrigerants. For each refrigerant, a limit is set on the amount that may be present in equipment located in areas where humans might be present. These limitations are shown in Table 15–11.

Minimum design pressures for equipment containing the refrigerants are also specified in the Standard (Table 15–12). Pressures in the low-pressure side of operating systems are based on the vapor pressure of the refrigerant at 80°F. On the high-pressure side, pressures are based on the vapor pressure at 105°F for evaporative or water-cooled condensers and at 125°F for air-cooled condensers. If the critical temperature of the refrigerant is lower than 105°F, the critical pressure is used for the minimum high-side design pressure. If the critical temperature is below 80°F, the critical pressure is also used for low-side design. Note that the pressures are in psig.

TABLE 15–11 MAXIMUM PERMISSIBLE QUANTITIES OF GROUP 1 REFRIGERANTS FOR DIRECT SYSTEMS[a]

Refrigerant	Maximum quantity in human-occupied space (lb/1000 ft^3)
R-11	35
R-12	31
R-13	27
R-13B1	38
R-14	23
R-21	21
R-22	22
R-30	6
R-113	24
R-114	44
R-115	40
R-C318	50
R-500	26
R-502	30
R-503	22
R-744	11

[a]These limits are equivalent to about 10 vol % at 1 atm (see UL classification in Table 15–12).

THRESHOLD LIMIT VALUES

For a number of years the American Conference of Governmental Industrial Hygienists [15–26] has published a booklet listing threshold limit values (TLVs) for many chemical products. These values are concentrations by volume in air believed to be safe for daily exposure. They are based on the best available information from industrial experience and from experimental human and animal studies. The limits are continually reviewed and may be changed as new developments occur. The latest issue of the publication should be consulted.

The threshold limits are recommendations and should be used only as guidelines and criteria for good practice. However, they may be included in codes or regulations issued by federal, state, or municipal authorities, in which case they become legally binding.

Three categories of threshold limit values are recognized:

1. *Time-weighted average* (*TWA*). The average concentration for a normal 8-hr workday and a 40-hr workweek to which nearly all people may be exposed without adverse effect.

TABLE 15–12 MINIMUM DESIGN PRESSURES[a]

Refrigerant	Minimum design pressures (psig)		
		High side	
	Low side	Evap. or water cooled	Air cooled
R-11	15	15	21
R-12	85	127	169
R-13	521	547	547
R-13B1	230	321	410
R-14	529	529	529
R-21	15	29	46
R-22	144	211	278
R-30	15	15	15
R-40	72	112	151
R-113	15	15	15
R-114	18	35	53
R-115	123	181	238
R-170	618	695	695
R-290	129	188	244
R-C318	34	59	85
R-500	102	153	203
R-502	162	232	302
R-503	617	617	617
R-600	23	42	61
R-600a	39	63	88
R-611	15	15	15
R-717	139	215	293
R-744	955	1058	1058
R-764	45	78	115
R-1150	732	732	732
R-1270	160	228	294

[a]Selection of higher design pressures may be required to satisfy actual shipping, operating, or standby conditions.

2. *Short-term exposure limit (STEL).* The concentration limit for exposures of about 15 min even though the average limit described above is not exceeded.
3. *Ceiling limit.* A concentration that should not be exceeded at any time (not given for all products).

Threshold limit values for some fluorinated refrigerants are given in Table 15–13. The maximum concentration is 1000 ppm by volume for any foreign substance in air. The only exception is carbon dioxide—set at 5000 ppm since that is about the concentration exhaled in normal breathing.

The ACGIH publication [15–26] should be consulted for more information about threshold limit values, and all questions should be referred to the Conference.

TABLE 15–13 THRESHOLD LIMIT VALUES

	TWA		STEL	
	ppm	lb/ft^3	ppm	lb/ft^3
R-11	1000	3.81×10^{-4}	—	—
R-12	1000	3.15×10^{-4}	1250	3.94×10^{-4}
R-13B1	1000	3.85×10^{-4}	1200	4.81×10^{-4}
R-22	1000	2.24×10^{-4}	1250	2.80×10^{-4}
R-113	1000	2.14×10^{-4}	1250	2.68×10^{-4}
R-114	1000	4.50×10^{-4}	1250	5.62×10^{-4}
R-115	1000	4.00×10^{-4}	—	—

MATERIAL SAFETY DATA SHEETS

The Occupational Safety and Health Administration (OSHA) of the U.S. Department of Labor requires that employers keep on file, and available to employees, information about chemicals employees might handle. This requirement has led to the issuance of "Material Safety Data Sheets" for the fluorinated refrigerants, available from DuPont and from other refrigerant manufacturers.

FLAMMABILITY

Completely halogenated and many partially halogenated compounds are not flammable in any mixtures with air at temperatures up to at least 100°C (212°F). Tests for some refrigerants were performed at the Underwriters' Laboratories [15–1] and repeated by the DuPont Company [15–27], including a number of other halogenated products. Nonflammable compounds and mixtures of compounds are listed in Table 15–14.

TABLE 15–14. NONFLAMMABLE HALOGENATED COMPOUNDS[a]

R	Formula	R	Formula[b]
12	CCl_2F_2	502	$CHClF_2/CClF_2CF_3$ (48.8/51.2)
13B1	$CBrF_3$	11/isopentane	$CCl_3F/(CH_3)_2CHCH_2CH_3$ (93/7)
22	$CHClF_2$	12/152a	CCl_2F_2/CH_3CHF_2 (90/10)
23	CHF_3	13B1/152a	$CBrF_3/CH_3CHF_2$ (75/25)
113	CCl_2FCClF_2	13B1/32	$CBrF_3/CH_2F_2$ (15/85)
114	$CClF_2CClF_2$	32/22	$CH_2F_2/CHClF_2$ (50/50)
115	$CClF_2CF_3$	32/115	$CH_2F_2/CClF_2CF_3$ (40/60)
123	$CHCl_2CF_3$	115/152a	$CClF_2CF_3/CH_3CHF_2$ (84/16)
134	CHF_2CHF_2	143a/115	$CH_3CF_3/CClF_2CF_3$ (60/40)

[a] Tested at 100°C (212°F).

[b] Percentages by weight are shown.

Some flammable fluorocarbons are listed in Table 15–15.

TABLE 15–15. FLAMMABLE FLUOROCARBONS

	Lower flammable limit (vol % in air)	
	25°C (77°F)	100°C (212°F)
R-152a, CH_3CHF_2	8	2[a]
R-143a, CH_3CF_3	13	6[a]
R-32, CH_2F_2	22	22

[a] Approximate composition; probably low.

Source: Ref. 15–27.

Autoignition Temperature

The autoignition temperature was determined by heating various mixtures of air and a variety of organic compounds in a glass flask held at a constant temperature. The temperature was slowly raised or lowered in steps of about 5°C until a flame or other reaction was observed or until the upper temperature limit of the apparatus was reached. The observed flames varied in intensity. The significance of these tests, at least in refrigeration applications, is open to question. The temperatures of autoignition are so high and in most cases the flame development so weak that there would seem to be little, if any, hazard. Some measurements from the UL [15–1] and from DuPont [15–27] are listed in Table 15–16.

It is noted that the results given here apply only to the UL test method described in Ref. 1 and may differ from the results of other tests.

Mixtures

Mixtures of flammable and nonflammable refrigerants may or may not be flammable depending on the relative amounts of each present. Determining the possible hazard depends on how the mixture is admitted to the atmosphere. If leakage, venting, spraying, and so on, occur in the liquid phase, the composition in the air will be essentially the total composition of the mixture. Measuring the flammable limits of that single mixture in air will be sufficient to define the hazard.

On the other hand, if vapor is removed, fractionation will occur and the composition of the mixture in the air will continually change. At some point during the evaporation it may be possible for a flammable mixture to form.

Test Methods

Measuring the flammability and flammable limits of mixtures is usually laborious and may be difficult. Different methods of testing may give different results and different sources of ignition may influence the outcome of the test. Some test methods are described briefly below. In many cases, modifications or refinements may be added by the person doing the test, so the test method should be described.

TABLE 15–16 AUTOIGNITION TEMPERATURES

Product[a]	Autoignition temperature °C	Autoignition temperature °F
Ammonia	>850	>1562
Ammonia	651	1204
Butane	430	806
Butane	435	815
Propane	466	871
Ethane	510	950
Ethyl bromide	511	952
Ethyl chloride	519	966
Gasoline	280	536
Illuminating gas	590	1094
Methyl bromide	537	999
Methyl chloride	632	1170
Methyl formate	456	853
Methylene chloride	662	1224
Dimethyl ether	<322	<612
R-12/dimethyl ether (90/10)	435	815
R-152a	455	851
R-11/isopentane (93/7)	482	900
R-115/R-152a (84/16)	543	1009
R-13B1/R-152a (75/25)	560	1040
R-32/R-22 (50/50)	612	1134
R-13B1/R-32 (35/65)	621	1150
R-13B1/R-32 (15/85)	625	1157
R-134	630	1166
R-22	635	1175
R-22	632	1170
R-32	648	1198
R-32/R-115 (40/60)	655	1211
R-13B1/R-32 (80/20)	680	1256
R-502	704	1299
R-143a	750	1382
R-143a/R-115	750	1382
R-23	765	1409
R-123	770	1418
R-13B1	>850	>1562
R-11/R-123 (78/22)	>855	>1571
R-116	870	1598
R-115	880	1616
R-12 R-11 R-114	Above maximum temperature attainable in test equipment	

[a] In mixtures, the percentage by weight is shown.

Explosion buret or eudiometer tube. This method usually consists of a glass tube from 2 to 4 in. in diameter and 3 to 4 ft long. Various mixtures of the test gas and air are put in the tube and ignited using either a match or electric spark. Flammability is observed visually by the development of a flame. A number of tests must be made to define the boundaries of the flammable area, if any. Procedures and various limitations of this test method have been discussed by Coward and Jones of the Bureau of Mines [15–28]. The test is also used at the Underwriters' Laboratories in flammability studies [15–1].

Explosion drum [15–3, 15–27]. The eudiometer tube may tend to give erratic or erroneous results. To increase the volume and, more realistically, represent possible hazardous conditions, a modified stainless steel 55-gal drum was developed by Sanders and de Brabander at the Freon Products Laboratory of the DuPont Company. The drum was placed on its side and the open end was covered by a clear polyethylene film. A glass pressure buret attached to copper tubing was inserted through the wall at the top equidistant from the ends and extending into the drum about 4 in. A small electric fan was located at the rear of the drum to ensure good mixing of the gases and air.

Two different ignition sources were used.

1. A spiral of Nichrome wire was designed to hold two kitchen matches that could be electrically ignited and located at the bottom of the drum midway between the ends.
2. An electric arc could be generated between two $\frac{1}{4}$-in. copper rods inserted in one side of the drum at about the midpoint. One rod was insulated from the drum while the other was directly connected. The length of the arc could be adjusted by moving the rods. For mixtures of a flammable gas and air, the gap was about $\frac{1}{2}$ in. When a fluorocarbon was also present, the gap was reduced to $\frac{1}{16}$ in. in order to get a satisfactory arc. A 15,000-V transformer was used to energize the arc.

A copper tube could be inserted at the top of the drum, extending to within an inch of the ignition source to withdraw a sample of the gas mixture for analysis.

If one or more of the following occurred on ignition, the test for flammability was considered positive.

1. An audible sound was produced.
2. The drum became perceptibly hot.
3. The polyethylene cover on the end was ruptured by an explosion.
4. The flame traveled from the ignition source at the bottom to the top of the drum.

This modified drum test was found to be the most accurate and reproducible test of flammability of any of the various methods examined.

Flash point. A modified Tagliabue Open Cup method (ASTM D1510–55T, "Flash Point of Volatile Flammable Materials by Tag Open Cup Apparatus") is used for testing mixtures of low-boiling gases. The liquid mixture is cooled to about −50°F or lower and allowed to warm up at a rate of about 2°F/min. A flame is moved across the surface of the liquid at 2°F intervals until a flash is observed or until the temperature reaches about 20°F or the sample has completely evaporated [15–29].

Evaporation. Liquefied gas mixtures are evaporated under controlled conditions and the vapors collected and analyzed at frequent intervals. The vapor samples are tested for flammability or compared with existing data if the system has been studied earlier. If the flammability limits have been determined for mixtures of two or more gases in air, the composition of vapors from an evaporating liquid mixture can be estimated with sufficient accuracy for most practical purposes as outlined below.

Department of Transportation [15–30]. A compressed gas is defined by the Department of Transportation as flammable [15–30] if one of the following occurs:

1. Either a mixture of 13 vol % or less with air forms a flammable mixture or the flammable range with air is wider than 12 vol % regardless of the lower limit. These limits shall be determined at atmospheric temperature and pressure. The method of sampling and test procedure shall be acceptable to the Bureau of Explosives and approved by the Associate Director for HMR.
2. Using the Bureau of Explosives Flame Projection apparatus, the flame projects more than 18 in. beyond the ignition source with valve opened fully or the flame flashes back and burns at the valve with any degree of valve opening. The liquid mixture under pressure is placed 6 in. behind a flame (laboratory burner, plumber's candle, etc.). The spray is directed toward the top half of the flame and the length of the flame extension (if any) is measured.
3. Using the Bureau of Explosives Open Drum Apparatus, there is any significant propagation of flame away from the ignition source. The apparatus consists of a 55-gal, open-head drum fitted with a hinged cover or plastic film over the open end. The closed end of the drum is equipped with a shuttered opening 1 in. in diameter for introduction of the sample. A plumber's candle is placed an equal distance from the ends on the bottom of the drum.
4. Using the Bureau of Explosives Closed Drum Test, there is any explosion of the vapor–air mixture in the drum. The closed drum test is conducted either with the open end covered by a thin film (containing a slit) or the hinged cover dropped into place. The liquid is sprayed into the drum for 1 min. Any explosion or rapid burning of the vapor/air mixture is a positive test.

Ignition Source

Among others, Coward and Jones [15–28] and Sanders [15–3] have studied the effect of the ignition on the results of flammability tests. If the ignition source is not sufficiently strong, some flammable mixtures will not be ignited, especially at compositions near the flammability limit. A flame—whether from match, burner, or candle—seems a better source of energy than an electric arc. In running tests in the explosion drum described above, the electric arc was used for preliminary runs because new matches did not need to be installed and a number of runs could be made without opening the drum. When compositions appeared to be near the flammability limit, matches were used as the ignition source.

A few comparisons of flammability using an electric arc versus a match are shown in Table 15–17. The flammable range is generally wider when a match is used for ignition. The tests in Table 15–17 were all done in a 4-ft x 4-in. eudiometer tube.

TABLE 15–17 DETERMINATION OF FLAMMABILITY BY MATCH AND ELECTRIC ARC IGNITION

Product	Ignition source	Flammability
R-152a	Match	4.0–13.2[a]
	Electric arc	5.5–12.5[a]
Isobutane	Match	1.6–7.2[a]
	Electric arc	2.4–6.4[a]
Mixtures of R-133a and isobutane[b]		
86/14	Electric arc	Negative
85/15	Electric arc	Borderline
84/16	Electric arc	Positive
96/4	Match	Negative
94/6	Match	Positive

[a] Flammability limits in air, vol %.

[b] Ratio of R-133a (CH_2ClCF_3) to isobutane, wt %.

Source: Ref. 15–31.

Comparison of Methods

Flammability tests on several mixtures by different methods are summarized in Table 15–18. Not all methods and modifications of methods have been tested with the same mixture, but some comparisons can be made. Mixtures of R-152a (CH_3CHF_2) and air were tested in the stainless steel explosion drum described above with a match ignition source and the flammable limits were found to be 3.9 to 16.9 vol % R-152a in air. The limits were 4.0 to 13.2 in a 4-ft x 4-in. eudiometer tube with match ignition [15–3]. It seems to be generally true that for these two methods the lower flammable limits are about the same but the upper limit is higher for the drum than for the tube. Apparently, the lower volume to wall area for the tube has a dampening effect on the flammability. The open cup flash point seems to give reasonable results and is in fairly good agreement with the other methods. Speed is one of its virtues and it can give a good qualitative answer, but for actual compositions, analytical or mathematical procedures will be necessary.

The compositions in Table 15–18 are in percent by weight in the liquid phase and show the highest concentration of flammable material that could be used and

TABLE 15–18 COMPARISON OF FLAMMABILITY BY DIFFERENT METHODS

Mixture	Flash point		Eudiometer	Modified closed drum	Evaporation	Reference
	Composition (wt %)	°F				
R-12/propane	91/9	−70		90/10	95/5	15–27
R-12/vinyl chloride	91/9	No flash	68/32 (35)	75/25	91/9	15–27
	80/20	No flash	72/28			15–31
	65/35	−2				15–31
R-12/R-11/vinyl chloride	35/35/30	No flash	72/28 (35)	77/23	85/15	15–27
	39/39/22	No flash				15–31
	32.5/32.5/35	18	72/28 (34)			15–27
	32.5/32.5/35	7				15–31
R-12/R-11/isobutane	45/45/10	No flash (to dryness)		89.5/10.5	92.5/7.5	15–27
	45/45/10	No flash				15–31
R-12/R-11/dimethyl ether	75/10/15	5		78/10/12	80/10/10	15–27
	75/10/15	−6				15–31

still give a negative flammability test. Exceptions are the flash point and evaporation tests. These methods reflect the changing composition as the liquid evaporates and the possibility of flammable fractions being formed even though the original total composition was not flammable. For example, the R-12/propane (91/9) mixture gave a flash point of −70°F because the lower boiling propane would be higher in concentration in the first fractions evaporating from the liquid. From the evaporation test, the original liquid mixture would need to contain 95% R-12 to ensure that no flammable compositions containing two fluorocarbon products would be formed. When two fluorocarbon products are present, the composition is sometimes described by their combined concentrations.

Calculated Compositions

The composition of liquid mixtures is generally given in terms of percent by weight of each component, as R-12/R-11 (50/50 by weight). The lower-boiling component is usually listed first. An exception is a mixture of a fluorocarbon with a flammable or otherwise distinctive refrigerant. In this case, the fluorocarbon is often given first even though the other component is lower boiling: as, R-12/propane.

In flammability studies of gaseous mixtures, compositions are usually given as percent by volume. The exact relationship between volume percent and weight percent can be obtained by actual analysis or by the use of tables (or equations) of properties for each component that are based on real measurements. However, for most purposes the use of Raoult's law and the assumption of ideal gases gives sufficiently reliable conversions.

Raoult's Law. At a given temperature, for each component the mole fraction in the liquid multiplied by its vapor pressure equals the partial pressure of the component in the vapor phase above and in equilibrium with the liquid.

Ideal gas. For an ideal gas, mole percent = volume percent = pressure percent. These relationships apply very well for simple gases with small molecules such as oxygen, nitrogen, hydrogen, helium, methane, and so on. For heavy gases such as the fluorocarbon refrigerants the relationships also apply very well at low pressures, but deviations may become evident at higher pressures. Compounds not containing hydrogen, such as R-12, CCl_2F_2, generally follow the rules better than those containing hydrogen, such as R-22, $CHClF_2$. However, the latter type are sufficiently regular to give good results in conversions for flammability studies and also for most refrigeration applications.

Total evaporation

Weight-to-Volume Percentage. When the total composition as percent by weight is known, the volume relationship in a gaseous mixture can be calculated.

R-12/R-11 (50/50 Percent by Weight)

Molecular weights: R-12 = 120.93
R-11 = 137.37

$$\frac{\text{weight}}{\text{molecular weight}} = \text{mol}$$

	Weight	Mol	Mole fraction	Vol %
R-12	50	0.41346	0.5318	53.2
R-11	50	0.36398	0.4682	46.8
		0.7744		

A more exact calculation can be made using tables of thermodynamic properties. Assuming a temperature of 80°F and a pressure of 14.7 psia for each gas, the specific volume of each can be obtained. The volume of the mixture will be the sum of the specific volumes for each gas at the same pressure.

R-12: 3.194 ft^3/lb = 53.6 vol %

R-11: 2.761 ft^3/lb = 46.6 vol %

R-12/Propane (90/10 Percent by Weight)

Molecular weights: R-12 = 120.93
Propane = 44.097

	Weight	Mol	Mole fraction	Vol %
R-12	90	0.74423	0.7665	76.7
Propane	10	0.22677	0.2335	23.3

From tables at 80°F and a pressure of 14.7 psia:

R-12: 3.1942 ft^3/lb = 2.8748 ft^3 per 0.9 lb = 76.4 vol %

Propane: 8.8739 ft^3/lb = 0.8874 ft^3 per 0.1 lb = 23.6 vol %

R-22/R-12 (80/20 Percent by Weight)

Note: Even though R-22 and R-12 form an azeotrope in the liquid phase, this calculation can be made.

Molecular weights: R-22 = 86.47
R-12 = 120.93

	Weight	Mol	Mole fraction	Vol %
R-22	80	0.92518	0.8483	84.8
R-12	20	0.16538	0.1517	15.2
		1.09056		

From tables at 80°F and a pressure of 14.7 psia,

R-22: 4.4912 ft^3/lb = 3.5930 ft^3 per 0.8 lb = 84.9 vol %

R-12: 3.1942 ft^3/lb = 0.6388 ft^3 per 0.2 lb = 15.1 vol %

4.2318

The temperature and pressure are not critical at moderate conditions.

1. At 0°F and 14.7 psia

   R-22: 3.7824 ft^3/lb = 3.0259 ft^3/0.8 lb = 84.97 vol %

   R-12: 2.6764 ft^3/lb = 0.5353 ft^3/0.2 lb = 15.03 vol %

2. At 200°F and 14.7 psia

   R-22: 5.5303 ft^3/lb = 4.4242 ft^3/0.8 lb = 84.87 vol %

   R-12: 3.9439 ft^3/lb = 0.7888 ft^3/0.2 lb = 15.13 vol %

   5.2130

3. At 200°F and 100 psia

R-22: 0.77712 ft^3/lb = 0.6217 ft^3/0.8 lb = 85.1 vol %

R-12: 0.54413 ft^3/lb = 0.1088 ft^3/0.2 lb = 14.9 vol %

0.7305

Volume-to-weight percentage. Assume that volume fraction is equal to mole fraction. For example, if a gaseous mixture contains 75% R-12 and 25% propane:

	Mole fraction	Molecular weight	Weight	Percent by weight
R-12	0.75	120.93	90.6975	89.2
Propane	0.25	44.097	11.0242	10.8
			101.7217	

Check: From tables at 80°F and 14.7 psia,

$$\frac{\text{volume fraction (ft}^3\text{)}}{\text{specific volume (ft}^3\text{)}} = \text{weight}$$

R-12: 3.1942 ft^3/lb = 0.2348 lb = 89.3 wt %

Propane: 8.8739 ft^3/lb = 0.0282 lb = 10.7 wt %

0.2630

Fractionation. Evaporation is a continuous process of liquid changing to vapor and the vapor moving away from the liquid. With a mixture, the composition is changing all the time. The more volatile component will be in higher concentration in the vapor than in the liquid and will be continuously decreasing in the remaining liquid. The atmospheric boiling point of a compound is an indication of its volatility. For example, in a mixture of R-12 (boiling point −21.6°F) and propane (boiling point −43.7°F), the propane is more volatile than R-12 and the first portions that evaporate will be richer in propane than the original mixture. If a flammable mixture develops, it will be at the beginning of the evaporation. On the other hand, if the flammable component has a higher boiling point, the hazard will develop toward the end of the evaporation. This is the case with R-12 and isobutane (boiling point 10.9°F).

The composition of the evaporating vapor can be calculated in step-wise increments to give a reasonable account of the changes during evaporation. The calculation is somewhat laborious. A computer program could be developed to calculate a large number of steps. However, 5 to 10 or 20 steps or more can be calculated by hand and give a very useful description of the vapor.

The following assumptions are made:

1. Raoult's law applies.
2. The vapor and liquid are in equilibrium.
3. A concrete amount of vapor is removed and is replaced with new vapor from the liquid.
4. Equilibrium is again established.
5. Ideal gases are involved.

Example

a. R-12/propane (90/10 wt %)
b. Assume that all of the product is in the liquid phase and weighs 1 lb.
c. Evaporation takes place until the vapor weighs 0.05 lb.
d. The molecular weight of R-12 is 120.93 and propane 44.097 psia.
e. The vapor pressure of R-12 at 80°F is 98.87 psia and of propane is 143.77 psia.
f. Composition of the vapor
   (1) Raoult's law: mole fraction in the liquid multiplied by the vapor pressure of the pure component equals the partial pressure in the vapor
   (2) Mole fraction in the liquid

$$\text{R-12:} \quad \frac{0.9}{120.03} = 0.0074423 \text{ mol} = 0.76646 \text{ mole fraction}$$

$$\text{Propane:} \quad \frac{0.1}{44.097} = 0.0022677 \text{ mol} = 0.23354 \text{ mole fraction}$$

   (3) Mole fraction in the vapor

		Partial pressure	Pressure fraction or mole fraction in vapor
R-12:	(0.76646)(98.87) =	75.78	0.69297
Propane:	(0.23354)(143.77) =	33.576	0.30703
		109.356	

   (4) Weight fraction in the vapor

mole fraction times molecular weight = weight

		Weight	Weight fraction
R-12:	(0.69297)(120.93) =	83.8009	0.86092
Propane:	(0.30703)(44.097) =	13.5378	0.13908
		97.3387	

(5) Assume that 0.05 lb of vapor is removed.

$$(0.8609)(0.05) = 0.04305 \text{ lb of R-12}$$

$$(0.1391)(0.05) = 0.00695 \text{ lb of propane}$$

g. Remaining in liquid

R-12: $0.9 - 0.04305 = 0.85695$ lb

Propane: $0.1 - 0.00695 = 0.09305$ lb

(1) Mole fraction in liquid

	Mol	Mole fraction
R-12:	$\frac{0.85695}{120.93} = 0.0070863$	0.77055
Propane:	$\frac{0.09305}{44.097} = 0.0021101$	0.22945

h. Remove a second 0.05 lb of vapor and repeat the calculation.

After 20 such calculations the curve shown in Figure 15–1 can be obtained, where percent evaporation is plotted against the ratio of the mole fraction of R-12 in the vapor to the mole fraction of propane.

Removal of 0.05 lb of the mixture equals 5% evaporation:

$$\text{ratio} = \frac{\text{mole fraction R-12} = 0.69297}{\text{mole fraction propane} = 0.30703} = 2.257$$

From a study of the flammability of R-12, propane, and air (Figure 15–4) it is found that mixtures must contain at least 13.4% R-12 to be nonflammable in

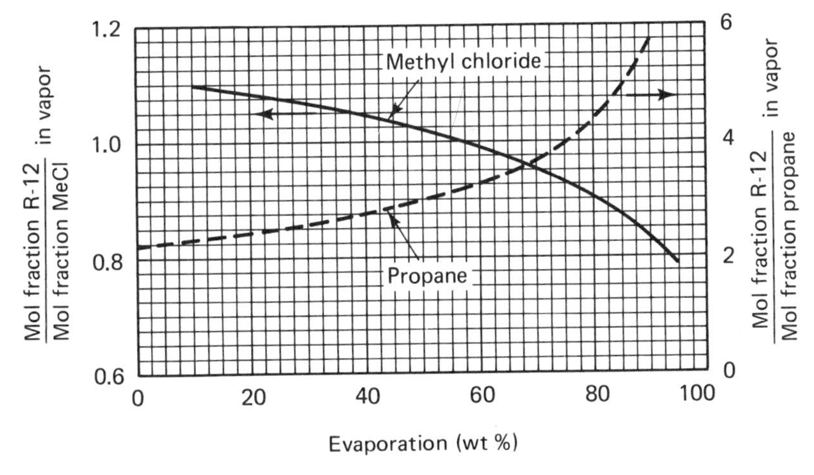

Figure 15–1 Ratio in vapor phase: mole fraction R-12/mole fraction methyl chloride (left) and R-12/propane (right).

any mixture with air. The ratio of R-12 to propane at this point is 13.4/4.1 = 3.268. When the ratio is less than 3.268, flammable mixtures with air can be obtained. Assuming that volume percent is equal to mole percent and referring to Figure 15–1, it can be seen that flammable fractions may be obtained until about two-thirds of the liquid has evaporated when the original composition contains 90 Wt % R-12 and evaporation takes place at 80°F. At this point the liquid contains about 93.2 wt % R-12 (see Figure 15–2).

An evaporation curve for mixtures of R-12 and methyl chloride is also shown in Figure 15–1. From Figure 15–3, the composition must be at least 12 vol % R-12 in air for complete nonflammability. The ratio of R-12 to methyl chloride is 12/13 = 0.923 when the original mixture is R-12/methyl chloride (70/30 by weight). From Figure 15–1 the ratio does not fall below 0.923 until about 75% of the liquid has evaporated and from that point on, flammable vapors might be expected. At this point the liquid contains about 65.6 wt % R-12 (see Figure 15–2).

In all cases, potentially flammable mixtures must be mixed with the right amounts of air to be flammable.

Flammability limits. Mixtures of flammable gases and air are hazardous only within certain ranges of concentrations bounded by the lower flammability limit and the upper flammability limit. When an inert gas is added to the mixture the flammable range is narrowed and at some concentration the mixture becomes nonflammable. Halogenated compounds are very good at suppressing flammability

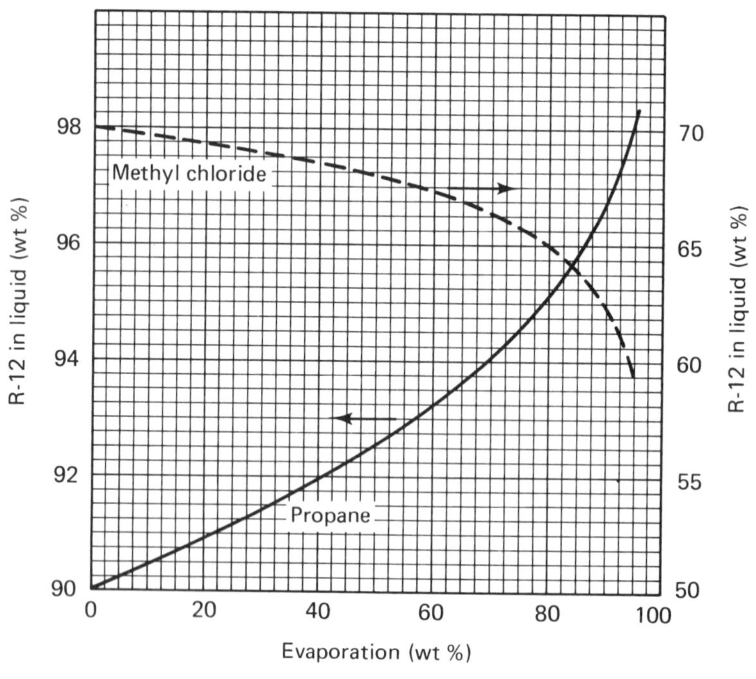

Figure 15–2 R-12 remaining in liquid phase during evaporation of mixtures with methyl chloride and propane.

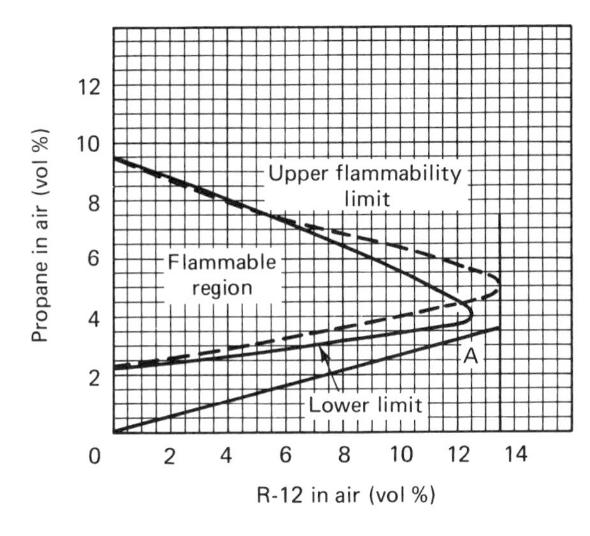

Figure 15–3 Flammability of R-12/ methyl chloride mixtures in air.

and usually produce nonflammable mixtures at relatively low concentrations. For example, the flammable range for propane in air is about 2.2 to 9.5 vol %. To eliminate the flammability of propane, an additional 43% of nitrogen must be added to the mixture. With carbon dioxide, an additional 30% must be added. With R-12 only about 10 to 13% is needed. Some flammability limits with and without added refrigerant are summarized in Table 15–19.

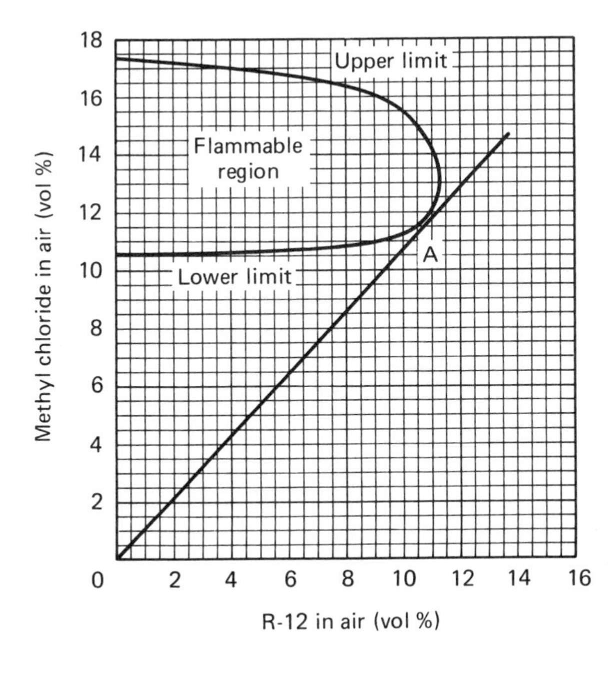

Figure 15–4 Flammability of R-12/ propane mixtures in air.

TABLE 15–19 SOME FLAMMABILITY LIMITS

Flammable range in air			Concentration of fluorocarbon producing nonflammability		
Product	Vol %	R	Vol %[a]	Wt %[b]	Reference
Methyl chloride	10–18	12	12	70	15–27
Butane	2–9	12	13.2		15–28
Propane	2–9	12	13.4	93.2	15–28
Propane	2–9	14	16	92	15–27
Ethylene	3–29	14	21	93	15–28
R-32, CH_2F_2	14–31	22	7		15–27
R-142b, CH_3CClF_2	9–15	114	16	92	15–3

[a] Concentration of fluorocarbon in a gaseous mixture (including air) necessary to produce nonflammability.

[b] Concentration of fluorocarbon in the liquid phase necessary to eliminate flammable fractions during evaporation.

R-12/Propane. In Figure 15–4 the region of flammability for mixtures of R-12, propane, and air is outlined. the concentration of propane in percent by volume is shown on the vertical axis and R-12 on the horizontal axis. The concentration of air is the difference between the sum of the R-12 and propane concentrations and 100. The solid curve is from the Bureau of Mines [15–28] and the dotted curve from Sanders (DuPont) [15–3]. When R-12 is present at concentrations of about 13.4 vol % or more, no mixture is flammable regardless of the concentration of propane or air. At 13.4% R-12, the propane concentration is about 4.1 on a straight line from zero concentration that does not intercept the flammable area.

R-12/Methyl Chloride. The flammable region for mixtures of R-12, methyl chloride, and air is shown in Figure 15–3. About 12 vol % R-12 is required to make all mixtures nonflammable. On a straight line from the origin at an R-12 concentration of 12% the methyl chloride concentration is about 13 vol % and the ratio is 0.923.

R-11/n-Pentane. A mixture of R-11/*n*-pentane (80/20 wt %) was judged flammable as determined by a flash point of 4°F measured by the Tagliabue Open Cup method. Vapors from this liquid mixture ignited and burned when tested 20 sec after evaporation began at 75°F [15–27].

R-12/n-Pentane. For a mixture containing 79 wt % R-12 and 21 wt % normal pentane, the flammability limits in air are 5 to 18 vol %.

R-114/R-142b [15–3]. A mixture containing 25 wt % R-114 and 75 wt % R-142b (CH_3CClF_2) was not flammable in any mixture with air. At concentrations of 76 wt % of R-142b and higher, flammable compositions could form.

R-152a Blends [15–3]. The maximum concentration in percent by weight of R-152a that can be present without forming flammable mixtures is

R-12 = 34% R-152a

R-22 = 24% R-152a

R-114 = 30% R-152a

R-14 Mixtures. *Propane.* The maximum amount of propane that can be present with R-14 and still be nonflammable is 8 wt %. This corresponds to 85.2 vol % of R-14 and 14.8 vol % propane, for a ratio of 5.76. With lower ratios of R-14 to propane, flammable mixtures are possible.

Ethylene. The maximum concentration of ethylene for nonflammability is 7 wt %. This corresponds to 80.9 vol % R-14 and 19.1 vol % ethylene.

FOOD FREEZING

In 1967, after 2 years of intensive testing, R-12 was approved for direct contact freezing of food. The following excerpt from the Federal Register [15–32] is included here to illustrate the high degree of purity and label identification required for this application.

Title 21–FOOD AND DRUGS

Chapter 1—Food and Drug Administration, Department of Health, Education, and Welfare.

Subchapter B—Food and Food Products

Part 121—FOOD ADDITIVES

Subpart D—Food Additives Permitted in Food for Human Consumption.

Dichlorodifluoromethane

The Commissioner of Food and Drugs, having evaluated the data in a petition (FAP 7A2123) filed by E. I. du Pont de Nemours & Co., Wilmington, Del, 19898 and other relevant material has concluded that a food additive regulation should be issued to provide for the safe use of dichlorodifluoromethane as a direct-contact freezing agent for foods. Therefore, pursuant to the provisions of the Federal Food, Drug, and Cosmetic Act (sec. 409(c)(1), 72 Stat. 1786; 21 U.S.C. 348(c)(1) and under the authority delegated to the Commissioner by the Secretary of Health, Education, and Welfare (21 CFR. 2.120), Part 121 is amended by adding to Subpart D the following new section:

121.1209 Dichlorodifluoromethane.

The food additive dichlorodifluoromethane may be safely used in food in accordance with the following prescribed conditions:

(a) The additive has a purity of not less than 99.97 percent.

(b) It is used or intended for use, in accordance with good manufacturing practice, as a direct-contact freezing agent for foods.

(c) To assure safe use of the additive:

(1) The label of its container shall bear, in addition to the other information required by the act, the following:

(i) The name of the additive, dichlorodifluoromethane, with or without the parenthetical name "Food Freezant 12."

(ii) The designation "food grade."

(2) The label or labeling of the food additive container shall bear adequate directions for use.

REUSE OF CYLINDERS

The fluorocarbon refrigerants are shipped in cylinders designed and tested to withstand pressures that may develop during transportation. The amount of refrigerant that can safely be stored in each cylinder is specified by Department of Transportation regulations. The Compressed Gas Association has issued a safety bulletin, SB-1, describing the hazards of refilling cylinders [15–33]. A copy of this bulletin is reproduced here.

The fluorocarbon refrigerants are usually shipped in small quantities (up to about 50 to 60 lb) in lightweight nonreusable cylinders. These cylinders have been approved (Department of Transportation container 39) for one-time use only and should be discarded when empty. However, they are very handy cylinders and the temptation is great to use them again—not only for refrigerant but for other purposes, especially for carrying compressed air. The repeated use with air or other gases is very hazardous, as outlined by the following warning taken from a safety bulletin published by a manufacturer of these lightweight cylinders [15–34].

AIR TANK . . . SAFETY BULLETIN

Air tank conversions of used disposable refrigerant cylinders are unsafe, dangerous, and potentially illegal.

These DOT 39 type cylinders are designed only for single trip service for refrigerant gases. The used tanks must not be adopted for reuse as compressed air tanks.

The disposable cylinder is intended for use one-time only in a corrosion free application. If used as air tanks, the cylinders are most likely filled with service station air. Compressors at these stations are seldom equipped with dryers. The humidity-rich air is compressed in the tank and the collected moisture forms a puddle. The contact between water, oxygen-rich air, and the unpainted interior initiates a rapid corrosion process.

The cylinder is soon weakened and cannot contain even a low pressure charge of air. The tank will subsequently explode. These low-pressure explosions can cause serious injuries, even death.

Refrigerant tank conversions for compressed air are hazardous. Only cylinders built and used in accordance with applicable DOT regulations and ASME codes can be considered safe for holding compressed air. Any other air tank conversions are dangerous.

These comments about air apply equally as well to other gases and refrigerants. In addition, repeated use of the cylinders may cause gradual weakening of the cylinder walls—hastened by bumps and scratches that accompany continued use.

The following safety bulletin information from the Compressed Gas Association details hazards of refilling refrigerant gas cylinders.

HAZARDS OF REFILLING COMPRESSED REFRIGERANT (HALOGENATED HYDROCARBONS) GAS CYLINDERS

Injuries and Property Damage Result from Improperly Filled Cylinders

Each year the Compressed Gas Association receives reports of serious personal injury and property damage resulting from improperly filled refrigerant gas cylinders. The incident rate of these accidents is fortunately extremely low and the industry as a whole is proud of its safety record. The alarming feature is that many of the accidents involve cylinders filled in the field by unqualified persons without proper cylinder charging equipment. Filling small cylinders from larger ones is sometimes practiced by distributors and service organizations in the refrigeration and air-conditioning field who are not equipped to perform this job safely. The hazard of overfilling small cylinders is always present when the filling is done by inexperienced operators who lack adequate knowledge of proper filling densities and the properties of the gases being handled. Safe filling densities vary for each refrigerant and overfilling through lack of knowledge or inadequate equipment may result in cylinder rupture and resulting damage. All operations involved in the transfer of refrigerant gases from one container to another require experienced personnel and a thorough knowledge of equipment, operation and maintenance in order to avoid personal injury, property damage or contamination of the refrigerant.

Federal and Local Authorities Recognize Hazard

Basic safety measures for handling and transporting compressed gases are included in the regulations of the Interstate Commerce Commission. Some state and local governments have either adopted these regulations or have issued similar ones to protect the public. Important among these regulations are:

1. the specification for the required type of cylinder, valve and safety device based on the service pressures likely to be encountered for each particular refrigerant
2. stipulation of a maximum filling limit by cylinder type for each refrigerant gas
3. the requirement of periodic re-inspection or hydrostatic testing of cylinders
4. the frequent inspection, repair or replacement of safety relief devices, valves and other components to insure safe operation at all times

Dangers of Cylinder Filling by Untrained Persons

1. *Overfilling*. Faulty loading equipment and inadequate technical knowledge of the proper filling densities and properties of the refrigerant being handled may lead to filling a cylinder beyond its rated capacity. This may not be evident at the time of filling, but a change in conditions such as an increase in temperature may lead to cylinder failure. The consequences can be personal casualty, severe property damage or both.
2. *Use of improper containers*. Minimum cylinder test criteria, materials and fabrication techniques are published in the Interstate Commerce Commission's specifications and regulations. Only containers and auxiliary equipment meeting these requirements should be used.
3. *Loss of product quality*. Each time a refrigerant is transferred, precautions must be taken to preserve the quality of the product. Exclusion of contamination in the refrigerant is essential for high performance and trouble-free operation

of mechanical refrigeration and air-conditioning equipment. Inexpert handling of a refrigerant can introduce harmful contamination and render the gas unfit for use.

Manufacturers of refrigerant gases are thoroughly familiar with the safety requirements for their products. They are equipped to perform the required testing of cylinders and auxiliary equipment and are qualified to perform essential maintenance work where necessary. As a final quality control measure, all refrigerant manufacturers analyze their products to make sure the packaged refrigerant meets acceptable purity standards. Chemists and laboratory facilities are not normally available in the field.

Recommendation. The risk of personal injury and property damage is sufficient reason to recommend the discontinuance of transfilling of refrigerant gases by unqualified persons. The possibility of the deterioration of product quality to levels below industry standards only adds strength to this recommendation.

The safety and quality record of the refrigerant gas industry as it applies to production, transportation, storage and handling of refrigerant gases is an enviable one. The Compressed Gas Association earnestly solicits the cooperation of the refrigerant distributors and service trades to support the above recommendation and hopes through this cooperation to eliminate accidents that may occur as a result of transferring refrigerant gases by persons not adequately trained to perform the operation safely.

REFERENCES

15–1. A. H. Nuckolls, "The Comparative Life, Fire, and Explosion Hazards of Common Refrigerants," Underwriters' Laboratories, Miscellaneous Hazard No. 2375 (Nov. 1933). Other UL reports: MH-3134 (R-22), MH-2256 (R-115), MH-3072 (R-113), MH-2256 (R-502), and MH-3135 (R-124a).

15–2. American Society of Heating, Refrigerating, and Air-Conditioning Engineers, *Safety Code for Mechanical Refrigeration*, ANSI/ASHRAE 15–1978, Atlanta, Ga.: ASHRAE (1978).

15–3. P. A. Sanders, *Handbook of Aerosol Technology*, 2nd ed. New York: Van Nostrand Reinhold (1979), Chap. 25.

15–4. T. Midgley and A. L. Henne, "Organic Fluorides as Refrigerants," *Ind. Eng. Chem.*, 24 (June 1932), 641.

15–5. DuPont Company, Freon Products Division, "Freon Compounds and Safety," Bulletin S-16 (1969).

15–6. D. Lester and L. A. Greenberg, "Acute and Chronic Toxicity of Some Halogenated Derivatives of Methane and Ethane," *Ind. Hyg. Occup. Med.*, 11 (July 1950), 335.

15–7. R. R. Sayers, et al., "Toxicity of Dichlorodifluoromethane," U.S. Bureau of Mines, Report R.I. 3013 (May 1930).

15–8. W. P. Yant, et al., "Toxicity of Dichlorotetrafluoroethane," U.S. Bureau of Mines, Report R.I. 3185 (Oct. 1932).

15–9. J. A. Pendergast, R. A. Jones, L. J. Jenkins, Jr., and J. Siegel, "Effects on Experimental Animals of Long Term Inhalation of Trichloroethane, Carbon Tetrachloride, 1,1,1-Trichloroethane, Dichlorodifluoromethane, and 1,1-Dichloroethylene," *Toxicol. Appl. Pharmacol.*, 10 (1967), 270.

15–10. DuPont Company, Freon Products Division, "Freon Fluorocarbons, Properties and Applications," Bulletin G-1 (Mar. 1984).

15–11. R. C. Downing and D. Madinabeitia, "The Toxicity of Fluorinated Hydrocarbon Aerosol Propellants," *Aerosol Age*, 5 (Sept. 1960), 25.

15–12. R. A. Kehoe, "Report on Human Exposure to Dichlorodifluoromethane Air," The Kettering Laboratory, University of Cincinnati, Cincinnati, Ohio (July 1943).

15–13. R. T. Johnstone and S. E. Miller, *Occupational Diseases and Industrial Medicine*. Philadelphia: W. B. Saunders Company (1961).

15–14. American Society of Heating, Refrigerating and Air-Conditioning Engineers, *Handbook and Product Directory, Fundamentals*. Atlanta, Ga.: ASHRAE (1977).

15–15. DuPont Company, Haskell Laboratory, unpublished data.

15–16. T. F. Slater, "A Note on the Relative Toxic Activities of Tetrachloromethane and Trichlorofluoromethane on the Rat," *Biochem. Pharmacol.*, 14 (1965), 178.

15–17. H. Sherman et al., "The Oral Toxicity of Dichlorodifluoromethane," *Toxicol Appl. Pharm.*, 29 (1974), 152.

15–18. M. C. Quevauviller, "Hygiène et securité des pulseurs pour aerosols médicamenteux," *Prod. Pharm.*, 20 (1965), 14.

15–19. G. D. Clayton and F. E. Clayton, eds., *Patty's Industrial Hygiene and Toxicology*, rev. 3rd ed. New York: John Wiley & Sons, Inc. (1981).

15–20. C. F. Reinhardt, A. Azar, M. E. Maxfield, E. Smith, Jr., and L. S. Mullin, "Cardiac Arrhythmia and Aerosol Sniffing," *Arch. Environ. Health*, 22 (1971), 265.

15–21. C. F. Reinhardt, L. S. Mullin, and M. E. Maxfield, "Epinephrine-Induced Cardiac Arrhythmia Potential of Some Common Industrial Solvents," *J. Occup. Med.*, 15 (1973), 953.

15–22. L. S. Mullin, A. Azar, C. F. Reinhardt, P. E. Smith, Jr., and E. F. Fabryka, "Halogenated Hydrocarbon-Induced Cardiac Arrhythmias Associated with Release of Endogenous Epinephrine," *Am. Ind. Hyg. Assoc. J.*, 33 (1972), 389.

15–23. J. W. Clayton, Jr., "The Toxicity of Fluorocarbons with Special Reference to Chemical Constitution," *J. Occup. Med.*, 4 (May 1962), 262.

15–24. J. W. Clayton, Jr., "Fluorocarbon Toxicity and Biological Action," *Fluo. Chem. Rev.*, 1 (1967), 197.

15–25. J. W. Clayton, Jr., "Fluorocarbon Toxicity: Past, Present, Future," *J. Soc. Cosmet. Chem.*, 18 (1967), 333.

15–26. American Conference of Governmental Industrial Hygienists, *TLV'S, Threshold Limit Values for Chemical Substances and Physical Agents in the Work Environment with Intended Changes for 1983–84*. Cincinnati, Ohio: ACGIH (1983–84).

15–27. DuPont Company, Freon Products Division, unpublished information.

15–28. H. F. Coward and G. W. Jones, "Limits of Flammability of Gases and Vapors," Bureau of Mines, Bulletin 503 (1953).

15–29. Underwriters' Laboratories, "Standard for Safety, Tests for Comparative Flammability of Liquids," UL 340 (1972).

15–30. *Hazardous Materials Regulations of the Department of Transportation*. Washington,

D.C.: Bureau of Explosives, Association of American Railroads, T. A. Phemister, Agent (1984).

15–31. L. T. Flanner and J. E. Van der Mey, ''Aerosol Flash Points,'' *Aerosol Age*, 7 (Oct. 1962), 91.

15–32. *Federal Register*, Vol. 32, No. 174, Friday, Sept. 8, 1967, Title 21—Food and Drugs, Part 121—Food Additives.

15–33. Compressed Gas Association, ''Hazards of Refilling Compressed Refrigerant (Halogenated Hydrocarbons) Gas Cylinders,'' Pamphlet SB-1 (1965).

15–34. Worthington Industries, ''Air Tank—Safety Bulletin,'' Columbus, Ohio: Worthington Industries, Inc.

16

THERMODYNAMIC PROPERTIES

In most cases, properties of refrigerants are based on direct, experimental measurements at different temperatures and pressures. The experimental data can be represented graphically or by equations. Graphs are useful for finding information quickly but equations are better suited for obtaining more precise values and for use with computers.

The thermodynamic properties of refrigerants are useful for comparing actual with theoretical performance and for comparing one refrigerant with another.

The same experimental data can be represented by different equations and many different forms have been suggested. For the most part, the equations given here were developed by J. J. Martin at the University of Michigan [16–1].

EQUATIONS

Equations based on measured data (or sometimes on estimations) for the following four basic properties of the liquid or gas are developed:

1. Liquid density
2. Vapor pressure
3. Pressure–volume–temperature relationships of the gas
4. Heat capacity of the gas

Other properties, such as enthalpy, entropy, velocity of sound, latent heat, and so on, can be calculated by thermodynamic relationships from these equations [16–2].

It is possible to obtain real values for the enthalpy and entropy by assuming that they are zero for the solid at absolute zero. Data for the heat capacity of the solid, the latent heat of melting from solid to liquid, and the heat capacity of the liquid from the melting point to the temperature range of interest in refrigeration can be measured or estimated. However, good measurements at these low temperatures are difficult to obtain and are not really necessary. It is sufficient in refrigeration applications to know the change in enthalpy and entropy as the refrigerant changes from one state to another or from one condition of temperature and pressure to another. So it is customary to assign arbitrary values for the enthalpy and entropy at a given reference point and develop thermodynamically valid properties for other conditions. In the past, values of zero for the liquid enthalpy and liquid entropy at $-40°$ have been selected for the reference points, possibly because this temperature is the same on both the Fahrenheit and Celsius scales. One disadvantage is that values for the liquid become negative at temperatures below $-40°$. Although this change of sign does not create a problem with numerical calculations, it is sometimes confusing. More recently, assigned values for the liquid enthalpy and entropy at the reference point are of sufficient magnitude that negative numbers are not likely.

The widespread use of computers has greatly increased interest in using the equations for the direct calculation of refrigerant properties [16–3,16–4]. This method is especially useful when the form of the equation is the same for various refrigerants, so that one program can be used by changing the numerical value of the constants. The equations are also used for the preparation of tables of properties and charts or diagrams. The charts show five properties of a refrigerant, including pressure, temperature, volume, enthalpy, and entropy. Any two properties can be used for the axes and the other three shown as lines of constant value. In refrigeration applications these charts are often called pressure–enthalpy diagrams, with pressure on the vertical axis and enthalpy on the horizontal axis. Other common arrangements use temperature versus entropy or enthalpy versus entropy. All of these arrangements are known as Mollier diagrams, named after Richard Mollier, a teacher in Germany especially interested in displaying properties of gases and liquids in chart form.

Pressure–enthalpy diagrams for a number of refrigerants (Figures 16–1 to 16–13) are included here. Extensive tables and charts are available from ASHRAE [16–5].

Liquid Density

$$d_L = A_L + B_L\left(1 - \frac{T}{T_c}\right)^{1/3} + C_L\left(1 - \frac{T}{T_c}\right)^{2/3} + D_L\left(1 - \frac{T}{T_c}\right) + E_L\left(1 - \frac{T}{T_c}\right)^{4/3} + F_L\left(1 - \frac{T}{T_c}\right)^{1/2} + G_L\left(1 - \frac{T}{T_c}\right)^{2}$$

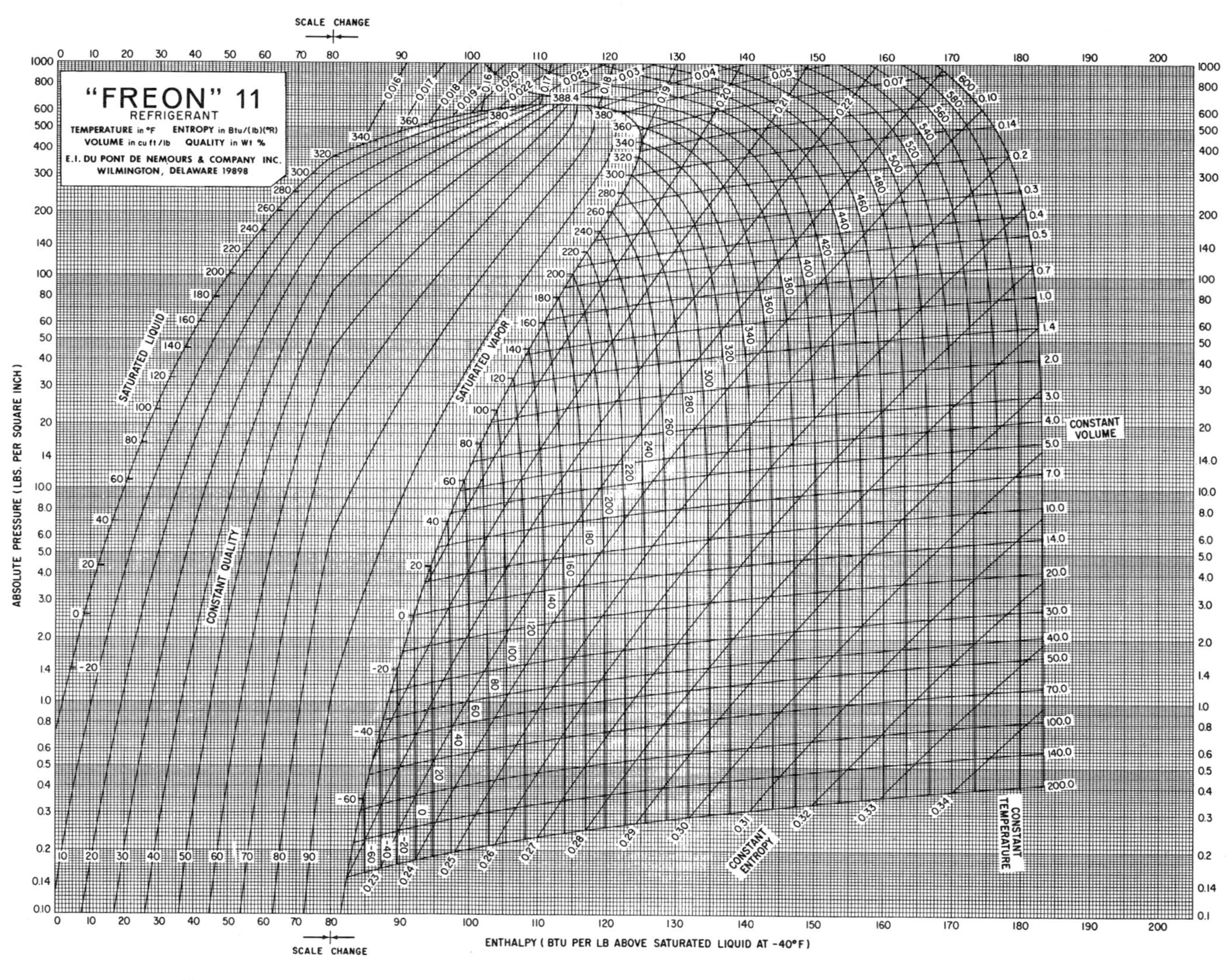

Figure 16–1 Pressure–enthalpy diagram for R-11. (Reprinted by permission of the DuPont Company.)

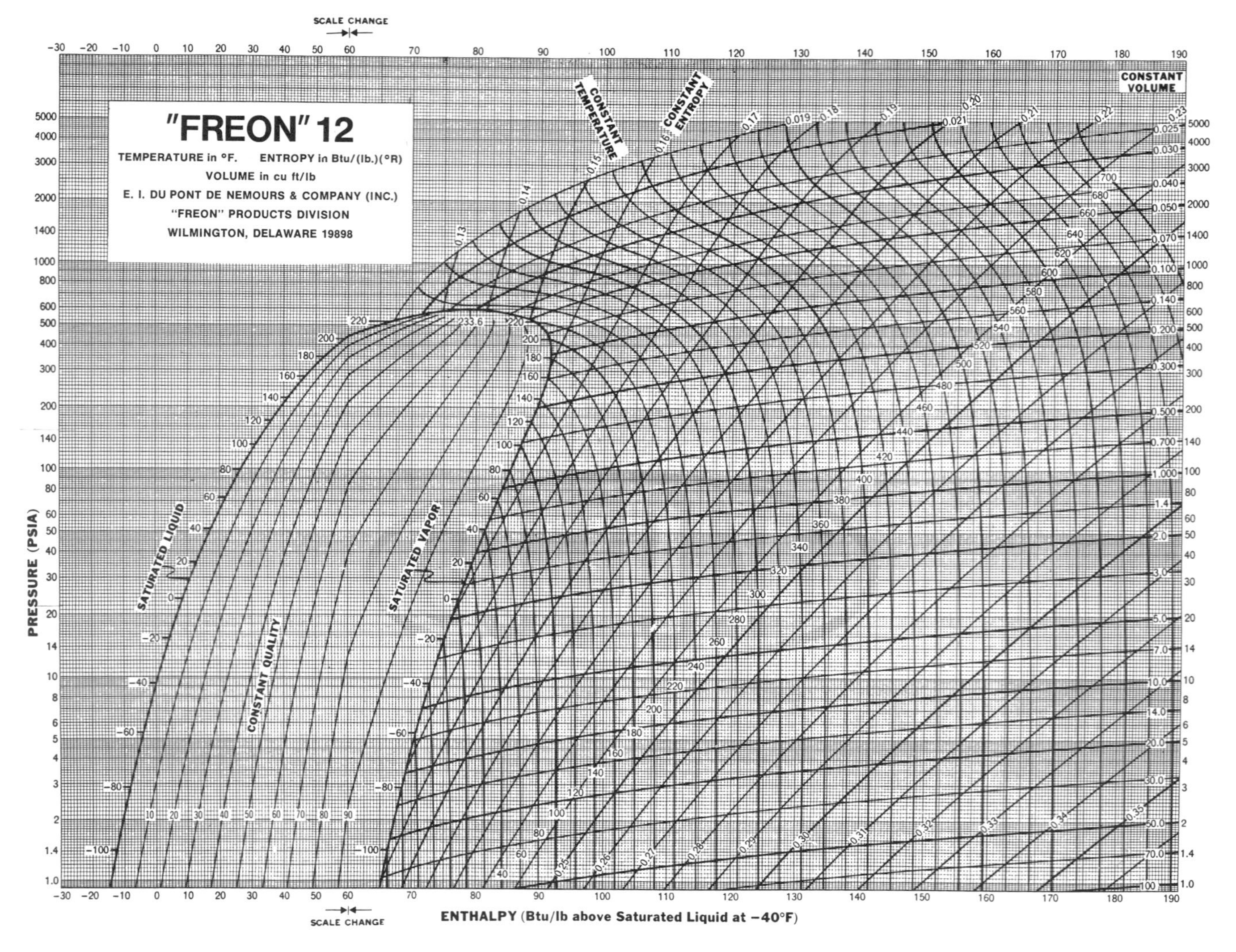

Figure 16–2 Pressure–enthalpy diagram for R-12. (Reprinted by permission of the DuPont Company.)

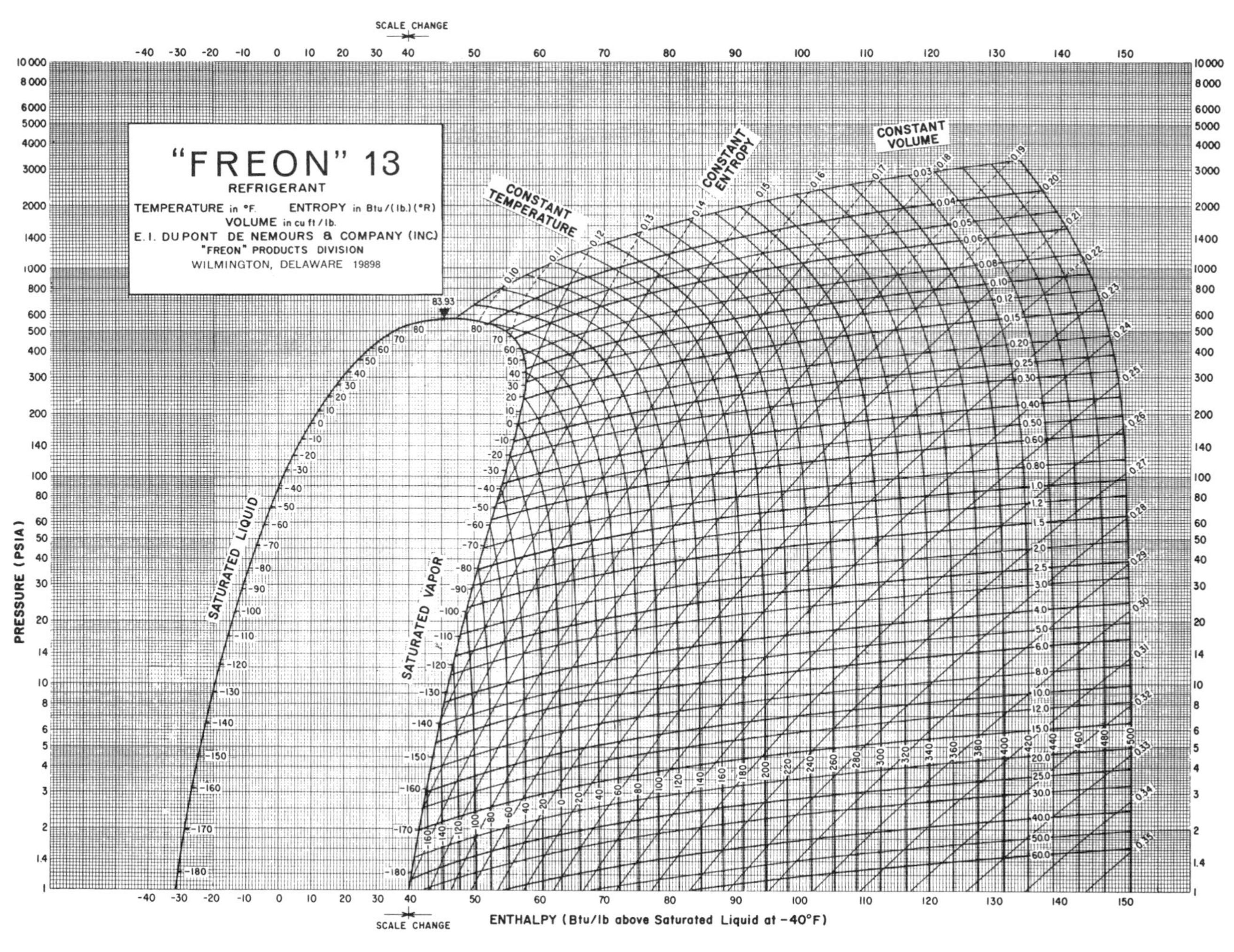

Figure 16–3 Pressure–enthalpy diagram for R-13. (Reprinted by permission of the DuPont Company.)

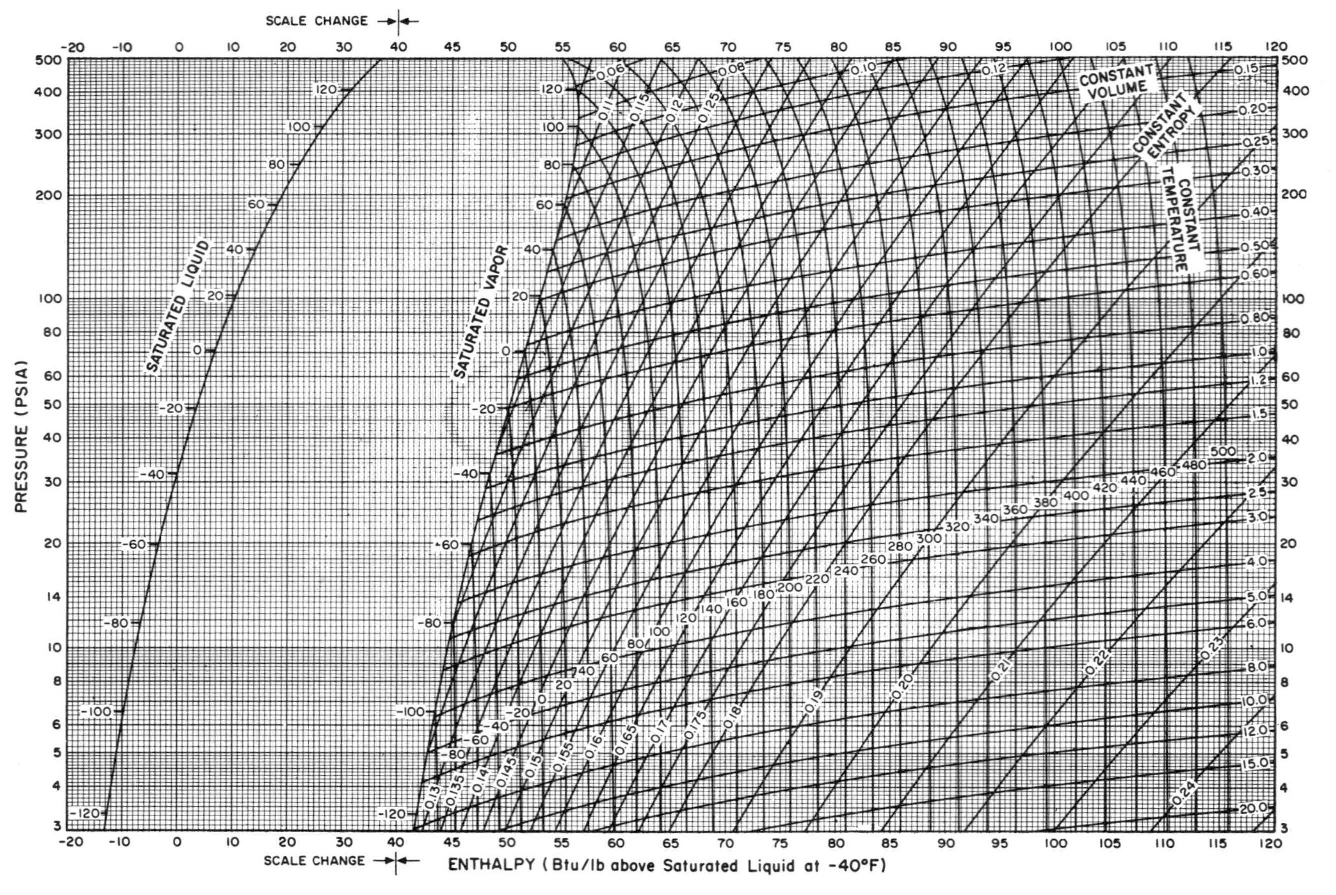

Figure 16–4 Pressure–enthalpy diagram for R-13Bl. (Reprinted by permission of the DuPont Company.)

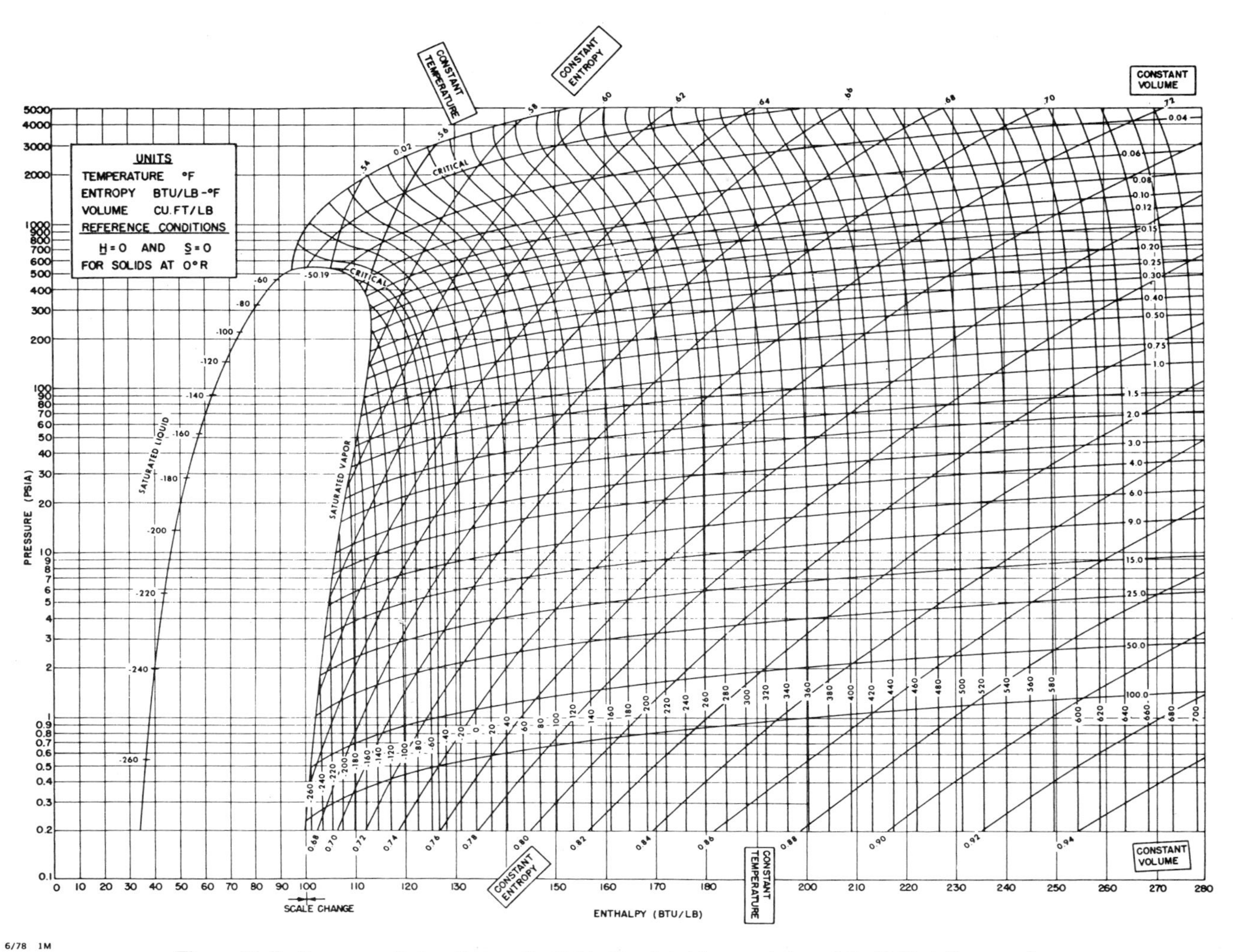

Figure 16–5 Pressure–enthalpy diagram for R-14. (Reprinted by permission of the DuPont Company.)

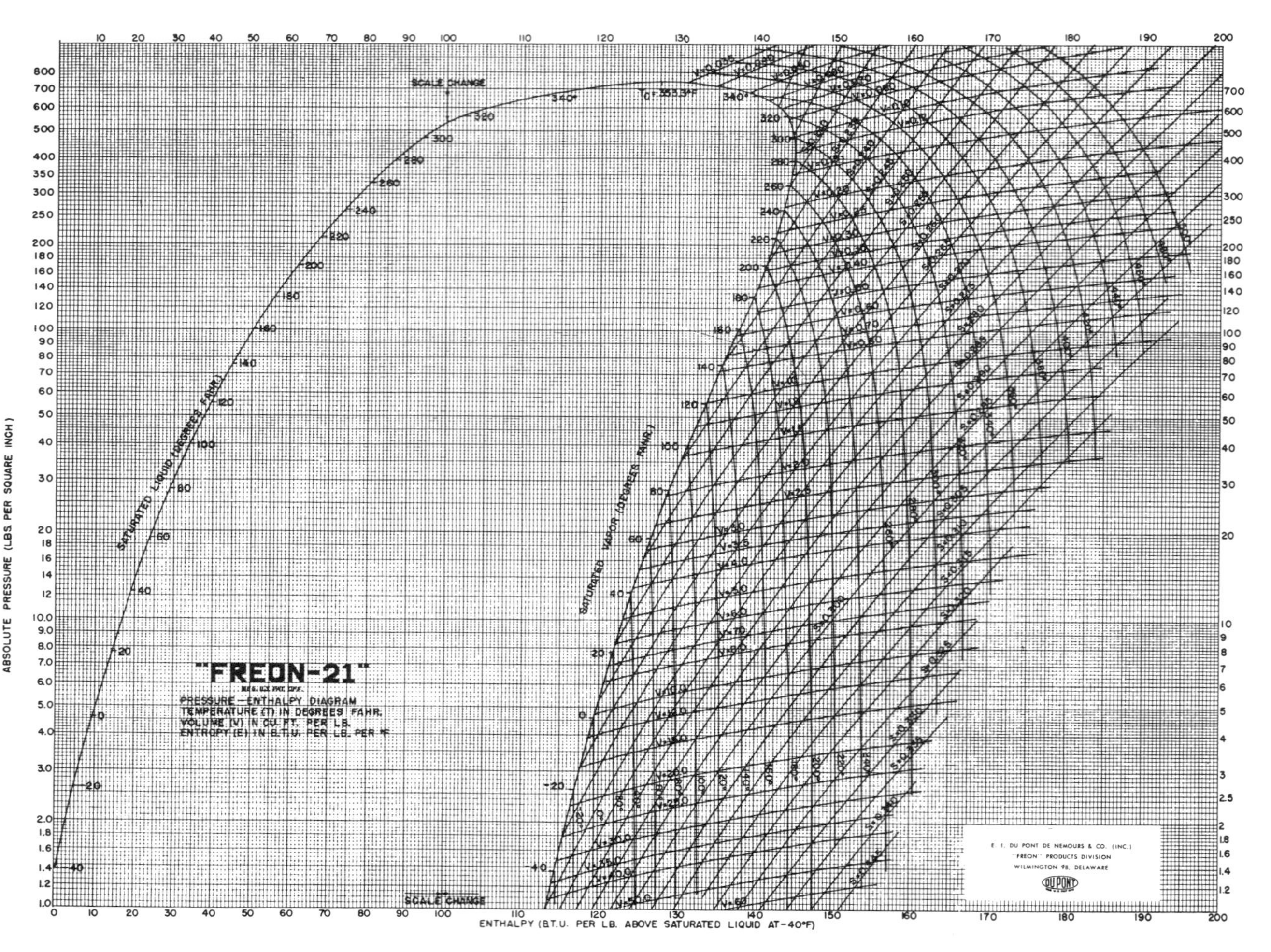

Figure 16–6 Pressure–enthalpy diagram for R-21. (Reprinted by permission of the DuPont Company.)

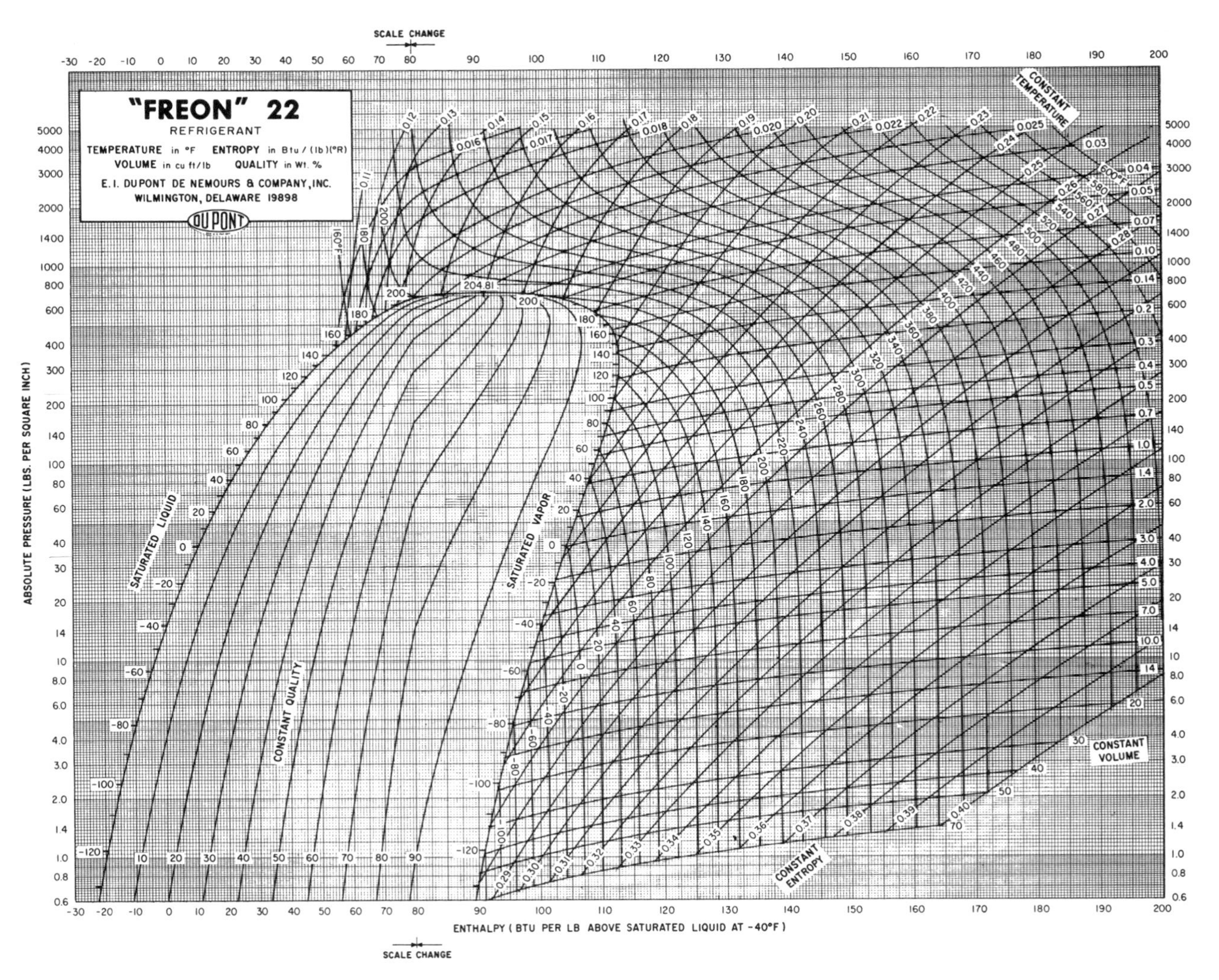

Figure 16–7 Pressure–enthalpy diagram for R-22. (Reprinted by permission of the DuPont Company.)

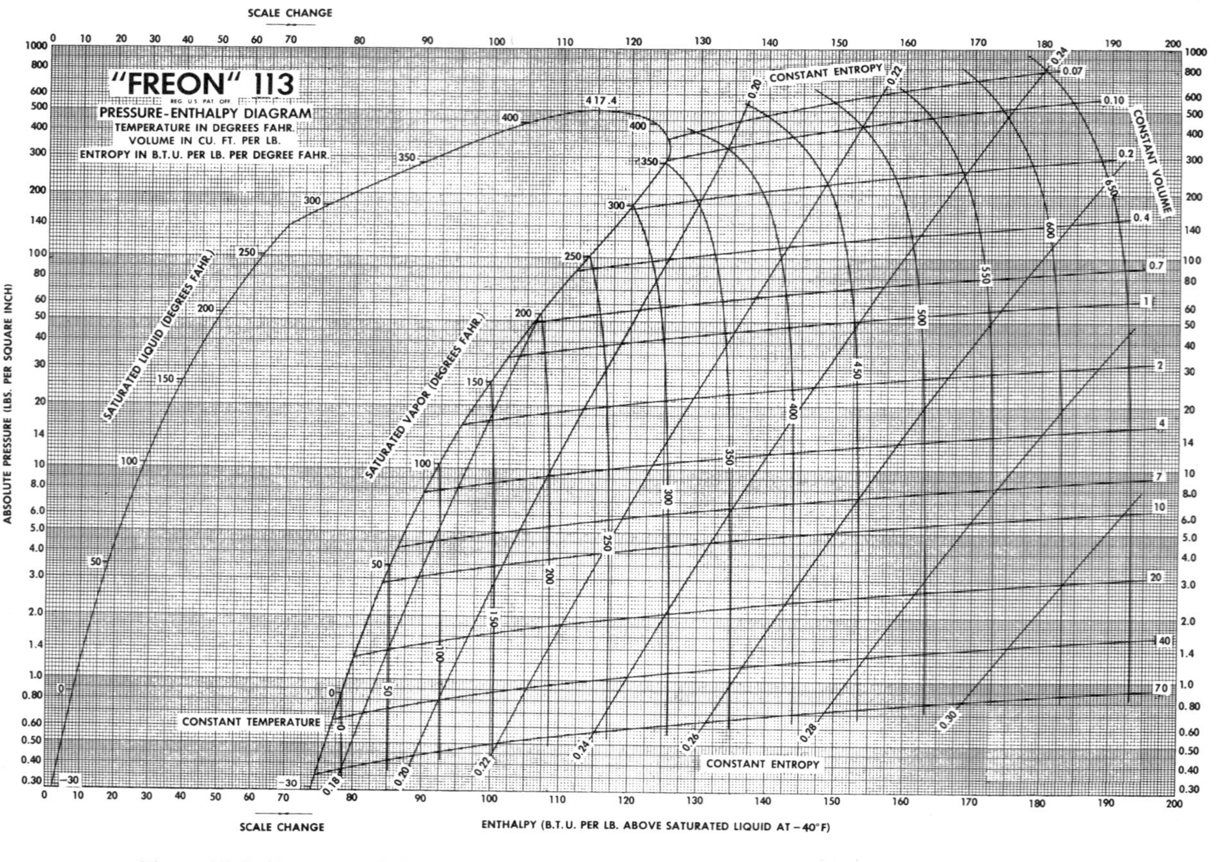

Figure 16–8 Pressure–enthalpy diagram for R-113. (Reprinted by permission of the DuPont Company.)

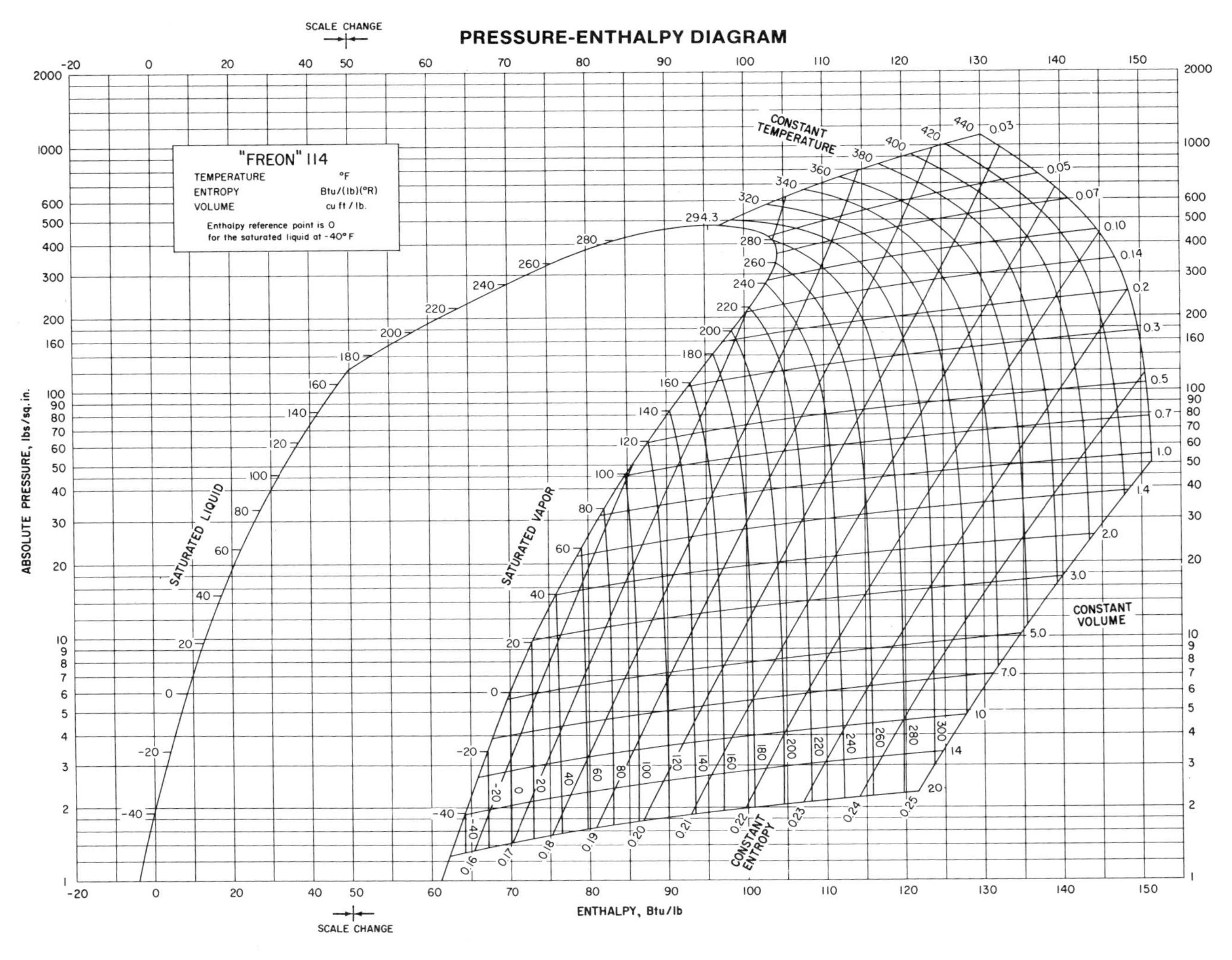

Figure 16–9 Pressure–enthalpy diagram for R-114. (Reprinted by permission of the DuPont Company.)

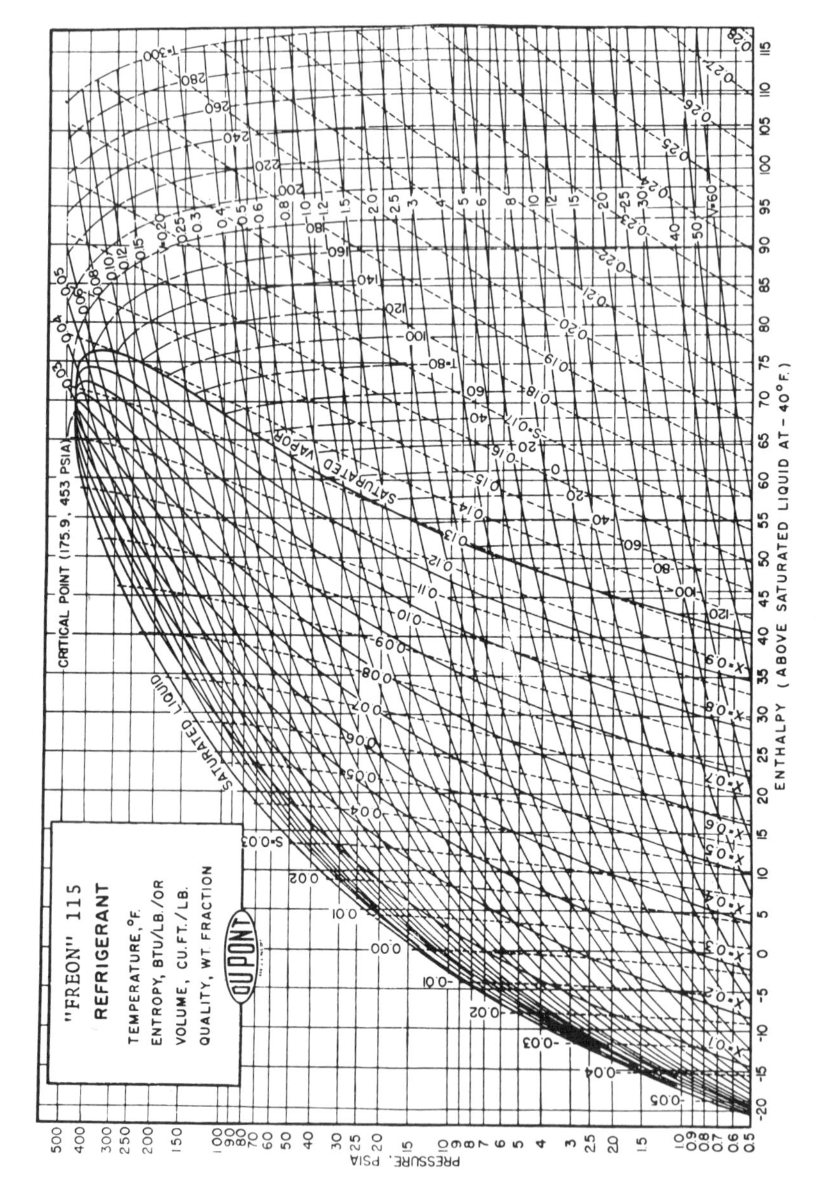

Figure 16–10 Pressure–enthalpy diagram for R-115. (Reprinted by permission of the DuPont Company.)

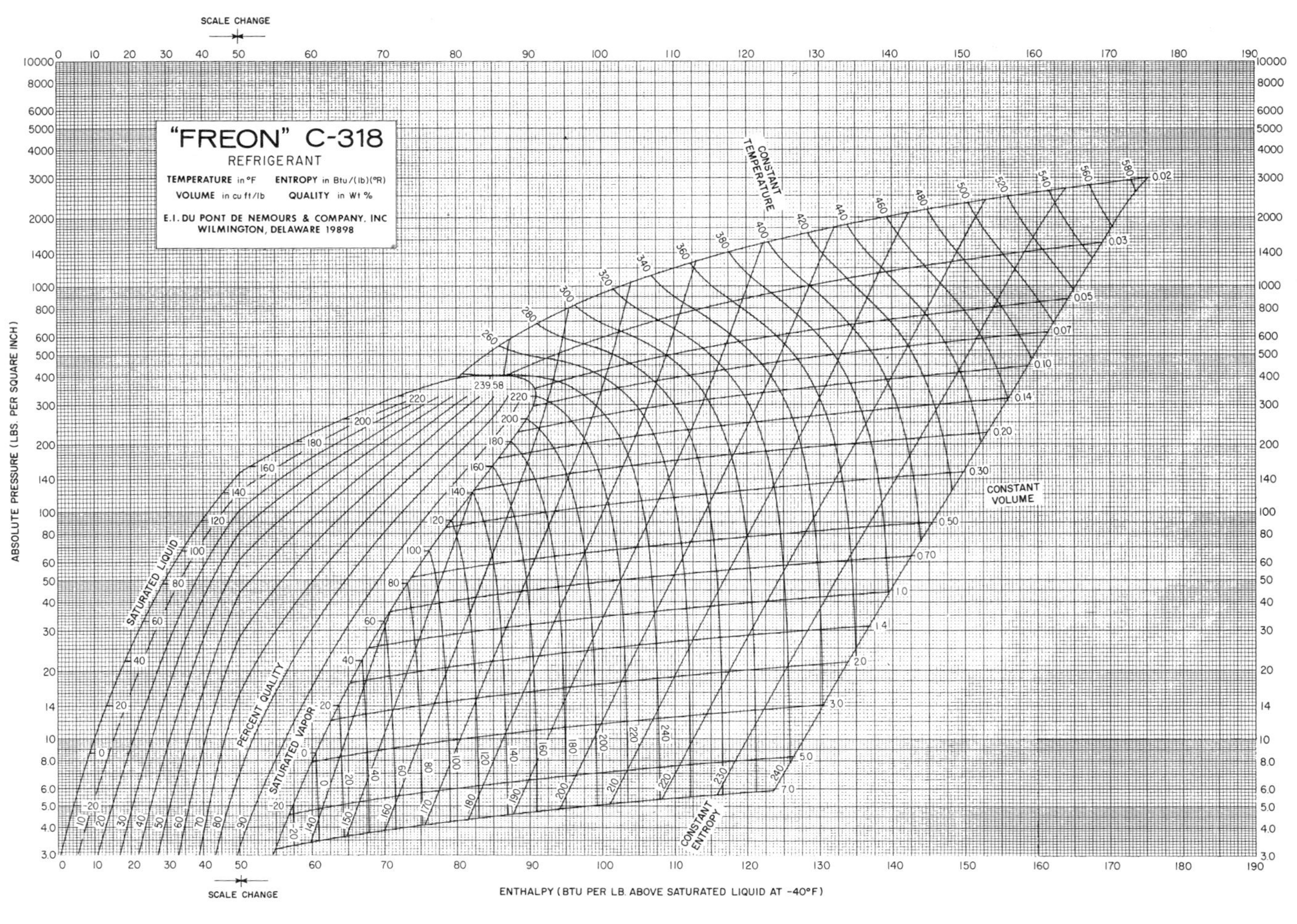

Figure 16–11 Pressure–enthalpy diagram for R-C318. (Reprinted by permission of the DuPont Company.)

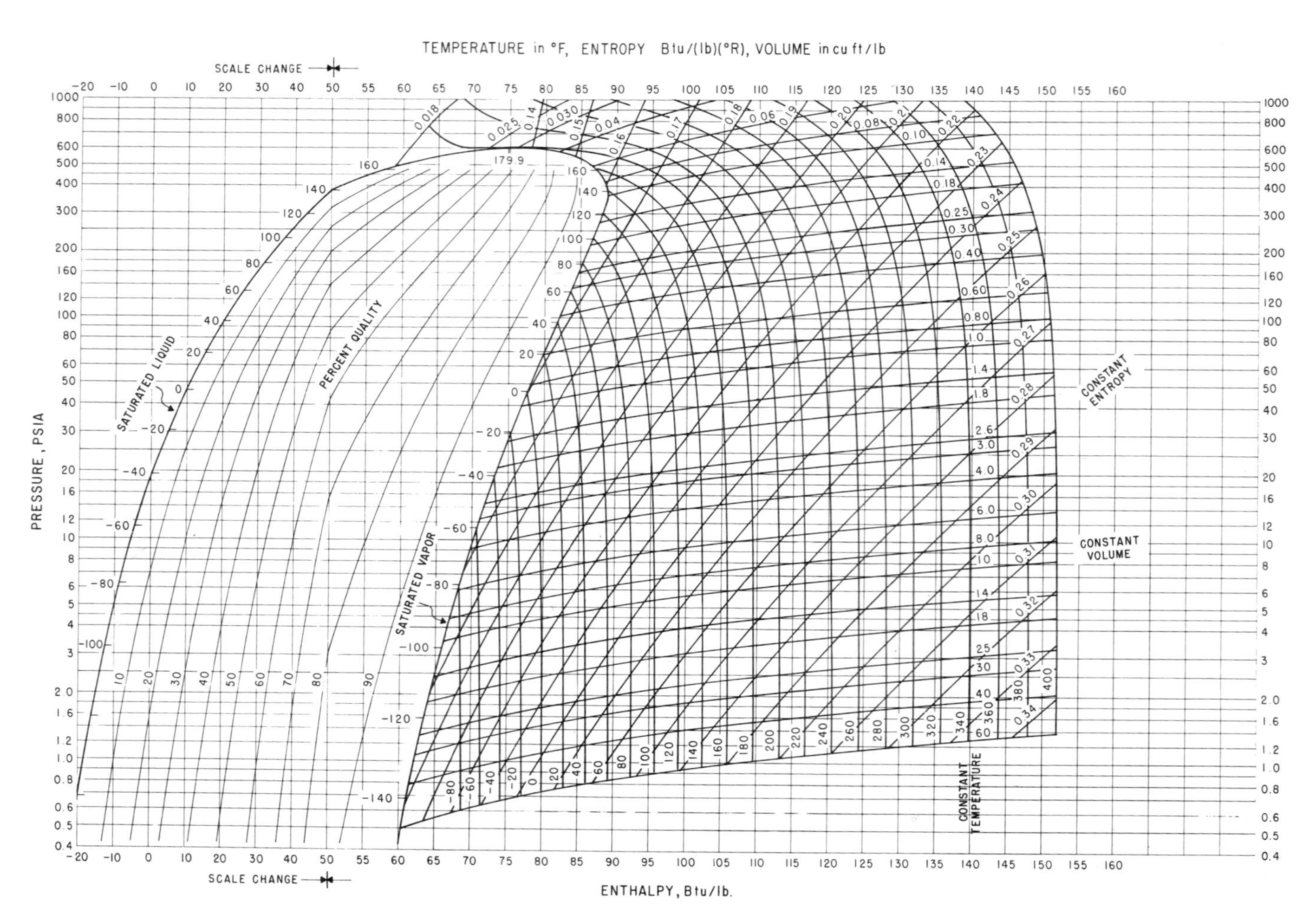

Figure 16–12 Pressure–enthalpy diagram for R-502. (Reprinted by permission of the DuPont Company.)

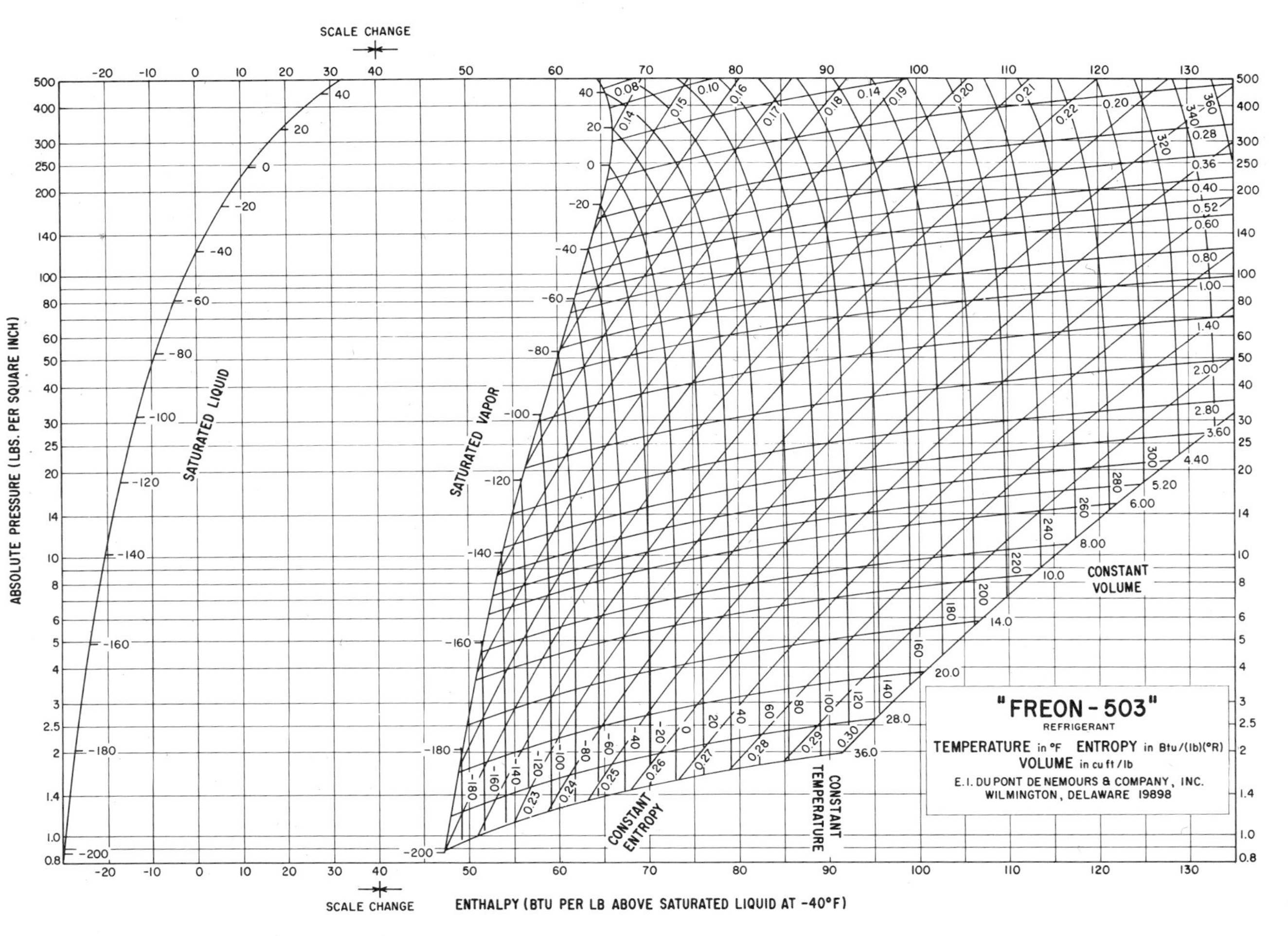

Figure 16–13 Pressure–enthalpy diagram for R-503. (Reprinted by permission of the DuPont Company.)

TABLE 16–1 CONSTANTS FOR LIQUID DENSITY EQUATION

Refrigerant	A_L	B_L	C_L	D_L	E_L	F_L	G_L	T_C (°R)	Add to °F	Reference
11	34.57	57.63811	43.6322	−42.82356	36.70663	0	0	848.07	459.67	16–6
12	34.84	53.341187	0	18.69137	0	21.98396	−3.150994	693.3	459.7	16–7
13	36.06996	54.395124	0	8.512776	0	25.879906	9.589006	543.60	459.69	16–8
14	39.06	69.568489	4.5866114	36.1716662	−8.058986	0	0	409.50		16–9
21[a]	116.37962	−0.03105808	−0.0000501	0	0	0	0	812.9	459.6	16–10
22	32.76	54.634409	36.74892	−22.2925657	20.4732886	0	0	664.50	459.69	16–11
23	32.7758	63.37784	−25.30533	144.16182	−106.1328	0	0	538.33	459.69	16–12
113[a]	122.872	−0.0128	0.0000636	0	0	0	0	877.0	459.6	16–10
114	36.32	61.146414	0	16.418015	0	17.476838	1.119828	753.95	459.69	16–13
500	31.00	43.562	74.709	−87.583	56.483	0	0	681.59	459.69	16–14
502	35.0	53.48437	63.86417	−70.08066	48.47901	0	0	639.56	459.67	16–15
C318	38.70	70.858318	23.609759	15.989182	−8.9243856	0	0	699.27	459.69	16–16

[a]The form of the equation is $d_L = A_L + B_L T + C_L T^2$.

Vapor Pressure

$$\log_{10}P = A + \frac{B}{T} + C\log_{10}T + DT + E\left(\frac{F-T}{T}\right)\log_{10}(F-T)$$

$$\frac{dP}{dT} = P\left(\frac{-\ln 10[B + EF\log_{10}(F-T)]}{T^2} + \frac{C-E}{T} + D\ln 10\right)$$

$$\frac{d(\log_{10}P)}{dT} = \frac{1}{P}\log_{10}e\,\frac{dP}{dT}$$

$$\frac{dP}{dT} = (P)(\ln 10)\left[\frac{-B}{T^2} + \frac{C}{(\ln 10)T} + D - E\left(\frac{\log_{10}e}{T} + \frac{F\log_{10}(F-T)}{T^2}\right)\right]$$

Equation of State

$$P = \frac{RT}{V-b} + \frac{A_2 + B_2T + C_2e^{-KT/T_c}}{(V-b)^2} + \frac{A_3 + B_3T + C_3e^{-KT/T_c}}{(V-b)^3} + \frac{A_4 + B_4T + C_4e^{-KT/T_c}}{(V-b)^4} + \frac{A_5 + B_5T + C_5e^{-KT/T_c}}{(V-b)^5} + \frac{A_6 + B_6T + C_6e^{-KT/T_c}}{e^{\alpha v}(1 + C'e^{\alpha v})}$$

$$\left(\frac{dP}{dV}\right)_T = -\frac{RT}{(V-b)^2} - \frac{2(A_2 + B_2T + C_2e^{-KT/T_c})}{(V-b)^3} - \frac{3(A_3 + B_3T + C_3e^{-KT/T_c})}{(V-b)^4} - \frac{4(A_4 + B_4T + C_4e^{-KT/T_c})}{(V-b)^5} - \frac{5(A_5 + B_5T + C_5e^{-KT/T_c})}{(V-b)^6} + (A_6 + B_6T + C_6e^{-KT/T_c})\left[-\frac{\alpha e^{\alpha v} + 2\alpha c'e^{2\alpha v}}{(e^{\alpha v} + c'e^{2\alpha v})^2}\right]$$

$$\left(\frac{dP}{dT}\right)_v = \frac{R}{V-b} + \frac{B_2 - \dfrac{KC_2e^{-KT/T_c}}{T_c}}{(V-b)^2} + \frac{B_3 - \dfrac{KC_3e^{-KT/T_c}}{T_c}}{(V-b)^3} + \frac{B_4 - \dfrac{KC_4e^{-KT/T_c}}{T_c}}{(V-b)^4} + \frac{B_5 - \dfrac{KC_5e^{-KT/T_c}}{T_c}}{(V-b)^5} + \frac{B_6 - \dfrac{KC_6e^{-KT/T_c}}{T_c}}{e^{\alpha v}(1 + C'e^{\alpha v})}$$

$$\left(\frac{d^2P}{dT^2}\right)_v = \frac{K^2e^{-KT/T_c}}{T_c^2}\left[\frac{C_2}{(V-b)^2} + \frac{C_3}{(V-b)^3} + \frac{C_4}{(V-b)^4} + \frac{C_5}{(V-b)^5} + \frac{C_6}{e^{\alpha v}(1 + C'e^{\alpha v})}\right)$$

TABLE 16–2 CONSTANTS FOR VAPOR-PRESSURE EQUATION

Refrigerant	A	B	C	D	E	F	Add to °F	Reference
11	42.14702865	−4344.343807	−12.84596753	4.0083725×10^{-3}	0.0313605356	852.07	459.67	16–6
12	39.88381727	−3436.632228	−12.47152228	$4.73044244 \times 10^{-3}$	0	0	459.7	16–7
13	25.967975	−2709.538217	−7.17234391	2.545154×10^{-3}	0.280301091	546.00	459.67	16–17
14	20.71545389	−2467.505285	−4.69017025	6.4798076×10^{-4}	0.770707795	424	459.69	16–18
21	42.7908	−4261.34	−13.0295	3.9851×10^{-3}	0	0	459.6	16–19
22	29.35754453	−3845.193152	−7.86103122	2.1909390×10^{-3}	0.445746703	686.1	459.69	16–11
23[a]	328.90853	−7952.76913	−144.5142304	0.24211502	$-2.1280665 \times 10^{-4}$	9.434955×10^{-8}	459.69	16–12
113	33.0655	−4330.98	−9.2635	2.0539×10^{-3}	0	0	459.6	16–19
114	27.071306	−5113.7021	−6.3086761	6.913003×10^{-4}	0.78142111	768.35	459.69	16–13
500	17.780935	−3422.69717	−3.63691	5.0272207×10^{-4}	0.4629401	695.57	459.67	16–20
502	10.644955	−3671.153813	−0.369835	-1.746352×10^{-3}	0.8161139	654	459.67	16–15
C318	15.63242	−4301.063	−2.128401	-1.19759×10^{-3}	0.6625898	714	459.69	16–16

[a]Form of equation is: $\log_{10}P = A + B/T + C\log_{10}T + DT + ET^2 + FT^3$.

TABLE 16–3 CONSTANTS FOR EQUATION OF STATE[a]

	Refrigerant					
	11	12	13	14	21	22
R	0.078117	0.088734	0.102728	0.1219336	0.10427	0.124098
b	0.00190	0.0065093886	0.0048	0.0015	0	0.002
A_2	−3.126759	−3.40972713	−3.083417	−2.162959	−7.316	−4.353547
B_2	1.318523×10^{-3}	$1.59434848 \times 10^{-3}$	2.341695×10^{-3}	2.135114×10^{-3}	4.6421×10^{-3}	2.407252×10^{-3}
C_2	−35.76999	−56.7627671	−18.212643	−18.941131	0	−44.066868
A_3	−0.025341	0.0602394465	0.058854	4.404057×10^{-3}	−0.20382376	−0.017464
B_3	4.875121×10^{-5}	$-1.87961843 \times 10^{-5}$	-5.671268×10^{-5}	1.282818×10^{-5}	3.593×10^{-4}	7.62789×10^{-5}
C_3	1.220367	1.31139908	0.571958	0.539776	0	1.483763
A_4	1.687277×10^{-3}	$-5.4873701 \times 10^{-4}$	-1.026061×10^{-3}	1.921072×10^{-4}	0	2.310142×10^{-3}
B_4	-1.805062×10^{-6}	0	1.338679×10^{-6}	-3.918263×10^{-7}	0	-3.605723×10^{-6}
C_4	0	0	0	0	0	0
A_5	-2.358930×10^{-5}	0	5.290649×10^{-6}	-4.481049×10^{-6}	0	-3.724044×10^{-5}
B_5	2.448303×10^{-8}	3.468834×10^{-9}	-7.395111×10^{-9}	9.062318×10^{-9}	0	5.355465×10^{-8}
C_5	-1.478379×10^{-4}	$-2.54390678 \times 10^{-5}$	-3.874233×10^{-5}	-4.836678×10^{-5}	0	-1.845051×10^{-4}
A_6	1.057504×10^{8}	0	7.378601×10^{7}	5.838823×10^{7}	0	1.363387×10^{8}
B_6	-9.472103×10^{4}	0	-7.435565×10^{4}	-9.263923×10^{4}	0	-1.672612×10^{5}
C_6	0	0	0	0	0	0
K	4.50	5.475	4.00	4.00	0	4.2
a[b]	580	0	625	661.199997	0	548.2
c'	0	0	0	0	0	0
T_C(°R)	848.07	693.3	543.60	409.50	812.9	664.50
Add to °F	459.67	459.7	459.67	459.69	459.69	459.69
Reference	16–6	16–7	16–18	16–18	16–21	16–11

[a]For refrigeration applications, the sixth term in the equation of state and other equations where it or derivatives occur may be omitted without loss of accuracy.

[b]When alpha, α, is zero, any term in which it appears in the denominator should be omitted.

TABLE 16–3 (Continued)

Refrigerant					
23	113	114	500	502	C-318
0.15327	0.05728	0.062780807	0.10805000	0.096125	0.053645698
0.00125	0	0.005914907	0.006034229	0.00167	0.0060114165
−4.679499	−4.035	−2.3856704	−4.549888	−3.2613344	−1.8947274
3.472778×10^{-3}	2.618×10^{-3}	1.0801207×10^{-3}	2.308415×10^{-3}	2.0576287×10^{-3}	9.8484745×10^{-4}
−159.775232	0	−6.5643648	−92.90748	−24.24879	−28.542156
−0.012475	−0.0214	0.034055687	0.08660634	0.034866748	0.026479892
7.733388×10^{-5}	5.00×10^{-5}	$-5.3336494 \times 10^{-6}$	-3.141665×10^{-5}	$-8.6791313 \times 10^{-6}$	-6.862101×10^{-6}
5.941212	0	0.16366057	2.742282	0.33274779	0.66384636
2.068042×10^{-3}	0	-3.857481×10^{-4}	-8.726016×10^{-4}	$-8.5765677 \times 10^{-4}$	$-2.4565234 \times 10^{-4}$
-3.684238×10^{-6}	0	0	0	7.0240549×10^{-7}	0
0	0	0	0	0.022412368	0
-3.868546×10^{-5}	0	1.6017659×10^{-6}	-1.375958×10^{-6}	8.8368967×10^{-6}	6.0887086×10^{-7}
6.455643×10^{-8}	0	$6.2632341 \times 10^{-10}$	9.149570×10^{-9}	$-7.9168095 \times 10^{-9}$	8.269634×10^{-10}
-7.394214×10^{-4}	0	$-1.0165314 \times 10^{-5}$	-2.102661×10^{-4}	$-3.7167231 \times 10^{-4}$	-3.849145×10^{-5}
7.502357×10^{7}	0	0	0	-3.8257766×10^{7}	0
-1.114202×10^{5}	0	0	0	5.5816094×10^{4}	0
0	0	0	0	1.5378377×10^{9}	0
5.50	0	3.0	5.475	4.2	5
520.0	0	0	0	609	0
0	0	0	0	7×10^{-7}	0
538.33	877.0	753.95	681.59	639.56	699.27
459.69	459.69	459.69	459.69	459.67	459.69
16–18	16–21	16–13	16–14	16–15	16–16

Heat Capacity of the Vapor

$$C_v = a + bT + cT^2 + dT^3 + \frac{f}{T^2} + \int_{\infty}^{V} JT\left(\frac{d_2P}{dT^2}\right)_v dV$$

$$C_v = a + bT + cT^2 + dT^3 + \frac{f}{T^2} - \frac{JK^2Te^{-KT/T_c}}{T_c^2}\left[\frac{C_2}{V-b} + \frac{C_3}{2(V-b)^2}\right.$$

$$\left. + \frac{C_4}{3(V-b)^3} + \frac{C_5}{4(V-b)^4} + \frac{C_6}{\alpha e^{\alpha v}} - \frac{C_6C'}{\alpha}(\ln 10)\log\left(1 + \frac{1}{C'e^{\alpha v}}\right)\right]$$

$$C_P = C_P^\circ + JT\int_P^\circ \left(\frac{d^2V}{dT^2}\right)_P dP_T$$

$$C_P = C_v - \frac{JT\left(\frac{dP}{dT}\right)_v^2}{\left(\frac{dP}{dV}\right)_T}$$

Enthalpy of the Vapor

$$H = aT + \frac{bT^2}{2} + \frac{cT^3}{3} + \frac{dT^4}{4} - \frac{f}{T} + JPV + J\left[\frac{A_2}{V-b} + \frac{A_3}{2(V-b)^2}\right.$$

$$\left. + \frac{A_4}{3(V-b)^3} + \frac{A_5}{4(V-b)^4} + \frac{A_6}{\alpha}\left(\frac{1}{e^{\alpha v}} - C'[\ln 10]\log\left[1 + \frac{1}{C'e^{\alpha v}}\right]\right)\right]$$

$$+ Je^{-KT/T_c}\left(1 + \frac{KT}{T_c}\right)\left[\frac{C_2}{V-b} + \frac{C_3}{2(V-b)^2} + \frac{C_4}{3(V-b)^3}\right.$$

$$\left. + \frac{C_5}{4(V-b)^4} + \frac{C_6}{\alpha e^{\alpha v}} - \frac{C_6C'(\ln 10)\log}{\alpha}\left(1 + \frac{1}{C'e^{\alpha v}}\right)\right] + X$$

$X^* = \Delta H$ (latent at T_1) $- H$ (at T_1V_1)

where T_1 = reference temperature (usually $-40°F$)
V_1 = saturated vapor volume at T_1

*Assuming the enthalpy of the liquid is zero at the reference temperature.

Entropy of the Vapor

$$S = a(\ln 10)\log T + bT + \frac{cT^2}{2} + \frac{dT^3}{3} - \frac{f}{2T^2} + JR(\ln 10)\log(V-b)$$

$$- J\left[\frac{B_2}{V-b} + \frac{B_3}{2(V-b)^2} + \frac{B_4}{3(V-b)^3} + \frac{B_5}{4(V-b)^4}\right.$$

$$\left. + \frac{B_6}{\alpha}\left(\frac{1}{e^{\alpha v}} - C'[\ln 10]\log\left[1 + \frac{1}{C'e^{\alpha v}}\right]\right)\right]$$

TABLE 16–4 CONSTANTS FOR HEAT CAPACITY EQUATION

Refrigerant	a	b	c	d	f	Add to °F	Reference
11	0.023815	2.798823×10^{-4}	-2.123734×10^{-7}	5.999018×10^{-11}	−336.80703	459.67	16–6
12	8.0945×10^{-3}	3.32662×10^{-4}	-2.413896×10^{-7}	6.72363×10^{-11}	0	459.7	16–7
13	0.01602	2.823×10^{-4}	-1.159×10^{-7}	0	0	459.69	16–8
14	0.0300559282	2.3704335×10^{-4}	$-2.85660077 \times 10^{-8}$	$-2.95338805 \times 10^{-11}$	0	—	16–9
21	0.0427	1.40×10^{-4}	0	0	0	459.6	16–22
22	0.02812836	2.255408×10^{-4}	-6.509607×10^{-8}	0	257.341	459.69	16–11
23	0.07628087	-7.561805×10^{-6}	3.9065696×10^{-7}	$-2.454905 \times 10^{-10}$	0	459.69	16–12
113	0.07963	1.159×10^{-4}	0	0	0	459.6	16–22
114	0.0175	3.49×10^{-4}	-1.67×10^{-7}	0	0	459.69	16–13
500	0.026803537	2.8373408×10^{-4}	$-9.7167893 \times 10^{-8}$	0	0	459.69	16–23
502	0.020419	2.996802×10^{-4}	-1.409043×10^{-7}	2.210861×10^{-11}	0	459.67	16–15
C318	0.0225178157	$3.69907814 \times 10^{-4}$	$-1.64842522 \times 10^{-7}$	$2.152780846 \times 10^{-11}$	0	459.69	16–16

$$+ \frac{JKe^{-KT/T_c}}{T_c} \left[\frac{C_2}{V-b} + \frac{C_3}{2(V-b)^2} + \frac{C_4}{3(V-b)^3} + \frac{C_5}{4(V-b)^4} \right.$$

$$\left. + \frac{C_6}{\alpha e^{\alpha v}} - \frac{C_6 C'(\ln 10)\log}{\alpha} \left(1 + \frac{1}{C' e^{\alpha v}}\right) \right] + Y$$

$$Y^* = \frac{\Delta H \text{ (latent at } T_1)}{T_1} - S \text{ (at } T_1 V_1)$$

where T_1 = reference temperature (usually −40°F).
V_1 = saturated vapor volume at T_1

*Assuming the entropy of the liquid is zero at reference temperature.

TABLE 16–5 CONSTANTS *X* AND *Y* FOR THE ENTHALPY AND ENTROPY EQUATIONS

Refrigerant	X	Y
11	50.5418	−0.0918395
12	39.556551	−0.016537936
13	20.911	−0.05676
14	86.102162	0.36172528
21	—	—
22	62.4009	−0.0453335
23	—	—
113	25.198	−0.40552
114	25.3396621	−0.11513718
500	46.4734	−0.09012707564
502	35.308	−0.07444
C318	12.19214242	−0.16828871

Use constants from the heat capacity equation and the equation of state.

Velocity of Sound

$$V_a = V \sqrt{\frac{857.3609T(dP/dT)_V^2}{C_v} - 4633.056 \left(\frac{dP}{dV}\right)_T}$$

Use constants from the equation of state.

Latent Heat of Vaporization

$$\Delta H_{lat} = JT(V_{\text{vap}} - V_{\text{liq}}) \frac{dP}{dT}$$

dP/dT is from the vapor-pressure equation.

NOMENCLATURE

P = pressure, psia

T = temperature, °R = °F + 459.67**

V = volume, ft^3/lb

d = density, lb/ft^3

R = gas constant ÷ molecular weight

V_a = velocity of sound, ft/sec

J = conversion factor from work units to heat units = 0.185053

C_v = heat capacity at constant volume, Btu/lb-°F

ΔH_{lat} = latent heat of vaporization, Btu/lb

H = enthalpy, Btu/lb

S = entropy, Btu/lb-°R

log = logarithm, base 10

ln = natural logarithm

a, b, etc. = constants in heat capacity equation

A, B, etc. = constants in vapor-pressure equation

A_2, B_2, etc. = constants in equation of state

$$e = 2.718281828$$

$$\ln 10 = 2.302585093 \qquad \ln x = \frac{\log x}{\log e}$$

$$\log e = 0.4342944819 \qquad \ln x = (\log x)(\ln 10)$$

$$\log e = \frac{1}{\ln 10}$$

REFERENCES

16–1. J. J. Martin, "Correlations and Equations Used in Calculating Thermodynamic Properties of Freon Refrigerants," in *Thermodynamic and Transport Properties of Gases, Liquids, and Solids*. New York: American Soceity of Mechanical Engineers (1959), p. 110.

16–2. R. C. Downing, "Refrigerant Equations," *ASHRAE Trans.*, 80 pt 2 (1974), 2313.

16–3. J. A. Schofield, "Computer Calculation of the Theoretical Performance Properties of Fluorocarbon Refrigerants," *ASHRAE Trans.*, 76, pt 1 (1970), 52.

16–4. G. T. Kartsounes and R. A. Erth, "Computer Calculations of the Thermodynamic Properties of Refrigerants 12, 22, and 502," *ASHRAE Trans.*, 77, pt 2 (1971), 88.

**In some cases 459.69 or 459.7 was used.

16–5. American Society of Heating, Refrigerating, and Air-Conditioning Engineers, *Thermodynamic Properties of Refrigerants.* Atlanta, Ga.: ASHRAE (1969).
16–6. DuPont Company, Freon Products Division, "Thermodynamic Properties of Freon 11," Bulletin T-11 (1965).
16–7. R. C. McHarness, B. J. Eiseman, Jr., and J. J. Martin, "Thermodynamic Properties of the Freon Refrigerants: I. Freon 12," *Refrig. Eng.*, 63 (1955), 31.
16–8. L. F. Albright and J. J. Martin, "Thermodynamic Properties of Chlorotrifluoromethane," *Ind. Eng. Chem.*, 44 (Jan. 1952), 188.
16–9. N. S. C. Chari, "Thermodynamic Properties of Carbontetrafluoride," dissertation, University of Michigan.
16–10. A. F. Benning and R. C. McHarness, "Thermodynamic Properties of Fluorochloromethanes and -Ethanes," *Ind. Eng. Chem.*, 32 (June 1940), 814.
16–11. DuPont Company, Freon Products Division, "Thermodynamic Properties of Freon 22," Bulletin T-22 (1964).
16–12. Y. C. Hou and J. J. Martin, "Physical and Thermodynamic Properties of Trifluoromethane," *AIChE J.*, 5 (Mar. 1959), 125.
16–13. J. J. Martin, "Thermodynamic Properties of Dichlorotetrafluoroethane," *J. Chem. Eng. Data*, 5 (July 1960), 334.
16–14. J. V. Sinka and K. P. Murphy, "Pressure–Volume–Temperature Relationship for a Mixture of Difluorodichloromethane and 1,1-Difluoroethane," *J. Chem. Eng. Data*, 12 (July 1967), 315.
16–15. J. J. Martin and R. C. Downing, "Thermodynamic Properties of Refrigerant 502," *ASHRAE Trans.*, 76, pt 2 (1970), 129.
16–16. J. J. Martin, "Thermodynamic Properties of Perfluorocyclobutane," *J. Chem. Eng. Data*, 7 (Jan. 1962), 68.
16–17. The equation is based on data from the following:
   a. L. Riedel, *Z. Kalte Ind.*, 48 (1941), 9.
   b. N. Thornton, A. Burg, and H. Schlesinger, *J. Am. Chem. Soc.*, 55 (1933), 3177.
   c. R. McNabney, thesis, Western Reserve University (1941).
   d. L. F. Albright and J. J. Martin, "Thermodynamic Properties of Chlorotrifluoromethane," *Ind. Eng. Chem.*, 44 (Jan. 1952), 188.
   e. DuPont Company, Freon Products Division, unpublished data.
16–18. J. J. Martin, University of Michigan, private communication.
16–19. A. F. Benning and R. C. McHarness, "Thermodynamic Properties of Fluorochloromethanes and -Ethanes," *Ind. Eng. Chem.*, 32 (Apr. 1940), 497.
16–20. DuPont Company, Freon Products Division, based on data in J. V. Sinka and K. P. Murphy, "Pressure–Volume–Temperature Relationship for a Mixture of Difluorodichloromethane and 1,1-Difluoroethane," *J. Chem. Eng. Data*, 12 (July 1967), 315.
16–21. A. F. Benning and R. C. McHarness, "Thermodynamic Properties of Fluorochloromethanes and -Ethanes," *Ind. Eng. Chem.*, 32 (May 1940), 698.
16–22. A. F. Benning et al., "Thermodynamic Properties of Fluorochloromethanes and -Ethanes," *Ind. Eng. Chem.*, 32 (July 1940), 976.
16–23. T. E. Morsey, "Thermodynamic Properties of Refrigerant 500," *Kaltetechnik*, 4 (1948), 94.

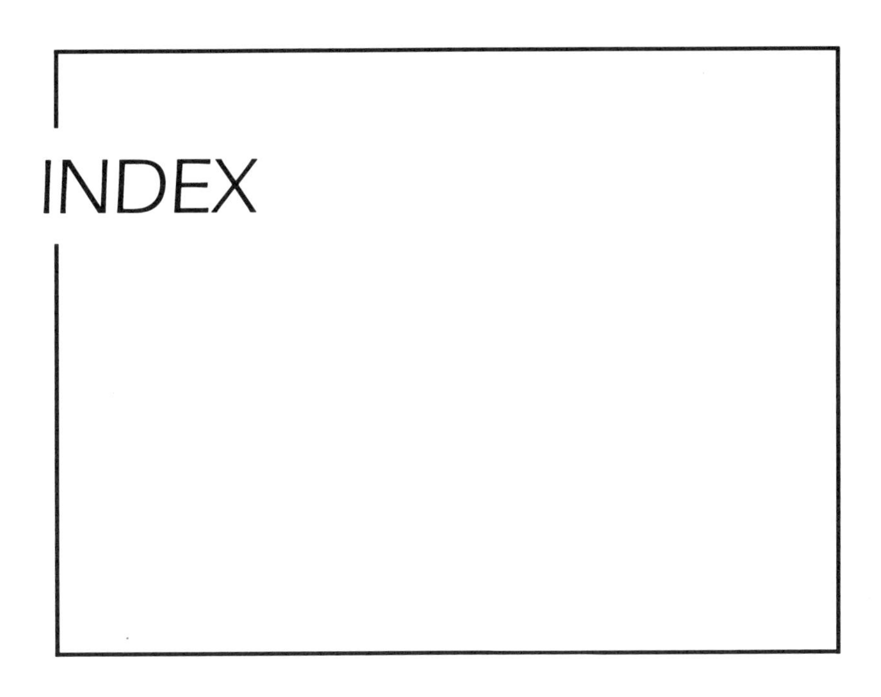

INDEX

A

B

C

D

E

F

P

R

S

T

U

V

W